Sergio Rodriguez
Department of Physics
Purdue University
West Lafayette, IN 47907-1396

Symmetry in Physics

Sergio Rodriguez

Symmetry in Physics

APPUNTI

SCUOLA NORMALE SUPERIORE
1999

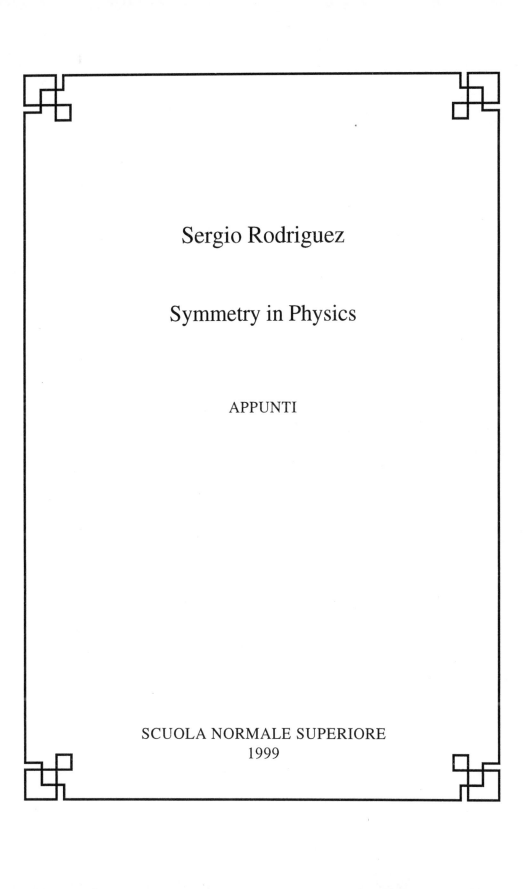

ISBN: 978-88-7642-254-6

For Katinka

Preface

These notes contain my lectures on the theory of group representations and its applications to the physics of atoms, molecules and crystals that I have given at Purdue University, Scuola Normale Superiore (SNS, Pisa, Italy) and Universidad Técnica Federico Santa María (UTFSM: Valparaiso, Chile) on and off over a period of more than 25 years. Invitations to lecture at UTFSM and at SNS (Pisa) during 1997–98 allowed me to have sufficient leisure to write the content of the lectures in the present form. The topics selected reflect my special interests and their scope is limited by the time available to the students. I am grateful to them for their interest in the subject and for requesting clarification of the more difficult parts of the course which, I am sure, led to improvements in the presentation. To what extent I have succeeded should be judged by the reader. The style is somewhat concise and will require careful attention on the part of the reader.

I owe a particular debt of gratitude to Dr. Zdenka Barticevic for inviting me to lecture at UTFSM and to Professor Franco Bassani and Professor Giuseppe La Rocca for inviting me to Pisa during a sabbatical leave from Purdue University in the school year 1997–98.

I am extremely grateful to Mrs. Virginia Messick who efficiently, patiently and cheerfully typed these notes, and to M.J. Seong for drawing the figures.

Contents

SYMMETRY IN PHYSICS

The concept of symmetry played a significant role in the development of the physical sciences since their very beginning. But its more systematic development did not take place until the mathematical theory of groups was created. The idea of symmetry can be traced back to the mathematics of the ancient Greeks. We mention, for example, the proof (which appears in Euclid's Elements) that there are only five regular polyhedra, the so-called Platonic solids, namely the regular tetrahedron, octahedron, cube, dodecahedron and icosahedron. The relation between these solids and the structure of molecules and crystals will be part of this course.

In describing the symmetry of physical systems we often consider geometrical transformations such as a rotation about a fixed axis, a mirror reflection in a plane, or a space inversion with respect to a fixed center (*i.e.*, a transformation of a point P defined by a vector $\boldsymbol{r} = \overrightarrow{OP}$ into $-\boldsymbol{r}$). If application of a geometrical transformation casts a physical system into a new position indistinguishable from its initial position, then we say that the system is invariant under the transformation.

The set of operations leaving a system invariant forms, what is called in mathematics, a group. The concept of group will be defined precisely in chapter 3.

The theory of groups owes its origin to the study of the roots of algebraic equations initiated by Lagrange, Ruffini, Gauss, Abel, Cauchy and culminating in the work of Galois. The definition of an abstract group was proposed by Cayley in 1854 but was ignored until, in 1870, Kronecker, independently, enunciated the postulates defining an abstract commutative group.

The study of permutation groups was greatly advanced by Jordan in his "Traité des substitutions". Finally, between 1890 and 1900, Frobenius, Schur and Burnside developed the theory of group representations. This theory and its applications in quantum mechanics will form the core of this course.

In parallel with these developments the theory of crystal symmetry also arose during the 19th century. It can be said to have been initiated by the Abbé Haüy at the beginning of the century. The derivation of the 32 crystallographic point groups was first published by Hessel in 1830 and rediscovered by Gadolin in 1869. The translation groups of crystals (lattices) were classified by Bravais who showed that there are 14 different possible lattices in three-dimensional space. Fedorov and Schönflies derived the 230 space groups of three-dimensional crystals.

The application of the theory of groups in quantum mechanics can be said to begin with the works of H. A. Bethe, H. Weyl and E. P. Wigner.

We begin our study with a description of orthogonal transformations in three-dimensional space (3-space for short).

Chapter 1

Introduction

1.1 Orthogonal transformations in three-dimensional space

The position of a point P in 3-space with respect to an arbitrarily selected origin O is specified by the vector $\overrightarrow{OP} = \boldsymbol{r}$. One could, of course, be content in specifying P by this displacement but it is often desirable to define a vector in terms of its components with respect to three non-coplanar vectors. For simplicity we choose three orthogonal unit vectors $\hat{e}_1$, $\hat{e}_2$ and $\hat{e}_3$ with $\hat{e}_3$ directed so that, when the $(\hat{e}_1, \hat{e}_2)$ plane is viewed from the direction into which $\hat{e}_3$ points, the rotation by $90°$ taking $\hat{e}_1$ into $\hat{e}_2$ is in the positive or counterclockwise sense. Thus

$$\hat{e}_3 = \hat{e}_1 \times \hat{e}_2 \tag{1.1}$$

and the triple product of $\hat{e}_1$, $\hat{e}_2$ and $\hat{e}_3$ is equal to unity:

$$(\hat{e}_1 \times \hat{e}_2) \cdot \hat{e}_3 = [\hat{e}_1 \hat{e}_2 \hat{e}_3] = 1 \ . \tag{1.2}$$

The frame of reference defined by $\hat{e}_1, \hat{e}_2$ and $\hat{e}_3$ is said to be a right-handed Cartesian system and the vectors themselves are said to form a right-handed triad of orthogonal unit vectors. In general

$$\boldsymbol{r} = x_1 \hat{e}_1 + x_2 \hat{e}_2 + x_3 \hat{e}_3 \tag{1.3}$$

where $x_i = \boldsymbol{r} \cdot \hat{e}_i$ $(i = 1, 2, 3)$ are the components of $\boldsymbol{r}$ in this frame.

Consider now a mapping $\boldsymbol{r} \to \boldsymbol{r}' \ : \ \boldsymbol{r}' = R\boldsymbol{r}$. The new vector $\boldsymbol{r}'$ is said to be the image of $\boldsymbol{r}$ under R. The transformation R is said to be linear if

$$(i) \ \ \boldsymbol{r}_1' = R\boldsymbol{r}_1 \ \text{ and } \boldsymbol{r}_2' = R\boldsymbol{r}_2 \Rightarrow \boldsymbol{r}_1' + \boldsymbol{r}_2' = R(\boldsymbol{r}_1 + \boldsymbol{r}_2) \ ,$$

and

$$(ii) \ \ \boldsymbol{r}' = R\boldsymbol{r} \Rightarrow a\boldsymbol{r}' = R(a\boldsymbol{r}) \ .$$

A linear transformation R is orthogonal if it preserves the lengths of all vectors, *i.e.*, if $\boldsymbol{r}' = R\boldsymbol{r}$, then $\boldsymbol{r}'\cdot\boldsymbol{r}' = \boldsymbol{r}\cdot\boldsymbol{r}$. The linearity of R implies that it also preserves angles. In fact, if $\hat{u}$ and $\hat{v}$ are any two unit vectors and $\hat{u}'$ and $\hat{v}'$ their images under R, then

$$(\hat{u} + \hat{v})^2 = (\hat{u}' + \hat{v}')^2$$

so that

$$\hat{u} \cdot \hat{v} = \hat{u}' \cdot \hat{v}' \ .$$

Let $\hat{e}_1', \hat{e}_2'$ and $\hat{e}_3'$ be the images of $\hat{e}_1, \hat{e}_2, \hat{e}_3$ under the orthogonal transformation R. These unit vectors are, necessarily, orthogonal. The image $\boldsymbol{r}'$ of the vector $\boldsymbol{r}$ in Eq. (1.4) has, with respect to the new vectors $\hat{e}_1', \hat{e}_2', \hat{e}_3'$, the same components x_1, x_2 and x_3 as $\boldsymbol{r}$ had with respect to $\hat{e}_1, \hat{e}_2, \hat{e}_3$, *i.e.*,

$$\boldsymbol{r}' = R\boldsymbol{r} = \sum_{j=1}^{3} x_j \hat{e}_j' = x_j \hat{e}_j' \tag{1.4}$$

where in the last equality we use the convention that a sum over a repeated index is implied. The components of $\boldsymbol{r}'$ with respect to $\hat{e}_1, \hat{e}_2, \hat{e}_3$ are

$$x_i' = \boldsymbol{r}' \cdot \hat{e}_i = \hat{e}_i \cdot \hat{e}_j' x_j \ . \tag{1.5}$$

We define R in terms of the nine numbers

$$R_{ij} = \hat{e}_i \cdot \hat{e}_j' \tag{1.6}$$

so that Eq. (1.5) reads

$$x_i' = R_{ij} x_j \ . \tag{1.7}$$

The transformation relating the triads $\{\hat{e}_i\}$ and $\{\hat{e}_i'\}$ $(i = 1, 2, 3)$ is

$$\hat{e}_i' = \hat{e}_j \hat{e}_j \cdot \hat{e}_i' = R_{ji} \hat{e}_j \ . \tag{1.8}$$

Thus the matrix associated with the transformation of the basis vectors is the transpose of that describing the transformation of the coordinates of a vector. The equality of $x_i x_i$ and $x_i' x_i'$ for all $\boldsymbol{r}$ implies that

$$R_{ij} R_{ik} = \delta_{jk} \ , \tag{1.9}$$

or, in matrix form, that

$$\widetilde{\boldsymbol{R}}\boldsymbol{R} = \boldsymbol{R}\widetilde{\boldsymbol{R}} = \boldsymbol{E} \tag{1.10}$$

where $\widetilde{\boldsymbol{R}}$ is the transpose of $\boldsymbol{R}$ and $\boldsymbol{E}$ the unit matrix. Denoting the determinant of $\boldsymbol{R}$ by $\parallel \boldsymbol{R} \parallel$, Eq. (1.10) shows that

$$\parallel \boldsymbol{R} \parallel = \pm 1 \ . \tag{1.11}$$

Furthermore, this shows that the matrix $\boldsymbol{R}$ has an inverse $\boldsymbol{R}^{-1}$ and that

$$\boldsymbol{R}^{-1} = \widetilde{\boldsymbol{R}} \tag{1.12}$$

The triple product of $\hat{e}'_1, \hat{e}'_2$ and $\hat{e}'_3$ is

$$\hat{e}'_1 \cdot (\hat{e}'_2 \times \hat{e}'_3) = R_{i1}R_{j2}R_{k3}\hat{e}_i \cdot (\hat{e}_j \times \hat{e}_k) = \epsilon_{ijk}R_{i1}R_{j2}R_{k3} \equiv \parallel \boldsymbol{R} \parallel \ . \tag{1.13}$$

Thus, the triad $\{\hat{e}'_i\}$ is right-handed if $\parallel \boldsymbol{R} \parallel = 1$ and left-handed if $\parallel \boldsymbol{R} \parallel = -1$. In equation (1.13) $\epsilon_{ijk} = 1$ if ijk is an even permutation of 123 (*i.e.*, 231 or 312), -1 if the permutation is odd and zero in all other cases. ϵ_{ijk} is called the Levi-Civita symbol (it is actually a tensor density).

1.2 Rotations in three-dimensional space. Parameters specifying a rotation. Euler angles. Composition of rotations.

A rotation by an angle φ about an axis through an origin O parallel to a unit vector $\hat{\boldsymbol{n}}$ transforms a vector $\boldsymbol{r} = \overrightarrow{OP}$ into $\boldsymbol{r}' = \overrightarrow{OP}'$ (see figure 1.1).

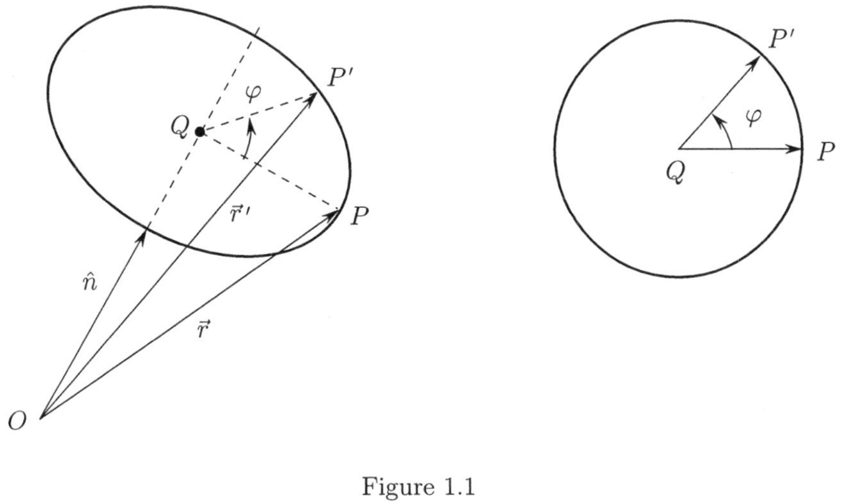

Figure 1.1

The projections of $\boldsymbol{r}$ and $\boldsymbol{r}'$ on $\hat{\boldsymbol{n}}$ are equal and

$$\overrightarrow{QP}' = \overrightarrow{QP} \cos\varphi + \hat{n} \times (\overrightarrow{QP}) \sin\varphi$$

so that

$$\begin{aligned} \boldsymbol{r}' &= \boldsymbol{r} \cdot \hat{n}\hat{n} + (\boldsymbol{r} - \boldsymbol{r} \cdot \hat{n}\hat{n}) \cos\varphi + \hat{n} \times (\boldsymbol{r} - \boldsymbol{r} \cdot \hat{n}\hat{n}) \sin\varphi \\ &= \boldsymbol{r} \cos\varphi + (1 - \cos\varphi)\hat{n}\hat{n} \cdot \boldsymbol{r} + \hat{n} \times \boldsymbol{r} \sin\varphi \ . \end{aligned} \tag{2.1}$$

This defines the matrix $\boldsymbol{R}$ of the transformation (2.2) as one whose components are

$$R_{ij} = \delta_{ij} \cos\varphi + n_i n_j (1 - \cos\varphi) - \epsilon_{ijk} n_k \sin\varphi \ . \tag{2.2}$$

The determinant of $\boldsymbol{R}$ is 1. In fact, $\| \boldsymbol{R} \|$ is a continuous function of φ but, when $\varphi = 0$, $R_{ij} = \delta_{ij}$ so that $\| \boldsymbol{R}(0) \| = 1$. If for some other value of φ, $\| \boldsymbol{R} \|$ were -1, the continuity of $\| \boldsymbol{R} \|$ implies the existence of at least one value of φ for which $\| R \|$ is, say, zero. But this is impossible since $\boldsymbol{R}$ is an orthogonal transformation.

The set of all orthogonal transformations in 3-space is denoted by $O(3)$. The subset of $O(3)$ consisting of those orthogonal transformations R such that $\| \boldsymbol{R} \| = 1$ is called SO(3). We shall see that O(3), the set of all orthogonal transformations in 3-space, forms a group called, appropriately, the orthogonal group in 3-space. SO(3) is also a group which receives the name the special orthogonal group in 3-space.

We have shown that every rotation is a member of SO(3). We now prove the converse of this statement. To demonstrate this we show first that there is a vector $\hat{e}_3$ such that $\boldsymbol{R}\hat{e}_3 = \hat{e}_3$. Since $\boldsymbol{R}$ is non-singular its null space consists of the zero vector alone. To prove the existence of $\hat{e}_3$ for $\boldsymbol{R} \neq \boldsymbol{E}$ it is enough to show that $\boldsymbol{R} - \boldsymbol{E}$ is singular. Now

$$\boldsymbol{R} - \boldsymbol{E} = \boldsymbol{R}(\boldsymbol{E} - \boldsymbol{R}^{-1}) = (-\boldsymbol{E})\boldsymbol{R}(\widetilde{\boldsymbol{R}} - \boldsymbol{E}) .$$

Since the determinants of $-\boldsymbol{E}$, $\boldsymbol{R}$ and $(\widetilde{\boldsymbol{R}} - \boldsymbol{E})$ are, respectively, -1, 1 and $\| \boldsymbol{R} - \boldsymbol{E} \|$ we obtain

$$\| \boldsymbol{R} - \boldsymbol{E} \| = - \| \boldsymbol{R} - \boldsymbol{E} \| = 0 .$$

Thus, $\hat{e}_3$ exists. Note that this proof holds for spaces of odd dimensionality. Clearly it is not true in two dimensions. Having established the existence of $\hat{e}_3$ we construct two vectors, $\hat{e}_1$ and $\hat{e}_2$ orthogonal to each other and to $\hat{e}_3$ and, forming with $\hat{e}_3$ a right-handed triad of unit vectors. Then the vectors $\hat{e}'_i = R\hat{e}_i$ also form a right-handed triad of orthogonal unit vectors. Let φ be the angle formed by $\hat{e}'_1$ and $\hat{e}_1$; then

$$\begin{aligned} \hat{e}'_1 &= R\hat{e}_1 = \hat{e}_1 \cos\varphi + \hat{e}_2 \sin\varphi , \\ \hat{e}'_2 &= R\hat{e}_2 = -\hat{e}_1 \sin\varphi + \hat{e}_2 \cos\varphi , \end{aligned}$$

and

$$\hat{e}'_3 = R\hat{e}_3 = \hat{e}_3 .$$

This represents a rotation by φ about $\hat{e}_3$.

If φ is infinitesimal, Eq. (2.2) reduces to

$$R_{ij} = \delta_{ij} - \epsilon_{ijk} n_k \varphi + O(\varphi^2) . \tag{2.3}$$

For rotations about the 3-axis, Eq. (2.3) is

$$\boldsymbol{R} = \boldsymbol{E} + \begin{pmatrix} 0 & -1 & 0 \\ 1 & 0 & 0 \\ 0 & 0 & 0 \end{pmatrix} \varphi + O(\varphi^2) .$$

Similar expressions are obtained for rotations by infinitesimal angles about the 1– and 2– axes. We express these rotations by means of the matrices

$$I_1 = \begin{pmatrix} 0 & 0 & 0 \\ 0 & 0 & -i \\ 0 & i & 0 \end{pmatrix} , \quad I_2 = \begin{pmatrix} 0 & 0 & i \\ 0 & 0 & 0 \\ -i & 0 & 0 \end{pmatrix} , \quad I_3 = \begin{pmatrix} 0 & -i & 0 \\ i & 0 & 0 \\ 0 & 0 & 0 \end{pmatrix} . \tag{2.4}$$

These matrices obey the commutation relations

$$[I_i, I_j] = I_i I_j - I_j I_j = i\epsilon_{ijk} I_k \tag{2.5}$$

often written in the form

$$\boldsymbol{I} \times \boldsymbol{I} = i\boldsymbol{I} , \tag{2.6}$$

with the notation $\boldsymbol{I} = I_i \hat{e}_i$. Defining the matrix

$$\boldsymbol{I} \cdot \hat{n} = I_i n_i \tag{2.7}$$

we find

$$(\boldsymbol{I} \cdot \hat{n})_{ij} = -i\epsilon_{ijk} n_k , \tag{2.8}$$

$$(\boldsymbol{I} \cdot \hat{n})^2 = \boldsymbol{E} - \hat{n}\hat{n} , \tag{2.9}$$

and

$$(\boldsymbol{I} \cdot \hat{n})^3 = \boldsymbol{I} \cdot \hat{n} . \tag{2.10}$$

This allows us to write

$$\boldsymbol{R} = \boldsymbol{E} - i\boldsymbol{I} \cdot \hat{n} \sin\varphi - (1 - \cos\varphi)(\boldsymbol{I} \cdot \hat{n})^2 . \tag{2.11}$$

This expression for $\boldsymbol{R}$ is identical to the power series expansion of the matrix

$$\begin{aligned} e^{-i\varphi \boldsymbol{I} \cdot \hat{n}} &= \sum_{\ell=0}^{\infty} \frac{(-i)^\ell}{\ell!} \varphi^\ell (\boldsymbol{I} \cdot \hat{n})^\ell \\ &= \boldsymbol{E} - i\boldsymbol{I} \cdot \hat{n} \sin\varphi - (\boldsymbol{I} \cdot \hat{n})^2 (1 - \cos\varphi) . \end{aligned}$$

Thus

$$\boldsymbol{R} = \exp(-i\varphi \boldsymbol{I} \cdot \hat{n}) . \tag{2.12}$$

The matrices I_1, I_2, I_3 allow us to express all elements of SO(3) in terms of them and are, therefore, called the generators of SO(3).

A second description of the rotation R consists in giving the vector

$$\boldsymbol{w} = 2\hat{n} \tan\left(\frac{1}{2}\varphi\right) . \tag{2.13}$$

The direction of the axis of rotation $\hat{n}$ is determined by two real numbers. As φ varies from $-\pi$ to π, $\tan(\frac{1}{2}\varphi)$ ranges from $-\infty$ to ∞. There is, therefore a one-to-one relation between the elements of SO(3) and the set of vectors of the form (2.13).

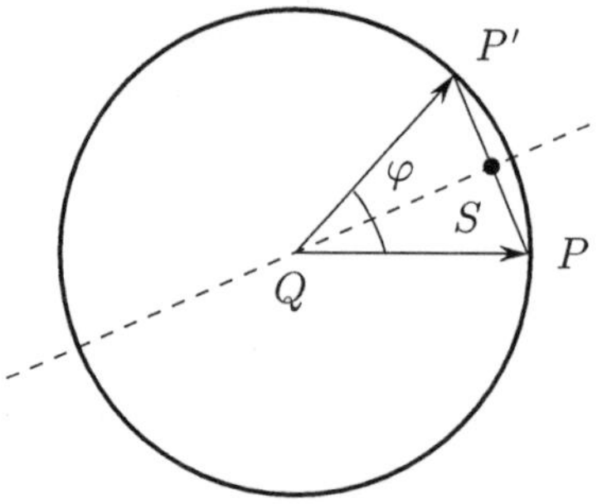

Figure 1.2

To obtain the form of the transformation $\boldsymbol{r} \to \boldsymbol{r}'$ associated with $\boldsymbol{w}$ we remark that the vectors $\boldsymbol{r}'-\boldsymbol{r}$ and $\frac{1}{2}\boldsymbol{w}\times(\boldsymbol{r}+\boldsymbol{r}')$ are equal. In fact, referring to figure 1.2, $\boldsymbol{r}'-\boldsymbol{r} = \overrightarrow{PP}'$; S is the tip of $\frac{1}{2}(\boldsymbol{r}+\boldsymbol{r}')$ so that $\frac{1}{2}\boldsymbol{w}\times(\boldsymbol{r}+\boldsymbol{r}')$ is parallel to $\overrightarrow{PP}'$. It is now enough to show that the magnitudes of these vectors are equal:

$$\left|\overrightarrow{PP}'\right| = 2\left|\overrightarrow{QP}\right|\sin\left(\frac{1}{2}\varphi\right) ,$$

and

$$\begin{aligned}\left|\frac{1}{2}\boldsymbol{w}\times(\boldsymbol{r}+\boldsymbol{r}')\right| &= \left|\boldsymbol{w}\times\overrightarrow{QS}\right| \\ &= 2\left|\overrightarrow{QP}\right|\sin\left(\frac{1}{2}\varphi\right) .\end{aligned}$$

Thus

$$\boldsymbol{r}'-\boldsymbol{r} = \frac{1}{2}\boldsymbol{w}\times(\boldsymbol{r}+\boldsymbol{r}') . \tag{2.14}$$

Equation (2.14) can be solved for $\boldsymbol{r}'$ by multiplying successively by $\boldsymbol{w}$ scalarly and vectorially to obtain

$$\boldsymbol{r}' = \boldsymbol{r} + \left(1+\frac{w^2}{4}\right)^{-1}\boldsymbol{w}\times\boldsymbol{r} + \frac{1}{2}\left(1+\frac{w^2}{4}\right)^{-1}\boldsymbol{w}\times(\boldsymbol{w}\times\boldsymbol{r}) . \tag{2.15}$$

Equation (2.14) allows us to find a convenient formula for the law of composition of rotations about intersecting axes. If a rotation $\boldsymbol{w}_1$ transforms $\boldsymbol{r}$ into $\boldsymbol{r}'$ and $\boldsymbol{w}_2$ transforms $\boldsymbol{r}'$ into $\boldsymbol{r}''$, then

$$\boldsymbol{w}_{21} = \frac{\boldsymbol{w}_1+\boldsymbol{w}_2-\frac{1}{2}\boldsymbol{w}_1\times\boldsymbol{w}_2}{1-(\boldsymbol{w}_1\cdot\boldsymbol{w}_2/4)} \tag{2.16}$$

describes the rotation that carries $\boldsymbol{r}$ directly into $\boldsymbol{r}''$. The verification of Eq. (2.16) is left to the reader. We note that $\boldsymbol{w}_{12}$, the rotation in which the order of the operations is reversed, is not identical to $\boldsymbol{w}_{21}$: rotations about intersecting axes do not commute in general. If the axes are coincident, then $\boldsymbol{w}_1$ is parallel to

$\boldsymbol{w}_2$ and, $\boldsymbol{w}_{21} = \boldsymbol{w}_{12}$. Also, two rotations by π about orthogonal axes commute (but see Sec. 6.10).

We consider now the composition of two infinitesimal rotations about orthogonal axes. For $\delta\varphi$ small, Eq. (2.13) gives

$$\boldsymbol{w} = \hat{n}\delta\varphi + \frac{1}{12}\hat{n}(\delta\varphi)^3 + \cdots \; . \tag{2.17}$$

The composition of a rotation by $\delta\varphi_1$ about $\hat{e}_1$ followed by a second by $\delta\varphi_2$ about $\hat{e}_2$, normal to $\hat{e}_1$, is described by

$$\boldsymbol{w}_{21} = \hat{e}_1\delta\varphi_1 + \hat{e}_2\delta\varphi_2 - \frac{1}{2}\hat{e}_3\delta\varphi_1\delta\varphi_2 + \cdots \; , \tag{2.18}$$

where $\hat{e}_3 = \hat{e}_1 \times \hat{e}_2$. Reversal of the order of the rotations gives

$$\boldsymbol{w}_{12} = \hat{e}_1\delta\varphi_1 + \hat{e}_2\delta\varphi_2 + \frac{1}{2}\hat{e}_3\delta\varphi_1\delta\varphi_2 + \cdots \; . \tag{2.19}$$

These results mean that, to second order in the infinitesimal angles $\delta\varphi_1$ and $\delta\varphi_2$, a rotation by $\delta\varphi_1$ about $\hat{e}_1$ followed by one by $\delta\varphi_2$ about $\hat{e}_2$ is equivalent to the operation obtained reversing the order of the rotations followed by a rotation by $-\delta\varphi_1\delta\varphi_2$ about $\hat{e}_3 = \hat{e}_1 \times \hat{e}_2$.

A third way to specify a rotation about a fixed point is by means of the so-called Euler angles. To define them we consider two arbitrary right-handed triads $(\hat{e}_1, \hat{e}_2, \hat{e}_3)$ and $(\hat{e}'_1, \hat{e}'_2, \hat{e}'_3)$ drawn from a common origin O. The rotation necessary to bring the first triad into coincidence with the second can be accomplished by three successive rotations by angles α, β and γ which we now define:

The planes $(\hat{e}_1, \hat{e}_2)$ and $(\hat{e}'_1, \hat{e}'_2)$, having the point O in common, must be either coincident or have a straight line intersection. In the first case the motion necessary to carry the first triad into coincidence with the second consists of a single rotation by an angle α about $\hat{e}_3$ if $\hat{e}'_3 = \hat{e}_3$ or by a rotation by π about $\hat{e}_1$ or $\hat{e}_2$ followed by a rotation by α about $\hat{e}'_3$ when $\hat{e}'_3 = -\hat{e}_3$. The second, more interesting case, requires some discussion. The line of intersection of the planes $(\hat{e}_1, \hat{e}_2)$ and $(\hat{e}'_1, \hat{e}'_2)$ is called the line of nodes. We select a positive sense on this line as that of $\hat{e}_3 \times \hat{e}'_3$. The angle between $\hat{e}_3$ and $\hat{e}'_3$, lying in the range $(0, \pi)$ is denoted by β; it is the polar angle of $\hat{e}'_3$, in the triad $\{\hat{e}_i\}$ with $\hat{e}_3$ as polar axis. The unit vector $\hat{\nu}$ defined by

$$\hat{e}_3 \times \hat{e}'_3 = \hat{\nu}\sin\beta \tag{2.20}$$

defines the sense of the line of nodes. The angle between $\hat{e}_2$ and $\hat{\nu}$, *i.e.*, the angle formed by the planes

$$(\hat{e}_3, \hat{e}_1) \text{ and } (\hat{e}_3, \hat{e}'_3)$$

is denoted by α. Finally the angle formed by the planes

$$(\hat{e}_3, \hat{e}'_3) \text{ and } (\hat{e}'_3, \hat{e}'_1)$$

is labeled γ.

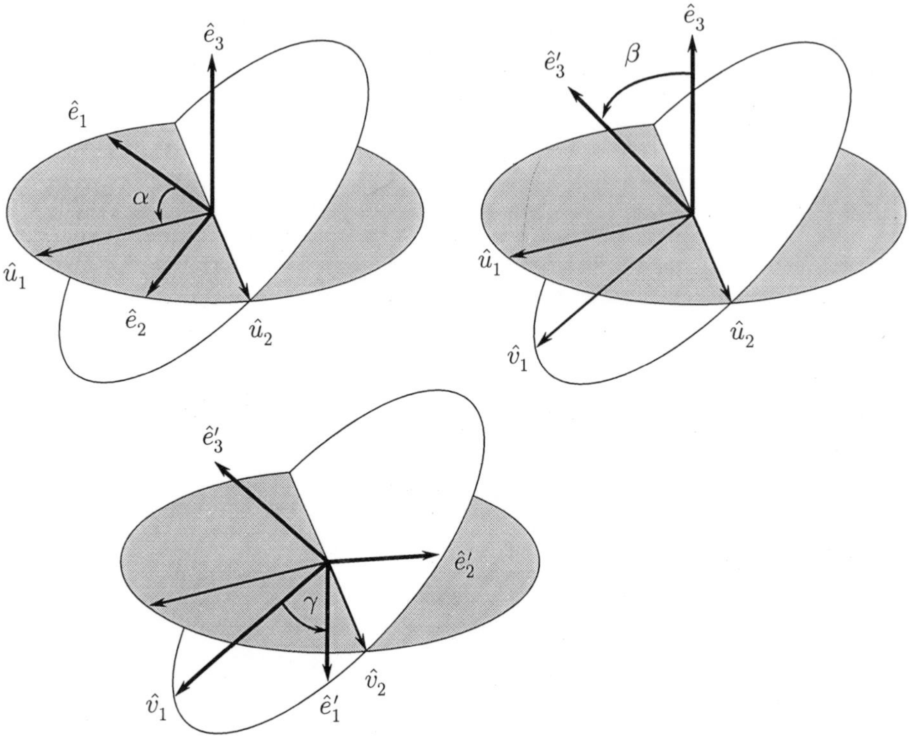

Figure 1.3

The rotation carrying the triad $\{\hat{e}_i\}$ onto $\{\hat{e}_i'\}$ is a composition of three successive rotations as follows:

(i) a positive rotation (*i.e.*, counterclockwise) by α about $\hat{e}_3$ until $\hat{e}_2$ is brought into coincidence with $\hat{\nu}$, the positive line of nodes. The range of α is, thus $0 \leq \alpha < 2\pi$,

(ii) a positive rotation by β about $\hat{\nu}$ until $\hat{e}_3$ comes into coincidence with $\hat{e}_3'$; $0 \leq \beta \leq \pi$,

(iii) a positive rotation by γ about $\hat{e}_3'$ until $\hat{e}_1$, in its new position after steps (i) and (ii) coincides with $\hat{e}_1'$. The range of γ is $0 \leq \gamma < 2\pi$.

We denote the intermediate triads after steps (i) and (ii) by $(\hat{u}_1, \hat{u}_2, \hat{u}_3)$ and $(\hat{v}_1, \hat{v}_2, \hat{v}_3)$, respectively. Then

$$\hat{u}_k = \hat{e}_j R^{(3)}_{jk}(\alpha) \tag{2.21}$$

where the $R^{(3)}_{jk}(\alpha)$ are the matrix elements of a rotation by α about the 3-axis, *i.e.*,

$$\boldsymbol{R}^{(3)}(\alpha) = \begin{pmatrix} \cos\alpha & -\sin\alpha & 0 \\ \sin\alpha & \cos\alpha & 0 \\ 0 & 0 & 1 \end{pmatrix} . \tag{2.22}$$

The triad $\{\hat{v}_i\}$ is given by

$$\hat{v}_\ell = \hat{u}_k R^{(2)}_{k\ell}(\beta) \tag{2.23}$$

with

$$\boldsymbol{R}^{(2)}(\beta) = \begin{pmatrix} \cos\beta & 0 & \sin\beta \\ 0 & 1 & 0 \\ -\sin\beta & 0 & \cos\beta \end{pmatrix} . \tag{2.24}$$

Finally, since

$$\hat{e}'_i = \hat{v}_\ell R^{(3)}_{\ell i}(\gamma) , \tag{2.25}$$

$$\hat{e}'_i = \hat{e}_j R^{(3)}_{jk}(\alpha) R^{(2)}_{k\ell}(\beta) R^{(3)}_{\ell i}(\gamma) = \hat{e}_j R_{ji}(\alpha, \beta, \gamma) \tag{2.26}$$

with

$$\boldsymbol{R}(\alpha,\beta,\gamma) = \boldsymbol{R}^{(3)}(\alpha)\boldsymbol{R}^{(2)}(\beta)\boldsymbol{R}^{(3)}(\gamma) . \tag{2.27}$$

The matrix $\boldsymbol{R}(\alpha,\beta,\gamma)$ is

$$\boldsymbol{R} = \begin{pmatrix} \cos\alpha\cos\beta\cos\gamma - \sin\alpha\sin\gamma & -\cos\alpha\cos\beta\sin\gamma - \sin\alpha\cos\gamma & \cos\alpha\sin\beta \\ \sin\alpha\cos\beta\cos\gamma + \cos\alpha\sin\gamma & -\sin\alpha\cos\beta\sin\gamma + \cos\alpha\cos\gamma & \sin\alpha\sin\beta \\ -\sin\beta\cos\gamma & \sin\beta\sin\gamma & \cos\beta \end{pmatrix} \tag{2.28}$$

The components of a vector $\boldsymbol{r} = x_i\hat{e}_i$ become $x'_i = \boldsymbol{r}' \cdot \hat{e}_i = x_j \hat{e}'_j \cdot \hat{e}_i$, or

$$x'_i = R_{ij}(\alpha,\beta,\gamma)x_j . \tag{2.29}$$

Rotations are usually labeled by the letter C to which we may add symbols giving the axis of rotation and the angle of rotation. Often these are implicit in the context in which they appear. A rotation by an integral sub-multiple of 2π, *e.g.*, $2\pi/n$ is denoted by C_n.

1.3 Vectors and pseudovectors

So far we have considered only rotations in 3-space, *i.e.*, elements of SO(3). We have seen that O(3) contains, in addition to rotations, transformations whose matrices have determinants equal to -1. An example is the inversion with respect to the origin:

$$\boldsymbol{r} \to \boldsymbol{r}' = -\boldsymbol{r} \ ; \ x'_i = -x_i \quad (i = 1,2,3) . \tag{3.1}$$

This operation is denoted by i whenever the context ensures that it will not be confused with the imaginary unit. It is also called the parity operation and is, sometimes, denoted by the symbol P. Its matrix form is

$$[i] = \begin{pmatrix} -1 & 0 & 0 \\ 0 & -1 & 0 \\ 0 & 0 & -1 \end{pmatrix} = -\boldsymbol{E} . \tag{3.2}$$

Its determinant is $(-1)^3 = -1$. Every element of O(3) can be expressed as an element of SO(3) times i unless it is, itself, a member of SO(3). We express this in symbols by O(3) = SO(3) $\times i$.

Other orthogonal operations having determinant -1 are reflections in planes and improper rotations. Reflections in mirror planes are denoted by σ. A reflection in the $(\hat{e}_1, \hat{e}_2)$ of a Cartesian system whose basis is $\{\hat{e}_i\}$ $(i = 1, 2, 3)$ transforms $\boldsymbol{r}$ into $\boldsymbol{r}'$ according to

$$\sigma_{12} \;:\; x_1' = x_1 \;,\; x_2' = x_2 \;,\; x_3' = -x_3$$

so that its corresponding matrix is

$$[\sigma_{12}] = \begin{pmatrix} 1 & 0 & 0 \\ 0 & 1 & 0 \\ 0 & 0 & -1 \end{pmatrix} \tag{3.3}$$

An improper rotation is a rotation followed by a reflection in a plane perpendicular to the axis of rotation. An improper rotation by α about the 3-axis is represented by

$$[\boldsymbol{S}_\alpha] = \begin{pmatrix} \cos\alpha & -\sin\alpha & 0 \\ \sin\alpha & \cos\alpha & 0 \\ 0 & 0 & -1 \end{pmatrix} . \tag{3.4}$$

If $\alpha = \pi/2$, this operation is denoted by S_4.

The velocity of a particle, its acceleration, a force, an electric current, etc. are vector quantities. They have the property that their components transform as the components of the position vector $\boldsymbol{r}$ under orthogonal transformations. Thus, if $\boldsymbol{V} = (V_1, V_2, V_3)$ is a vector, the quantities V_i transform according to

$$V_i' = R_{ij} V_j \tag{3.5}$$

under the operation $\boldsymbol{R}$. There exist other objects, *e.g.*, tensors, obeying more complicated laws of transformation. We shall be concerned with tensors later but now we shall study pseudovectors. We shall show later that pseudovectors are second rank antisymmetric tensors in 3-space.

The prototype pseudovector is the vector product of two vectors, say $\boldsymbol{u}$ and $\boldsymbol{v}$:

$$\boldsymbol{w} = \boldsymbol{u} \times \boldsymbol{v}; \quad w_i = \epsilon_{ijk} u_j v_k \;. \tag{3.6}$$

The components of $\boldsymbol{w}$ are

$$w_1 = u_2 v_3 - u_3 v_2 \;,\; w_2 = u_3 v_1 - u_1 v_3 \;,\; w_3 = u_1 v_2 - u_2 v_1 \;. \tag{3.7}$$

That $\boldsymbol{w}$ is not a vector is obvious because under inversion $u_i \to u_i' = -u_i$, $v_i \to v_i' = -v_i$ but $w_i \to w_i' = w_i$.

To obtain the transformation of the components of $\boldsymbol{w}$ under an orthogonal transformation R we make use of the sum

$$\epsilon_{\ell mn} R_{i\ell} R_{jm} R_{kn} \;.$$

If ijk is 123 or one of its even permutations this quantity equals the determinant $\| \boldsymbol{R} \|$ of $\boldsymbol{R}$. If ijk is an odd permutation of 123 it is $- \| \boldsymbol{R} \|$. If any two of the indices are equal, the expression is a determinant with two identical rows so that it vanishes. Hence

$$\epsilon_{\ell mn} R_{i\ell} R_{jm} R_{kn} = \| \boldsymbol{R} \| \epsilon_{ijk} \ . \tag{3.8}$$

Now

$$\begin{aligned} R_{i\ell} w_\ell &= \epsilon_{\ell mn} u_m v_n R_{i\ell} = \epsilon_{\ell mn} R_{i\ell} R_{jm} R_{kn} u'_j v'_k = \| \boldsymbol{R} \| \epsilon_{ijk} u'_j v'_k \\ &= \| \boldsymbol{R} \| w'_i \ . \end{aligned}$$

Thus, since $\| \boldsymbol{R} \|^2 = 1$,

$$w'_i = \| \boldsymbol{R} \| R_{il} w_\ell \ . \tag{3.9}$$

Therefore, pseudovectors transform as vectors under SO(3) but as $w'_i = -R_{i\ell} w\ell$ under improper operations. Thus, while under reflection in a plane the components of a vector parallel to the plane remain invariant and that normal to it changes sign, the opposite occurs for a pseudovector; the components of a pseudovector parallel to a reflection plane change sign under this operation and that normal to the plane remains invariant.

An identity which is often used is

$$\epsilon_{ijk} \epsilon_{i\ell m} = \delta_{j\ell} \delta_{km} - \delta_{jm} \delta_{k\ell} \ . \tag{3.10}$$

In fact, this quantity is equal to

$$\epsilon_{1jk} \epsilon_{1\ell m} + \epsilon_{2jk} \epsilon_{2\ell m} + \epsilon_{3jk} \epsilon_{3\ell m} \ .$$

At most one of the terms in this sum is different from zero for given j, k, ℓ and m and this occurs only if the pairs jk and ℓm are equal. The sum is equal to 1 if $j = \ell$ and $k = m$ and to -1 is $j = m$ and $k = \ell$.

1.4 Simple applications

It is instructive at this stage to make a few elementary applications of the transformations in the orthogonal group.

Let us consider a spherical distribution of charge around a center O, *i.e.*, a distribution of electric charge with a density which is invariant under all rotations about O. Let P be an arbitrary point distinct from O. The distribution of charge is invariant with respect to rotations about $\overrightarrow{OP}$. Thus, the electric field at P must also be invariant under these rotations. Therefore the component of the electric field, $\boldsymbol{E}$, normal to $\overrightarrow{OP}$ must vanish. Furthermore, the magnitude of $\boldsymbol{E}$ is a function of $r = |\overrightarrow{OP}|$ only because of the invariance with respect to rotations about axes normal to $\overrightarrow{OP}$.

A more interesting problem is that of calculating the magnetic induction $\boldsymbol{B}$ produced by an infinite straight wire carrying a constant electric current, I. We recall that $\boldsymbol{B}$ is a pseudovector as required by Ampère's law

$$\nabla \times \boldsymbol{B} = \frac{4\pi}{c}\boldsymbol{j} \ , \tag{4.1}$$

where $\boldsymbol{j}$ is the electric current density. Since $\boldsymbol{j}$ is proportional to the velocity of the charge carriers, it is a vector quantity. The components

$$\frac{\partial}{\partial x_i} = \partial_i \tag{4.2}$$

of the operator ∇ transform as a vector, *i.e.*, according to Eq. (3.5). In fact, if $x_i' = R_{ij}x_j$, $x_j = R_{ij}x_i'$ so that

$$\partial_i' = \frac{\partial x_j}{\partial x_i'}\partial_j = R_{ij}\partial_j \ . \tag{4.3}$$

Since $\boldsymbol{j}$ is a vector, Eq. (4.1) requires $\boldsymbol{B}$ to be pseudovector in character. Let $\overrightarrow{OP} = (\rho, \varphi, z)$ be the position of a point $P (\neq O)$ with respect to an origin O on the wire in cylindrical coordinates with the z-axis along the length of the wire. The system is invariant with respect to reflection in the plane through P containing the wire. Under this operation B_z and B_ρ change sign but on the other hand remain invariant: $B_z = -B_z$, $B_\rho = -B_\rho$. Thus only $B_\varphi \neq 0$. Since the system remains invariant under rotations about the wire and translations along it, B_φ depends only on ρ, the distance from P to the wire. Application of Ampère's law in its integral form yields

$$B_\varphi = \frac{2I}{c\rho} \ , \tag{4.4}$$

the law of Biot and Savart.

Another simple problem of interest is that of an infinite plane sheet carrying a constant current I per unit length normal to the direction of the current. We wish to obtain the magnetic induction $\boldsymbol{B}$ at a point P not on the plane. We drop the perpendicular straight line from P to the plane and denote by O its foot. With O as origin we select a Cartesian coordinate system with the z-axis perpendicular to the plane and the x-axis parallel to the current. The current distribution is invariant with respect to reflection in the x-z plane. Thus $B_x = B_z = 0$. For z outside the plane $(\nabla \times \boldsymbol{B})_x = -\partial B_y/\partial z = 0$ so that B_y is independent of z both above and below the plane. But this proof fails on the plane itself. The invariance of the current distribution under reflection in the x-y plane shows that

$$B_y(+0) = -B_y(-0) \ ,$$

i.e., $B_y(z)$ has a jump discontinuity at $z = 0$. Application of Ampère's law gives

$$B_y(z) = \begin{cases} -\dfrac{2\pi I}{c} & \text{if} \quad z > 0 \\ \dfrac{2\pi I}{c} & \text{if} \quad z < 0 \end{cases} \ .$$

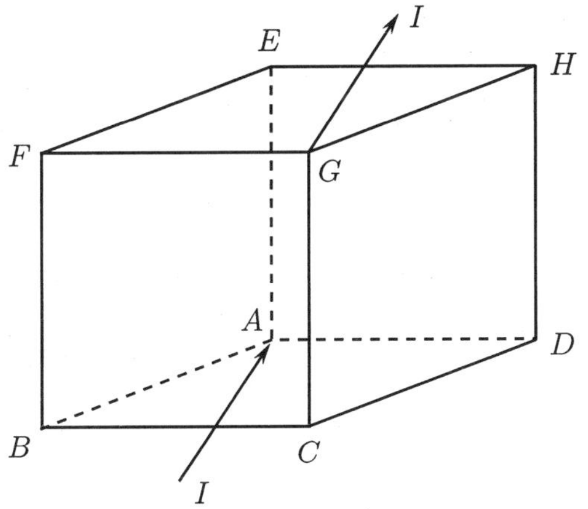

Figure 1.4

We can also deduce the existence of the discontinuity by integrating

$$-\frac{\partial B_y}{\partial z} = \frac{4\pi}{c} j_x$$

along z:

$$-(B_y(z))_{z>0} + (B_y(z))_{z<0} = \frac{4\pi}{c} I \ .$$

Finally we obtain the effective electrical resistance of 12 equal resistors of resistance R arranged in the form of a cube (see figure 1.4) when a potential difference V is applied between two opposite vertices (A and G in the figure). If I is the current, the effective resistance is

$$R_{eff} = \frac{V}{I} \ .$$

We wish to find R_{eff} in terms of R. We note that rotations by $2\pi/3$ about AG leave the arrangement invariant. Thus, the currents in AB, AD and AE are equal to $I/3$. Likewise the currents in CG, FH and HG are equal to $I/3$.

The system remains invariant upon reflection in the plane ACGE so that the currents in EF and EH are equal to $(I/6)$. Therefore,

$$V = \frac{1}{3} IR + \frac{1}{6} IR + \frac{1}{3} IR = \frac{5}{6} IR$$

and

$$R_{eff} = \frac{5R}{6} \ .$$

What is the effective electrical resistance if the potential source is applied between A and C? (Answer, $R_{eff} = 3R/4$).

Chapter 2

Symmetry in Classical Mechanics

2.1 Hamilton's principle and canonical transformations

We concentrate throughout on microscopic phenomena so that we shall deal only with conservative systems. A classical system with n degrees of freedom is characterized by n, suitably chosen, generalized coordinates $q_1, q_2, \ldots q_n$. The problem of mechanics is to find the trajectory $\{q_i(t)\}$ of the system in the configuration space $\{q_i\}$ for all times given the coordinates and their time derivatives $\{q_i(t_0)\}$ and $\{\dot{q}_i(t_0)\}$ at some initial time t_0. The equations satisfied by the coordinates are the second order equations

$$\frac{d}{dt}\frac{\partial L}{\partial \dot{q}_i} - \frac{\partial L}{\partial q_i} = 0 \ , \quad i = 1, 2, \ldots n \tag{1.1}$$

where L, called the Lagrangian, is a function of $q_1, q_2, \ldots q_n, \dot{q}_1, \dot{q}_2, \ldots \dot{q}_n$ and t given by

$$L = T - V \tag{1.2}$$

where T and V are the kinetic and potential energies of the system in terms of the q's and $\dot{q}$'s. The equations (1.1) are the familiar Lagrangian equations of motion. They can be deduced from a variational principle. Let us consider a motion from $q_i(t_0)$ to $q_i(t_1)$ $(i = 1, 2, \ldots n)$. The variational principle states that the action

$$S = \int_{t_0}^{t_1} L(q_1, q_2 \ldots q_n, \dot{q}_1, \dot{q}_2, \ldots \dot{q}_n, t)dt \tag{1.3}$$

is an extremum for the actual path with respect to variations $\delta q_i(t)$ such that the initial and final points are fixed, *i.e.*, such that $\delta q_i(t_0) = \delta q_i(t_1) = 0$. In fact,

$$\begin{aligned} \delta S &= \int_{t_0}^{t_1} dt \sum_{i=1}^{n} \left(\frac{\partial L}{\partial q_i}\delta q_i + \frac{\partial L}{\partial \dot{q}_i}\delta \dot{q}_i \right) \\ &= \sum_{i=1}^{n} \left[\frac{\partial L}{\partial \dot{q}_i}\delta q_i \right]_{t_0}^{t} + \int_{t_0}^{t_1} dt \sum_{i=1}^{n} \left(\frac{\partial L}{\partial q_i} - \frac{d}{dt}\frac{\partial L}{\partial \dot{q}_i} \right) \delta q_i \end{aligned}$$

The variation δS vanishes if Eqs. (1.1) are verified and conversely if $\delta S = 0$ for arbitrary variations $\delta q_i(t)$ subject to the restriction that the initial and final points are fixed, then Eqs. (1.1) are satisfied.

An important consequence of the variational principle is that it demonstrates the invariance of the form of the Lagrangian equations with respect to one-to-one point transformations $\{q_i\} \to \{Q_i\}$:

$$q_i = q_i(Q_1, Q_2, \ldots Q_n, t) \ , \ i = 1, 2, \ldots n \ .$$

In the new variables, as well as in the old, $\delta S = 0$ so that

$$\frac{d}{dt}\frac{\partial L}{\partial \dot{Q}_i} - \frac{\partial L}{\partial Q_i} = 0 \ , \ i = 1, 2, \ldots n \ . \tag{1.4}$$

The system of n second order equations (1.1) is equivalent to $2n$ first order equations most conveniently obtained by a Legendre transformation. We define a set of n generalized momenta p_i by

$$p_i = \frac{\partial L}{\partial \dot{q}_i} \ , \ i = 1, 2, \ldots n \tag{1.5}$$

and assume that Eqs. (1.5) can be solved for the $\dot{q}_i$ in terms of $q_1, q_2, \ldots q_n, p_1, p_2, \ldots p_n, t$. This is possible if the functional determinant whose elements are

$$\frac{\partial^2 L}{\partial \dot{q}_i \partial \dot{q}_j}$$

does not vanish. The Legendre transformation consists of forming the function

$$H(q_1, q_2, \ldots q_n, p_1, p_2, \ldots p_n, t) = \sum_{i=1} p_i \dot{q}_i(\boldsymbol{q}, \boldsymbol{p}, t) - L[\boldsymbol{q}, \dot{\boldsymbol{q}}(\boldsymbol{q}, \boldsymbol{p}, t), t] \tag{1.6}$$

which is called the Hamiltonian function. Here we denote by $\boldsymbol{q}$ the set of all $q_i (i = 1, 2, \ldots n)$. A similar convention is used for the generalized momenta.

Direct differentiation using Eq. (1.5) gives

$$\frac{\partial H}{\partial q_i} = -\frac{\partial L}{\partial q_i} \ , \ \frac{\partial H}{\partial p_i} = \dot{q}_i \ , \ \frac{\partial H}{\partial t} = -\frac{\partial L}{\partial t} \ . \tag{1.7}$$

So far, this is purely a change of variables; no physics is involved until we use the equations of motion (1.1) establishing that

$$\dot{p}_i = \frac{\partial L}{\partial q_i} \ . \tag{1.8}$$

Then, Eqs. (1.7) become

$$\dot{q}_i = \frac{\partial H}{\partial p_i} \ , \ \dot{p}_i = -\frac{\partial H}{\partial q_i} \ , \ \frac{\partial H}{\partial t} = -\frac{\partial L}{\partial t} \ , \tag{1.9}$$

the Hamiltonian equations of motion. The space $\{\boldsymbol{q},\boldsymbol{p}\} = \{q_1, q_2, \dots q_n, p_1 \dots p_n\}$ is called the phase space of the system, and the solution of Eqs. (1.9) for given initial conditions $\{\boldsymbol{q}(t_0), \boldsymbol{p}(t_0)\}$ is called the phase trajectory $\{\boldsymbol{q}(t), \boldsymbol{p}(t)\}$.

We saw that the Lagrangian equations are invariant with respect to point transformation $\{\boldsymbol{q}\} \to \{\boldsymbol{Q}\}$. However, we cannot expect the Hamiltonian equations to be invariant for arbitrary one-to-one transformations $\{\boldsymbol{q},\boldsymbol{p}\} \to \{\boldsymbol{Q},\boldsymbol{P}\}$ in phase space because, while the p_i's are related to the q's by Eq. (1.5), the corresponding relation between the P_i's and the Q_i's need not hold. To investigate under what conditions the transformation preserves the form of the Hamiltonian equations we cast the variational principle in the form

$$\delta \int_{t_0}^{t_1} \left(\sum_{i=1}^{n} p_i \dot{q}_i - H \right) dt = 0 \ . \tag{1.10}$$

This is equivalent to

$$\begin{aligned} \int_{t_0}^{t_1} \sum_{i=1}^{n} \Bigg(p_i \delta \dot{q}_i + \dot{q}_i \delta p_i & - \frac{\partial H}{\partial q_i} \delta q_i - \frac{\partial H}{\partial p_i} \delta p_i \Bigg) dt = \int_{t_0}^{t_1} \sum_{i=1}^{n} \Bigg[\delta p_i \left(\dot{q}_i - \frac{\partial H}{\partial p_i} \right) \\ & + \delta q_i \left(-\dot{p}_i - \frac{\partial H}{\partial q_i} \right) \Bigg] dt + \sum_{i=1}^{n} p_i \delta q_i \Bigg]_{t_0}^{t_1} = 0 \ . \end{aligned}$$

Thus, the Hamiltonian equations imply (1.10) and conversely, if (1.10) holds for arbitrary variations $\delta q_i, \delta p_i$ subject to the same restrictions as before on δq_i, then, Eqs. (1.9) follow. As a consequence we can state that a one-to-one transformation $\{\boldsymbol{q},\boldsymbol{p}\} \to \{\boldsymbol{Q},\boldsymbol{P}\}$ in phase space preserves the form of the Hamiltonian equations of motion if the differential forms

$$\sum_{i=1}^{n} p_i dq_i - H dt \ , \ \sum_{i=1}^{n} P_i dQ_i - K dt \tag{1.11}$$

differ by an exact differential dF. In fact, if this is verified

$$\delta \int_{t_0}^{t_1} \left(\sum_i p_i dq_i - H dt \right) = \delta \int_{t_0}^{t_1} \left(\sum_i P_i dQ_i - K dt \right)$$

so that

$$\dot{Q}_i = \frac{\partial K}{\partial P_i} \quad , \quad \dot{P}_i = -\frac{\partial K}{\partial Q_i} \tag{1.12}$$

Such transformations are called canonical. If $F(\boldsymbol{q}, \boldsymbol{Q}, t)$ is the function for which

$$\sum_{i=1}^{n} p_i dq_i - H dt = \sum_{i=1}^{n} P_i dQ_i - K dt + dF$$

we have

$$p_i = \frac{\partial F}{\partial q_i} \quad , \quad P_i = -\frac{\partial F}{\partial Q_i} \quad , \quad K = H + \frac{\partial F}{\partial t} \ . \tag{1.13}$$

The function F is the generating function of the transformation (1.13). The one-to-one transformation is obtained from the first two equations (1.13) by solving for the q's and p's as functions of the Q's and P's. Another form of generating function is

$$S(\boldsymbol{q}, \boldsymbol{P}, t) = F + \sum_{i=1}^{n} Q_i P_i \tag{1.14}$$

so that

$$\sum_{i=1}^{n} p_i dq_i - H dt = -\sum_{i=1}^{n} Q_i dP_i - K dt + dS \ . \tag{1.15}$$

The function S generates a transformation given by

$$p_i = \frac{\partial S}{\partial q_i} \quad , \quad Q_i = \frac{\partial S}{\partial P_i} \quad , \quad K = H + \frac{\partial S}{\partial t} \ . \tag{1.16}$$

The function

$$S_0 = \sum_i q_i P_i \tag{1.17}$$

generates the identity transformation

$$p_i = \frac{\partial S_0}{\partial q_i} = P_i \quad , \quad Q_i = \frac{\partial S_0}{\partial P_i} = q_i \quad , \quad K = H \ . \tag{1.18}$$

An infinitesimal transformation differs from the identity by small quantities proportional to an infinitesimal ϵ. Let

$$S(\boldsymbol{q}, \boldsymbol{P}, t) = S_0 + \epsilon S_1(\boldsymbol{q}, \boldsymbol{P}, t) \ . \tag{1.19}$$

Then

$$p_i = \frac{\partial S}{\partial q_i} = P_i + \epsilon \frac{\partial S_1}{\partial q_i} \quad , \quad Q_i = \frac{\partial S}{\partial P_i} = q_i + \epsilon \frac{\partial S_1}{\partial P_i} \quad , \quad K = H + \epsilon \frac{\partial S_1}{\partial t} \ . \tag{1.20}$$

To first order in ϵ

$$\delta q_i = Q_i - q_i \approx \epsilon \frac{\partial S_1}{\partial p_i} \quad , \quad \delta p_i = P_i - p_i = -\epsilon \frac{\partial S_1}{\partial q_i} \ . \tag{1.21}$$

Setting $\epsilon = \delta t$, a small increment in time, and $S_1 = H$ we obtain

$$\delta q_i \approx \delta t \frac{\partial H}{\partial p_i} \quad , \quad \delta p_i \approx -\delta t \frac{\partial H}{\partial q_i} \tag{1.22}$$

Equations (1.22) show that the evolution of the phase point of a conservative system can be regarded as the unfolding of a sequence of infinitesimal canonical transformations generated by the Hamiltonian function of the system.

2.2 Poisson brackets and constants of the motion

Let u and v be two functions of the coordinates, momenta and, possibly, of the time t. The Poisson bracket of u and v is defined by

$$[u, v] = \sum_{i=1}^{n} \left(\frac{\partial u}{\partial q_i} \frac{\partial v}{\partial p_i} - \frac{\partial u}{\partial p_i} \frac{\partial v}{\partial q_i} \right) = \sum_{i=1}^{n} \frac{\partial(u, v)}{\partial(q_i, p_i)} \ . \tag{2.1}$$

They satisfy the following identities

$$[u, v] = -[v, u] \ , \tag{2.2}$$

$$[u + v, w] = [u, w] + [v, w] \tag{2.3}$$

$$[uv, w] = u[v, w] + [u, w]v \tag{2.4}$$

and

$$[[u, v], w] + [[v, w], u] + [[w, u], v] = 0 \ . \tag{2.5}$$

Equations (2.2) –(2.4) are obvious; Eq. (2.5) is called the Jacobi identity. To prove Eq. (2.5) we regard $[u, w]$ as a differential operator D_u acting on w. To simplify the writing we use $q_i = x_i$ and $p_i = x_{i+n}$. In addition we write

$$\alpha_i = -\frac{\partial u}{\partial p_i} \quad , \quad \alpha_{i+n} = \frac{\partial u}{\partial q_i} \quad , \quad \beta_i = -\frac{\partial v}{\partial p_i} \quad , \quad \beta_{i+n} = \frac{\partial v}{\partial q_i} \ .$$

Then

$$D_u w = [u, w] = \sum_{\mu=1}^{2n} \alpha_\mu \frac{\partial w}{\partial x_\mu}$$

and a similar expression for $D_v w$,

$$D_v w = [v, w] = \sum_{\mu=1}^{2n} \beta_\mu \frac{\partial w}{\partial x_\mu} \ .$$

Now

$$\begin{aligned} D_u D_v w - D_v D_u w &= \sum_{\mu,\nu} \left[\alpha_\mu \frac{\partial}{\partial x_\mu} \left(\beta_\nu \frac{\partial w}{\partial x_\nu} \right) - \beta_\nu \frac{\partial}{\partial x_\nu} \left(\alpha_\mu \frac{\partial w}{\partial x_\mu} \right) \right] \\ &= \sum_{\mu,\nu} \left(\alpha_\mu \frac{\partial \beta_\nu}{\partial x_\mu} \frac{\partial w}{\partial x_\nu} - \beta_\nu \frac{\partial \alpha_\mu}{\partial x_\nu} \frac{\partial w}{\partial x_\mu} \right) \ . \end{aligned}$$

This shows that $(D_u D_v - D_v D_u)w$ is linear in the derivatives of w, the second derivatives of w having been canceled. Thus

$$[u,[v,w]] - [v,[u,w]] = \sum_i \left(A_i \frac{\partial w}{\partial p_i} + B_i \frac{\partial w}{\partial q_i} \right) .$$

The quantities A_i and B_i depend on u and v but not on w. Setting $w = p_i$ we find

$$A_i = \frac{\partial}{\partial q_i}[u,v]$$

while equating w to q_i gives

$$B_i = -\frac{\partial}{\partial p_i}[u,v] .$$

Thus

$$[u,[v,w]] - [v,[u,w]] = [[u,v],w]$$

which is equivalent to Eq. (2.5).

An important property of Poisson brackets is their invariance with respect to canonical transformation, *i.e.*, if $\{\boldsymbol{q},\boldsymbol{p}\} \to \{\boldsymbol{Q},\boldsymbol{P}\}$ is canonical, then

$$[u,v]_{Q,P} = [u,v]_{q,p} . \tag{2.6}$$

The time rate of change of a dynamical variable $u(\boldsymbol{q},\boldsymbol{p},t)$ is

$$\begin{aligned} \frac{du}{dt} &= \sum_i \left(\frac{\partial u}{\partial q_i}\dot{q}_i + \frac{\partial u}{\partial p_i}\dot{p}_i \right) + \frac{\partial u}{\partial t} = \sum_i \left(\frac{\partial u}{\partial q_i}\frac{\partial H}{\partial p_i} - \frac{\partial u}{\partial p_i}\frac{\partial H}{\partial q_i} \right) + \frac{\partial u}{\partial t} \\ &= [u,H] + \frac{\partial u}{\partial t} . \end{aligned} \tag{2.7}$$

Thus, a variable u is a constant of the motion if

$$[u,H] = -\frac{\partial u}{\partial t} . \tag{2.8}$$

If u does not depend explicitly on the time, the condition for u to be constant is

$$[u,H] = 0 . \tag{2.9}$$

If u and v are constants of the motion so is $[u,v]$ in addition to their sum and product. To prove that $[u,v]$ is a constant we use the Jacobi identity in the form

$$[[u,v],H] + [[v,H],u] + [[H,u],v] = 0$$

or

$$[[u,v],H] + \left[-\frac{\partial v}{\partial t}, u \right] + \left[\frac{\partial u}{\partial t}, v \right] = [[u,v],H] + \frac{\partial}{\partial t}[u,v] = 0. \qquad \text{Q.E.D.}$$

If H is independent of t, then, replacing u with H in Eq. (2.9) we deduce that H itself is a constant of the motion.

2.3 Symmetry and conservation laws

(i) Transformation of variables

Consider a canonical transformation $\{\boldsymbol{q}, \boldsymbol{p}\} \to \{\boldsymbol{Q}, \boldsymbol{P}\}$. To simplify the notation set $q_i = x_i, p_i = x_{i+n}$, $Q_i = X_i, P_i = X_{i+n}$. Let R be the transformation $\boldsymbol{x} \to \boldsymbol{X}$ which we assume to be one-to-one and onto. Thus R has an inverse R^{-1}. When a function $u(\boldsymbol{x})$ is transformed by R we define the transformed function as $P_R u(\boldsymbol{X})$ having the same numerical value at $\boldsymbol{X}$ as the function $u(\boldsymbol{x})$ had at the original point $\boldsymbol{x}$. Thus, the new function represents the same object as $u(\boldsymbol{x})$ represented in the original variables. Thus

$$P_R u(\boldsymbol{X}) = u(\boldsymbol{x}) = u(R^{-1}\boldsymbol{X}) \ . \tag{3.1}$$

We illustrate this definition by a simple example. Let the point (x, y) experience a rotation by an angle α to its new position (X, Y):

$$X = x\cos\alpha - y\sin\alpha \ , \quad Y = x\sin\alpha + y\cos\alpha$$

A function $u(x, y)$ is transformed into

$$P_R u(X, Y) = u(x, y) = u(X\cos\alpha + Y\sin\alpha, -X\sin\alpha + Y\cos\alpha) \ .$$

If $u = x^2 + y^2$, $P_R u(X, Y) = X^2 + Y^2$ but if $u = x^2 - y^2$,

$$P_R u(X, Y) = (X^2 - Y^2)\cos 2\alpha + 2XY\sin 2\alpha \ .$$

An infinitesimal transformation changes x_μ into $X_\mu = x_\mu + \delta x_\mu$ ($\mu = 1, 2, \ldots$ $\ldots 2n$). The change in $u(\boldsymbol{x})$ is

$$\begin{aligned} \delta u &= u(\boldsymbol{X}) - u(\boldsymbol{x}) = u(\boldsymbol{X}) - P_R u(\boldsymbol{X}) \\ &= \sum_{\mu=1}^{2n} \frac{\partial u}{\partial x_\mu}\delta x_\mu = \sum_{i=1}^{n}\left(\frac{\partial u}{\partial q_i}\delta q_i + \frac{\partial u}{\partial p_i}\delta p_i\right) \ . \end{aligned}$$

We suppose now that the canonical transformation has been generated by $S_0 + \epsilon S_1$ given in Eq. (1.19):

$$\delta u = \epsilon \sum_{i=1}^{n}\left(\frac{\partial u}{\partial q_i}\frac{\partial S_1}{\partial p_i} - \frac{\partial u}{\partial p_i}\frac{\partial S_1}{\partial q_i}\right) = \epsilon[u, S_1] \ . \tag{3.2}$$

Thus u remains invariant if $[u, S_1] = 0$.

(ii) Translational invariance and conservation of momentum

Consider N particles at positions $\boldsymbol{r}_i = (x_i, y_i, z_i)(i = 1, 2, \ldots N)$. Let $\boldsymbol{p}_i$ be the momentum of the ith particle and make an infinitesimal canonical transformation generated by $S_0 + \epsilon S_1$, with

$$S_1 = \sum_{i=1}^{N} p_{ix} \ . \tag{3.3}$$

This transformation gives

$$\delta x_i = \epsilon \ , \ \ \delta y_i = 0 \ , \ \ \delta z_i = 0 \ , \ \ \delta p_{ix} = \delta p_{iy} = \delta p_{iz} = 0 \ .$$

Thus S_1 generates a translation of all the particles by ϵ along the x-axis. According to Eq. (3.2)

$$\delta H = \epsilon \left[H, \sum_{i=1}^{N} p_{ix} \right] \ . \tag{3.4}$$

Thus, if H is invariant under translations parallel to the x-axis, $\delta H = 0$ and

$$\sum_{i=1}^{N} p_{ix} \ ,$$

the projection of the total momentum on the x-axis, is a constant of the motion.

(iii) Rotational invariance and conservation of angular momentum

Take

$$S_1 = L_z = \sum_{i=1}^{N} (x_i p_{iy} - y_i p_{ix}) \tag{3.5}$$

under the same conditions as in (ii). We obtain

$$\begin{aligned} \delta x_i &= -\epsilon y_i \ , \ \delta y_i &= \epsilon x_i \ , \ \delta z_i &= 0 \\ \delta p_{ix} &= -\epsilon p_{iy} \ , \ \delta p_{iy} &= \epsilon p_{ix} \ , \ \delta p_{iz} &= 0 \ . \end{aligned}$$

The changes in coordinates and momenta correspond to a rotation about the z-axis by ϵ of the positions and momenta of all particles in the system. Then,

$$\delta H = \epsilon [H, L_z] \ . \tag{3.6}$$

If H is invariant with respect to rotations about the z-axis, then L_z is a constant of the motion.

(iv) Transformation of a vector under rotations

Let $\boldsymbol{V}$ be a vector quantity. Under a rotation by an infinitesimal angle ϵ about the 3-axis of a Cartesian coordinate system, its components change according to

$$\delta V_1 = -\epsilon V_2 \ , \ \ \delta V_2 = \epsilon V_1 \ , \ \ \delta V_3 = 0 \ . \tag{3.7}$$

But, using Eq. (3.2) and the result of sub-section (iii)

$$\delta V_1 = \epsilon [V_1, L_3] \ , \ \ \delta V_2 = \epsilon [V_2, L_3] \ .$$

Thus

$$[V_1, L_3] = -V_2 \ , \ \ [V_2, L_3] = V_1 \ . \tag{3.8}$$

Furthermore

$$[V_3, L_3] = 0 \ . \tag{3.9}$$

In general we have

$$[L_i, V_j] = \epsilon_{ijk} V_k \ . \tag{3.10}$$

2.4 Time reversal symmetry in classical mechanics

Consider a particle of mass m in a potential field $\phi(\boldsymbol{r})$; its equation of motion is

$$m\frac{d^2\boldsymbol{r}}{dt^2} = -\nabla\phi(\boldsymbol{r}) \ . \tag{4.1}$$

if $\boldsymbol{r}(t)$ is a solution of the equation, then so is $\boldsymbol{r}(-t)$. The velocity at time t for the second solution is reversed with respect to that of the first. But the trajectories are the same. In fact, eliminating t from $x_1 = x_1(t), x_2 = x_2(t)$ we obtain a surface, a second surface, obtained eliminating t from $x_1(t)$ and $x_3(t)$ intersects the first in a curve, the trajectory of the particle. Clearly, the same surfaces are obtained eliminating t from $x_1(-t), x_2(-t)$ and $x_3(-t)$ and hence both solutions trace out the same trajectory but in opposite senses.

We define the time reversal on a mechanical system as the operation in which all positions are retained but all momenta are reversed. We may visualize the situation as follow: a particle starts its motion at time t_0; at time $t > t_0$ its momentum is abruptly reversed and is then allowed to proceed according to the laws of motion. Under the conditions of Eq. (4.1) the particle will then move along its initial trajectory passing through all of the points previously described but in reverse order with reversed momentum. At time $t_0 + 2(t - t_0) = 2t - t_0$ it passes through its initial position. Such a system is said to obey time reversal symmetry.

Not all physical systems obey time reversal symmetry. Certainly, if there is friction, time reversal fails. There are two cases of special interest and they both involve, like friction, velocity dependent forces. One such situation is provided by the motion of a charged particle in a magnetic field. For example, an electron moving in a plane normal to a constant, uniform magnetic field, $\boldsymbol{B}$, describes a circle. We imagine its motion to start at time t_0 from position P_0 and velocity $\boldsymbol{v}_0$ normal to $\boldsymbol{B}$. At $t > t_0$ the electron has reached the point P at which time its velocity is abruptly reversed. The subsequent path does not retrace the original circle but at time $2t - t_0$ reaches the point P' in the figure 2.1.

If, however, at the same time that $\boldsymbol{v}$ is reversed at P we also reverse the magnetic field, then time reversal symmetry is restored. This means that if we include in our system the source of the magnetic field as well as the electron itself,then the system as a whole obeys time reversal symmetry. In fact,the magnetic field arises from the electric currents within its source. If the velocities are reversed for every charged particle in the system so are the electric currents and the magnetic field. When studying atoms, molecules or nuclei the internal magnetic fields produced by the motion of the charged particles within the system will change sign under the operation of time reversal. Such systems will, therefore obey time reversal symmetry.

A second example of failure of time reversal symmetry occurs in descriptions of motion in a rotating frame of reference because of the presence of fictitious forces. In fact, the Coriolis force, being velocity-dependent, causes a breakdown of the invariance under $t \to -t$.

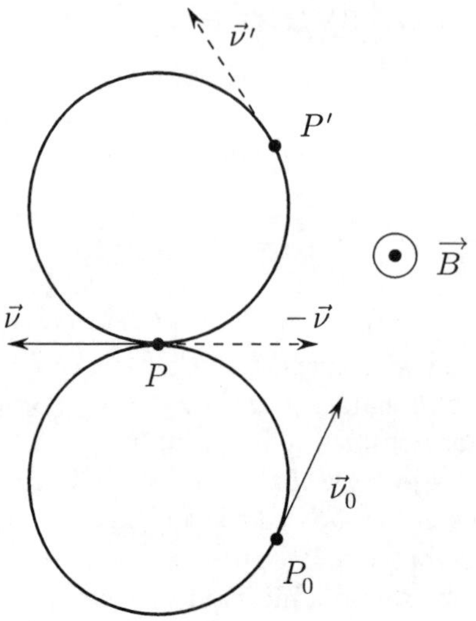

Figure 2.1

For a system described in phase space the operation of time reversal is effected by the transformation

$$q_i = Q_i \ , \ p_i = -P_i \ , \ t' = -t \ . \tag{4.2}$$

If H is an even function of the momenta, it is unchanged by this operation. The equations of motion are, then

$$\frac{d}{dt'}Q_i = \frac{\partial H}{\partial P_i} \ , \ \frac{d}{dt'}P_i = -\frac{\partial H}{\partial Q_i} \ . \tag{4.3}$$

Even though the transformation (4.2) preserves the canonical equations of motion in this case, it is, itself, not canonical. In fact

$$[Q_i, Q_j]_{q,p} = [P_i, P_j]_{q,p} = 0 \ , \ [Q_i, P_j]_{q,p} = -[q_i, p_j] = -\delta_{ij} \ . \tag{4.4}$$

In general

$$[u, v]_{Q,P} = -[u, v]_{q,p} \ . \tag{4.5}$$

Transformations which reverse the sign of the Poisson brackets are called anti-canonical.

Chapter 3

The Elements of Group Theory

3.1 Binary operations in abstract sets

Let $S = \{a, b, c, \ldots\}$ be a set of distinct objects. An internal binary operation in S is a mapping of $S \times S$ into S, *i.e.*, a function f is defined associating a unique element of S with every ordered pair (a, b) of elements of S. We call the element associated to (a, b) their product and denote it by ab.

Let R be the set of real numbers. Addition and multiplication are examples of internal binary operations in R.

An internal binary operation is said to be commutative if $ab = ba$ for every pair of elements $a, b \in S$. It is said to be associative if $a(bc) = (ab)c$ for any three elements a, b and c of S. In such cases we can dispense with the use of parentheses and simply write abc. The powers of an element a are defined by $a^2 = aa, a^3 = aaa, \ldots$.

If an element $e_R \in S$ exists such that $ae_R = a$ for all $a \in S$ we call it a right identity. Similarly we define a left identity, e_L, as an element, if it exists in S, such that $e_La = a$ for all $a \in S$. If both e_R and e_L exist they are equal since

$$e_Le_R = e_L \quad , \quad e_Le_R = e_R \ .$$

Furthermore this element, called e from now on, is unique being equal to any right or to any left identity.

If S contains an identity e and, for a given $a \in S$, an element a'_R exists such that

$$aa'_R = e$$

we say that a'_R is a right inverse of a. Similarly we define a left inverse as an element a'_L, if it exists in S, such that

$$a'_La = e \ .$$

If both a'_R and a'_L exist and the operation $S \times S \to S$ is associative, then $a'_R = a'_L$ and is unique for given a. In fact

$$a'_R = ea'_R = (a'_L a)a'_R = a'_L(aa'_R) = a'_L e = a'_L \ .$$

When this is the case we denote the element $a'_R = a'_L$ by a^{-1}, call it the inverse of a and write

$$a^{-1}a = aa^{-1} = e \ .$$

In a set S with an associative internal binary operation and an identity element e, if a and b in S possess inverses a^{-1} and b^{-1}, then ab also has an inverse and

$$(ab)^{-1} = b^{-1}a^{-1} \ .$$

The proof consists of verifying that $(ab)(b^{-1}a^{-1}) = (b^{-1}a^{-1})(ab) = e$:

$$\begin{aligned} (ab)(b^{-1}a^{-1}) &= a(bb^{-1})a^{-1} = aa^{-1} = e \\ (b^{-1}a^{-1})(ab) &= b^{-1}(a^{-1}a)b = b^{-1}b = e \ . \end{aligned}$$

Let us suppose that a relation, denoted by the symbol $\sim$, is established between elements of a set $S = \{a, b, c, \ldots\}$ such that
(i) $a \sim a$ for all $a \in S$ (reflexivity),
(ii) $a \sim b \Rightarrow b \sim a$ for any pair $a, b \in S$ (symmetry), and
(iii) $a \sim b$ and $b \sim c \Rightarrow a \sim c$ for $a, b, c \in S$ (transitivity).

A relation satisfying conditions (i),(ii) and (iii) is called an equivalence relation. If between the elements of a set S an equivalence relation is established, the set S is divided into disjoint sub-sets X_i of equivalent elements such that the union of the X_i is S itself. Thus

$$\begin{aligned} X_i \cap X_j &= \phi \ \text{if} \ i \neq j \quad (\phi \text{ is the empty set}) \\ \bigcup_i X_i &= S \ . \end{aligned}$$

The sets X_i are called the equivalence classes under the relation $\sim$.

A simple example of an equivalence relation in the set of integers

$$Z = \{\cdots - 3, -2, -1, 0, 1, 2, 3, \cdots\}$$

is congruence modulo p with p a positive integer:

Definition: m and $n \in Z$ are congruent modulo p if $m - n$ is divisible by p. We write

$$m \equiv n \ (\text{mod.} p)$$

Taking $p = 3$, the equivalence classes are

$$\{\cdots - 6, -3, 0, 3, 6 \cdot\} \ , \ \{\cdots - 5, -2, 1, 4, \cdots\} \ , \ \{\cdots - 4, -1, 2, 5, \cdots\} \ .$$

This definition appears so simple that we do not expect it to lead to interesting results, but that is not the case. It is easy to prove that

(i) $m \equiv n$ (mod. p) and $r \equiv s$ (mod. p) imply that

$$\begin{aligned} m+r &\equiv n+s \ (\text{mod.}p) \\ m-r &\equiv n-s \ (\text{mod.}p) \\ mr &\equiv ns \ (\text{mod.}p) \end{aligned}$$

Consider now the integer

$$N = a_n 10^n + a_{n-1}10^{n-1} + \cdots + a_1 10 + a_0$$

where $a_0, a_1, a_2 \ldots a_n$ are integers ranging from 0 to 9. We note that

$$10 \equiv 1 \ (\text{mod. } 3), 10^2 \equiv 1 \ (\text{mod.}3), \ldots \ .$$

Thus

$$N \equiv a_0 + a_1 + \cdots + a_n \ (\text{mod. } 3) \ .$$

This is just the test of divisibility of an integer by 3. The rule of divisibility of 11 is proved by noting that

$$10 \equiv -1 \ , \ 10^2 \equiv 1 \ , \ 10^3 \equiv -1 \ldots \ (\text{mod. } 11) \ .$$

Hence

$$N \equiv a_0 - a_1 + a_2 \mp \cdots + (-1)^n a_n \ (\text{mod.}11) \ .$$

For example 9867 is a multiple of 11 since $7 - 6 + 8 - 9 = 0$.

3.2 The concept of group

Definition: A set $G = \{a, b, c, \ldots\}$ is said to be a group under an associative internal binary operation if it contains an identity element and the inverse of each of its elements.

Hence, for G to be a group the following conditions must be verified

(i) $a, b \in G \Rightarrow ab \in G$ (closure)

(ii) $a, b, c \in G \Rightarrow a(bc) = (ab)c$

(iii) $\exists e \in G$ such that $ae = ea = a$ for all $a \in G$

(iv) $\forall a \in G$, $a^{-1} \in G$ exists such that $aa^{-1} = a^{-1}a = e$.

A group with a commutative binary operation, is said to be Abelian. Examples of groups are:

a. R under addition; the identity element is zero and the inverse of a is $-a$

b. R is not a group under multiplication because zero has no inverse. However, if we remove 0, then $R - \{0\}$ is a group. The identity is 1 and the inverse of a is $1/a$.

c. The set $\{1, -1\}$ is a group under multiplication.

d. Define $\omega = \exp(2\pi i/n)$ where n is a positive integer. The set $\{1, \omega, \omega^2, \ldots \ldots \omega^{n-1}\}$ is a group under multiplication. This group is similar (more about this later) to the group of rotations about a fixed axis by multiples of $2\pi/n$.

e. The permutation groups:

The symmetric group S_n is the set of permutations of n objects. Labeling the objects by $1, 2, \ldots n$, a permutation P consists of reordering the objects so that the one labeled i is replaced by one labeled p_i:

$$P = \begin{pmatrix} 1 & 2 & 3 & \cdots & n \\ p_1 & p_2 & p_3 & \cdots & p_n \end{pmatrix}.$$

The labels $p_1, p_2, \ldots p_n$ must all be distinct; the order of the columns is irrelevant for the specification of P. Multiplication of permutations is defined as simple successive application of them. For example the product of

$$P_1 = \begin{pmatrix} 1 & 2 & 3 & 4 \\ 3 & 4 & 2 & 1 \end{pmatrix}$$

and

$$P_2 = \begin{pmatrix} 1 & 2 & 3 & 4 \\ 1 & 3 & 4 & 2 \end{pmatrix}$$

is

$$P_1P_2 = \begin{pmatrix} 1 & 2 & 3 & 4 \\ 3 & 4 & 2 & 1 \end{pmatrix}\begin{pmatrix} 1 & 2 & 3 & 4 \\ 1 & 3 & 4 & 2 \end{pmatrix} = \begin{pmatrix} 1 & 2 & 3 & 4 \\ 3 & 2 & 1 & 4 \end{pmatrix}.$$

The product of P_2 and P_1 is

$$P_2P_1 = \begin{pmatrix} 1 & 2 & 3 & 4 \\ 1 & 3 & 4 & 2 \end{pmatrix}\begin{pmatrix} 1 & 2 & 3 & 4 \\ 3 & 4 & 2 & 1 \end{pmatrix} = \begin{pmatrix} 1 & 2 & 3 & 4 \\ 4 & 2 & 3 & 1 \end{pmatrix}.$$

Thus $P_1P_2 \neq P_2P_1$ in general. These operations form a group; the identity is

$$P_e = \begin{pmatrix} 1 & 2 & \cdots & n \\ 1 & 2 & \cdots & n \end{pmatrix}$$

and the inverse of

$$P = \begin{pmatrix} 1 & 2 & \cdots & n \\ p_1 & p_2 & \cdots & p_n \end{pmatrix}$$

is

$$P^{-1} = \begin{pmatrix} p_1 & p_2 & \cdots & p_n \\ 1 & 2 & \cdots & n \end{pmatrix}.$$

The number of elements in S_n is $n!$

Permutations can be expressed in the form of cycles. For example, instead of P_1 we consider the sequence

$$1 \to 3 \to 2 \to 4 \to 1 .$$

Thus we write

$$P_1 = (1324)$$

This is a cycle and P_1 consists of a single cycle. The permutation

$$\begin{pmatrix} 1 & 2 & 3 & 4 \\ 2 & 1 & 4 & 3 \end{pmatrix}$$

consists of two cycles $1 \to 2$, $3 \to 4$. Thus we write it in the form

$$(12)(34) .$$

We note that cycles can be written in any order and that any permutation can be expressed as the product of disjoint cycles. The simplest cycle is one in which a particular object is not changed, *e.g.*, P_2 above is

$$P_2 = (1)(234) .$$

The next simple cycle is an interchange of two objects. Any cycle can be obtained from a series of interchanges. For example

$$P_1 = (1324) = (13)(32)(24)$$

can be obtained by three interchanges. The number of interchanges associated with a given permutation has always the same parity. An even permutation is one that can be achieved by an even number of interchanges. An odd permutation is one which requires an odd number of interchanges.

f. As stated before $O(3)$ and $SO(3)$ are groups.

If the number of elements in a group G is finite we say G is a finite group. Otherwise it is infinite. The number of elements in a finite group is called its order. Thus S_n is of order $n!$ while $O(3)$ and $SO(3)$ are infinite groups.

3.3 Subgroups

Definition. Let $G = \{e, a, b, \ldots\}$ be a group under an operation $(a, b) \to ab$. A non-empty subset H of G is a subgroup of G under the same binary operation if

(i) $a \in H$ and $b \in H \Rightarrow ab \in H$, and
(ii) $a \in H \Rightarrow a^{-1} \in H$.

From (i) and (ii) it follows that $e \in H$. Thus H is, itself, a group. Clearly G itself and $\{e\}$ are subgroups of G. Every other subgroup of G is called a proper subgroup.

Example: $\{1, i, -1, -i\}$ is a group under multiplication. Its only proper subgroup is $\{1, -1\}$.

Consider the group Z of integers under addition and one of it proper subgroups H. There exists an integer $p > 1$ such that

$$H = \{\cdots - 3p, -2p, -p, 0, p, 2p, 3p, \cdots\} = pZ$$

Since H is a proper subgroup of Z it contains the identify 0 and, at least, one other element, say n. H must also contain $-n$ so that it contains positive numbers. Let p be the smallest positive integer in H. Then H contain $0, \pm p, \pm 2p, \ldots$; $p \neq 1$ because otherwise $H (= Z)$ would not be a proper subgroup. We must now show that all elements of H have been listed. Assume q, not a multiple of p is in H. Then, dividing q by p we have $q = kp + r$ where the residue r lies between 1 and $p - 1$. But the $q - kp = r$ would be an element of H in contradiction to the assumption that p is the smallest positive element of H.

Bezout's theorem. If a and b are non-zero elements of Z, then u and v in Z can be found so that

$$ua + vb = d$$

where d is the greatest common divisor of a and b.

To prove this theorem we consider the set $\{ua + vb\}$ where, for fixed a and b, u and v take all possible values in Z. This set is a non-empty subgroup H of Z (it could be Z itself). Thus, a positive integer n exists such that

$$H = nZ \ .$$

Now, if d divides a and b it divides every element of H and, in particular, its smallest positive element n. Conversely, every divisor of n divides every element of H and, in particular, a and b. Thus $n = d$.

Corollary. If a and b are relatively prime, then u and $v \in Z$ exist such that

$$ua + vb = 1 \ .$$

In this case $d = 1$ and $H \equiv Z$.

3.4 Cayley's rearrangement theorem. Multiplication Tables

Let G be a group $\{e, a, b, \ldots\}$ and m one of its elements and consider the mapping $a \to ma$ or $G \to mG$ where mG is the set of all products ma as a ranges over G. This mapping is onto and one-to-one (*i.e.*, bijective) In fact:

(i) $\{ma\}$ for fixed m contains all elements of G since $b \in G$ is the image of $m^{-1}b$ under this mapping,

(ii) $a \to ma$ is one-to-one because $ma = mb \Rightarrow a = b$. Different elements of G are mapped into different elements of mG.

Clearly a similar result holds for the mapping $a \to am$. Thus, Cayley's theorem is that mG and Gm are simple rearrangements of the elements of G.

Since each element m of a group of order n generates a permutation of its elements, Cayley's theorem can be stated that all groups of order n are subgroups of the symmetric group S_n.

The products of elements of a group $G = \{e, a, b, c, \ldots\}$ can be displayed in a multiplication table as shown below. Each row and each column contains all elements of G without repetition.

	e	a	b	c	...
e	e	a	b	c	...
a	a	a^2	ab	ac	...
b	b	ba	b^2	bc	...
c	c	ca	cb	c^2	...
⋮	⋮	⋮	⋮	⋮	

If the order of the group is either 2 or 3 only one form of the multiplication table is possible. The multiplication tables for groups of order 2 or 3 are

	e	a
e	e	a
a	a	e

	e	a	b
e	e	a	b
a	a	b	e
b	b	e	a

The group of order two is generated by the single element $a(a^2 = e)$. A realization of this group is $\{1, -1\}$ under multiplication. Another is the group consisting of the identity element E of $SO(3)$ and a rotation by π about a fixed axis ($C_2^2 = E$). The group of order three is generated by the element a: $b = a^2$ and $e = a^3$. An example is $\{1, \omega, \omega^2\}$ under multiplication where $\omega = \exp(2\pi i/3)$.

If $a \neq e$ is an element of a finite group, the set of distinct powers of a: $a, a^2, a^3, \ldots$ must be finite, *i.e.*, integers m and k ($m > k > 0$ without loss of generality) exist so that

$$a^m = a^k \ .$$

Thus positive numbers $n = m - k$ exist so that $a^n = e$. Let n be the smallest such number. It is called the order of a. The set

$$\{e, a, a^2, \ldots a^{n-1}\}$$

is then a subgroup.

A group of the form

$$H = \{e, a, a^2, \ldots a^{n-1}\}$$

with $a^n = e$ is called the cyclic group of order n assuming n is the smallest positive integer for which $a^n = e$. H is clearly Abelian.

3.5 Cosets

Let G be a group and H one of its proper subgroups. We define a relation between a and b of G as

$$a \sim b \text{ if } a^{-1}b \in H \ .$$

This is an equivalence relation. In fact

(i) $a \sim a$ since $a^{-1}a = e \in H$,

(ii) $a \sim b \Rightarrow b \sim a$ since $a^{-1}b \in H$ implies that $(a^{-1}b)^{-1} = b^{-1}a \in H$, and

(iii) $a \sim b$ and $b \sim c \Rightarrow a \sim c$ because $a^{-1}b \in H$ and $b^{-1}c \in H \Rightarrow a^{-1}c \in H$.

This relation, like every equivalence relation, classifies the elements of G into non-overlapping classes. For each element $a \in G$ we form

$$C_a = aH$$

and call it the left coset of a. Except for $C_e \equiv H$ the left cosets are not groups. If $a \in H$, $C_a = C_e = H$ by virtue of Cayley's theorem. If a is not in H, C_a does not contain the identity and is therefore not a group. We now prove that the left cosets are the equivalence classes of the relation $\sim$ just defined. In fact, if c is an element of C_a, then $a^{-1}c \in H$ and $a \sim c$. If C_a and C_b have an element in common they are identical because if c is such an element then all the elements of C_a as well as all the elements of C_b are equivalent to c.

The mapping

$$H \to aH$$

is one-to-one. The element c of C_a is the image of $a^{-1}c \in H$ under this mapping. Two distinct elements b and c of C_a are the images of $a^{-1}b$ and $a^{-1}c$, respectively, two distinct elements of H.

In an identical manner the equivalence relation,

$$a \sim b \text{ if } ba^{-1} \in H$$

leads to the definition of the right cosets of H: $\{Ha\}$.

Lagrange's theorem

If G is a finite group of order g and H one of its proper subgroups of order h, then h is a divisor of g.

Let n be the number of left cosets of H. Then $nh = g$. This proves the theorem.

Incidentally, the numbers of left and right cosets of H must be equal because Ha^{-1} contains the inverses of the elements of aH. As a corollary we can now state that groups of prime order cannot have proper subgroups. Furthermore, every element of such a group other than e must be of order equal to the order of the group. Hence, all groups of prime order are cyclic. If p is prime, all groups of order p are, then, of the form

$$\{e, a, a^2, \dots a^{p-1}\}$$

with $a^p = e$.

We are now in a position to find the possible multiplication tables of groups of order four. We denote the elements by $\{e, a, b, c\}$. Two cases need to be considered:

(i) There is only one element, say a, of order 2. Then the multiplication table is

	e	a	b	c
e	e	a	b	c
a	a	e	c	b
b	b	c	a	e
c	c	b	e	a

In this group $c = ab = ba$ and the group is generated by b only

$$b, b^2 = a, b^3 = c, b^4 = e \ .$$

It is the cyclic group of order four. It is Abelian.

(ii) There is more than one element of order 2. Let $a^2 = e$ and $b^2 = e$. The multiplication table is

	e	a	b	c
e	e	a	b	c
a	a	e	c	b
b	b	c	e	a
c	c	b	a	e

We note that $c^2 = e$ also. This group is Abelian and it is called the Vier-gruppe.

No other cases need to be considered. There can be no element of order 3 because if would generate a subgroup of order 3 which would be in contradiction with Lagrange's theorem. Thus, there are essentially two distinct groups of order four.

3.6 Homomorphisms and isomorphisms

A group G is said to be homomorphic to a group H if there exists a mapping of H onto G:

$$a \in H \to h(a) \in G$$

such that for any two elements a and b of H

$$h(a)h(b) = h(ab) \ ,$$

We emphasize that we require that the set $\{h(a)\}$ for all a in H spans G. In other words if $H = \{e, a, b, c, \ldots\}$ all elements of G appear in $\{h(e), h(a), h(b), \ldots\}$ but they need not, necessarily, be distinct. In other words, the image of H under h, $h(H)$ must be identical to G.

The simplest homomorphism is that between $H = \{e, a, b, \ldots\}$ and $G = \{e_G\}$ with $h(a) = e_G$ for all $a \in H$.

The group $\{1, -1\}$ under multiplication is homomorphic to Z under addition with

$$h(a) = \begin{cases} 1 & \text{if } a \text{ is even} \\ -1 & \text{if } a \text{ is odd} \end{cases}$$

Theorem. If G is homomorphic to H under the mapping $a \to h(a)$, then $h(e)$ is the unit element in G and $h(a^{-1}) = h^{-1}(a)$.
Proof:

(i) $h(a)h(e) = h(ae) = h(a)$; $h(e)h(a) = h(ea) = h(a)$

(ii) $h(a)h(a^{-1}) = h(aa^{-1}) = h(e); h(a^{-1})h(a) = h(a^{-1}) = h(e)$

The kernel K of a homomorphism h of H onto G is the set of all elements of H whose images under h are the identity in G. K is a subgroup of H. In fact

(i) $a \in K$ and $b \in K \Rightarrow h(a)$ and $h(b)$ are equal to e_G. Thus $h(a)h(b) = h(ab) = e_G$. Hence $ab \in K$.

(ii) $a \in K \Rightarrow h(a) = e_G$. But $h(a^{-1}) = h^{-1}(a) = e_G$ so that $a^{-1} \in K$.

Two groups G an H are said to be isomorphic if there is a one-to-one homomorphism between them. Then if $a \in H \to h(a) \in G$, h has an inverse h^{-1}.

3.7 Conjugate classes

Definition. Let $H = \{e, a, b, c, \ldots\}$ be a group. The element b is a conjugate of a if there exists an element u of H such that

$$b = u^{-1}au \ .$$

The relation of conjugation is an equivalence relation:

(i) $a \sim a$ because $e^{-1}ae = a$.

(ii) $a \sim b$ implies $b \sim a$ because $b = u^{-1}au \Rightarrow a = (u^{-1})^{-1}bu^{-1}$ and $u^{-1} \in H$

(iii) $a \sim b$ and $b \sim c$ then $a \sim c$ because the existence in H of u and v such that

$$b = u^{-1}au$$

and

$$c = v^{-1}bv$$

shows that

$$c = (uv)^{-1}a(uv)$$

and $uv \in H$.

Conjugation allows us to classify the elements of a group into non-overlapping *conjugate classes.* Often we shall use the term class when referring to a conjugate class to shorten the discourse when there is no danger of confusion. The class of elements conjugate to a is denoted by C_a. Since for every $u \in H$, $u^{-1}eu = e$, the unit element is in a class by itself, C_e. All other conjugate classes are not groups since they do not contain the identity.

If $u \in H$, the mapping

$$a \to u^{-1}au \ , \quad C_a \to u^{-1}C_a u$$

is a rearrangement of the class C_a. In fact $u^{-1}C_a u$ contains only elements conjugate to a and if a and b belong to C_a, $u^{-1}au$ and $u^{-1}bu$ are distinct if $a \neq b$.

Thus, for all $u \in H$,

$$u^{-1}C_a u = C_a \ .$$

An invariant subgroup or normal divisor of a group H is a subgroup of H which contains entire classes. Thus, if N is a normal divisor of H, then, for all $u \in H$,

$$u^{-1}Nu = N \ .$$

Thus

$$Nu = uN \ ,$$

in words, the left and right cosets of an element of a group with respect to an invariant subgroup are equal.

3.8 Class multiplication

Consider two subsets $S_1 = \{u, v, \ldots\}$ and $S_2 = \{a, b, \ldots\}$ of elements of a group H. We define two forms of products

$$S_1 S_2 = \{ua, ub, \ldots va, vb, \ldots\}$$

and

$$[S_1 S_2] = [ua, ub, \ldots va, vb, \ldots] \ .$$

In $S_1 S_2$ repeated elements are deleted while in $[S_1 S_2]$ they are kept.

Let $\mathcal{F}$ be the set of all cosets of an invariant subgroup of H,

$$\mathcal{F} = \{C_u\}$$

where

$$C_u = Nu \ .$$

Now

$$C_u C_v = C_{uv}$$

since

$$C_u C_v = (Nu)(Nv) = (Nu)(Nu^{-1}uv) = N(uNu^{-1})uv = NNuv = Nuv = C_{uv}$$

Thus under this operation $\mathcal{F}$ is a group called the factor group of H with respect to N. We denote it by

$$\mathcal{F} = H/N \ .$$

The kernel K of a homomorphism h of H onto G is an invariant subgroup of H since if $a \in K$, then

$$h(u^{-1}au) = h(u^{-1})h(a)h(u) = h(u^{-1})h(u) = h(e) = e_G$$

for all $u \in H$. Thus

$$u^{-1}Ku = K \ .$$

Let $\Re$ be a set of elements of H, some or all of which may be repeated any number of times. If $\Re$ consists of entire classes then, for all $u \in H$

$$u^{-1}\Re u = \Re \ .$$

The converse of this statement is true, *i.e.*, if $\Re$ is invariant under conjugation by all elements of H, then $\Re$ consists of entire classes. To prove this let $\Re'$ be the largest subset of $\Re$ containing entire classes. Then remove from $\Re$ the set $\Re'$. The residue, $\Re''$ remains invariant under conjugation but if there exists an element $a \in H$ in $\Re''$, an element conjugate to a other than itself is not in $\Re''$. But since $u^{-1}\Re''u = \Re''$ that element is in $\Re''$. Thus, the assumption of the presence of a in $\Re''$ leads to a contradiction and $\Re''$ is empty. We conclude that invariance of a collection of elements of a group under conjugation by all elements of the group, is a necessary and sufficient condition for that collection to consist of entire classes.

Let $H = \{e, a, b, \ldots\}$ be a finite group of order h containing N_c conjugate classes:

$$C_1 = \{e\} \ , \ C_2, \ldots C_{Nc} \ .$$

The number of elements in class C_μ is denoted by n_μ. Clearly

$$\sum_{\mu=1}^{Nc} n_\mu = h \ .$$

Consider the product $[C_\mu C_\nu]$ of two classes. Take an element $u \in H$ and form

$$u^{-1}[C_\mu C_\nu]u = [u^{-1}abu]$$

where $a \in C_\mu$ and $b \in C_\nu$. We can express

$$[u^{-1}abu] \ .$$

in the form

$$[u^{-1}auu^{-1}bu] \equiv [C_\mu C_\nu] \ .$$

Thus $[C_\mu C_\nu]$ consists of entire classes. Suppose the class C_λ appears in $[C_\mu C_\nu]$ $n_{\mu\nu\lambda}$ times ($n_{\mu\nu\lambda} = 0, 1, \ldots$). We write symbolically

$$[C_\mu C_\nu] = \sum_{\lambda=1}^{Nc} n_{\mu\nu\lambda} C_\lambda \ .$$

The products $[C_\mu C_\nu]$ and $[C_\nu C_\mu]$ are identical since

$$[C_\mu C_\nu] = [\{ab\}] = [\{aba^{-1}a\}] = [C_\nu C_\mu] \ .$$

This means that

$$n_{\mu\nu\lambda} = n_{\nu\mu\lambda} \ .$$

Since

$$\begin{aligned} [C_1 C_\nu] &= C_\nu \\ n_{1\nu\lambda} = n_{\nu 1\lambda} &= \delta_{\nu\lambda} \ . \end{aligned}$$

3.9 Direct product of two groups

A group H is said to be the direct product of two of its subgroups H_a and H_b if:

(i) Every element of H_a commutes with every element of H_b, and

(ii) Every element of H can be expressed in a unique way as a product ab of an element a of H_a and an element b of H_b. We denote this direct product by the notation

$$H = H_a \times H_b \ .$$

Both H_a and H_b are invariant subgroups of H. In fact, for every u of H

$$u^{-1} H_a u = H_a$$

since, if u is in H_a, $u^{-1}H_a u$ is a rearrangement of H_a. If u is in H but no in H_a we can write $u = u_a u_b$ with $u_a \in H_a, u_b \in H_b$ and we have

$$\begin{aligned} (u_a u_b)^{-1} H_a u_a u_b &= u_b^{-1}(u_a^{-1} H_a u_a) u_b \\ &= u_b^{-1} H_a u_b \\ &= H_a \ . \end{aligned}$$

The last equality follows from the fact that u_b commutes with all elements of H_a.

Clearly $O(3)$ is the direct product of $SO(3)$ and the group $C_i = \{E, i\}$:

$$O(3) = SO(3) \times C_i \ .$$

Chapter 4

The Linear Groups

4.1 Vector spaces

Definition. A vector space over a field F is a collection, $V = \{\xi, \eta \ldots\}$, of objects called vectors, among which we define operations of addition, $V \times V \to V$, and multiplication $F \times V \to V$. Addition associates a vector $\xi + \eta$ to every pair of vectors ξ and η of V. The second operation, scalar multiplication associates a vector $a\xi$ to a vector $\xi \in V$ and an element a of F. The following postulates are required to call V a vector space over the field F:

(i) V is an Abelian group under vector addition,
(ii) $(a+b)\xi = a\xi + b\xi \quad , \quad \xi \in V; a, b \in F$,
(iii) $a(\xi + \eta) = a\xi + a\eta \quad , \quad \xi, \eta \in V; a \in F$,
(iv) $a(b\xi) = (ab)\xi \quad , \quad \xi \in V; a, b \in F$, and
(v) $1\xi = \xi$ where 1 is the unit in F and $\xi \in V$.

The elements of F are called scalars. If $F = R$, the field of real numbers, then V is called a real vector space. If $F = C$, the field of complex numbers, it is a complex vector space.

From axioms (i)-(v) one can deduce a large number of results. The identity element in (i) is the zero vector denoted by 0 which is not to be confused with the zero element of the field which is written with the same symbol. The inverse of ξ is $(-1)\xi$ written as $-\xi$. We define $\eta - \xi = \eta + (-1)\xi$. We can show that $0\xi = 0$ and that if $a\xi = 0$, then $a = 0$ or $\xi = 0$.

The set of vectors $\{\xi_1, \xi_2, \ldots \xi_n\}$ is said to be linearly independent if

$$c_1\xi_1 + c_2\xi_2 + \cdots + c_n\xi_n = 0 \tag{1.1}$$

only if $c_1 = c_2 = \cdots = c_n = 0$. If $\xi_1, \xi_2, \cdots \xi_n$ are not linearly independent, then we say that they are linearly dependent, *i.e.*, in such a case Eq. (1.1) is verified by a set of $c_1, c_2, \cdots c_n$ not all of which are zero. Any set containing the zero vector is linearly dependent.

A vector space V_n is said to be of dimension n if it has n linearly independent vectors but any set of $n+1$ vectors is linearly dependent. If $\{\epsilon_1, \epsilon_2 \ldots \epsilon_n\}$ is a set of linearly independent vectors in V_n for any vector ξ of V_n, numbers $c_1, c_2 \dots c_n, c$, not all zero can be found such that

$$c_1\epsilon_1 + c_2\epsilon_2 + \cdots + c_n\epsilon_n + c\xi = 0 \ .$$

c cannot be zero because, then, $c_1\epsilon_1 + c_2\epsilon_2 + \cdots + c_n\epsilon_n = 0$ with not all c_i equal to zero, contrary to the hypothesis that the ϵ_i are linearly independent. Dividing by $-c$ we obtain

$$\xi = x_1\epsilon_1 + x_2\epsilon_2 + \cdots + x_n\epsilon_n \ ; \ x_i = -c_i/c \ . \tag{1.2}$$

We say that ξ is a linear combination of $\epsilon_1, \epsilon_2, \dots \epsilon_n$. It is easy to show that the coefficients x_i are uniquely determined by ξ. Then we say that $\{\epsilon_1, \epsilon_2, \dots \epsilon_n\}$ is a basis of V_n. All bases in V_n have n basis vectors (reader, prove this statement).

A vector space which is not of finite dimension is said to be of infinite dimension. An infinite set of vectors in such a space is said to be linearly independent if every one of its finite subsets is composed of linearly independent vectors.

An example of a vector space of dimension n is the set of all n-tuples of complex numbers such as $\xi = (x_1, x_2, \dots x_n)$, $\eta = (y_1, y_2, \cdots y_n)$, ... where we define

$$\xi + \eta = (x_1 + y_1, x_2 + y_2, \cdots x_n + y_n)$$

and

$$a\xi = (ax_1, ax_2, \dots ax_n) \ .$$

The vectors

$$\begin{aligned} \epsilon_1 &= (1, 0 \dots 0) \\ \epsilon_2 &= (0, 1, \dots 0) \\ &\vdots \\ \epsilon_n &= (0, 0, \dots 1) \end{aligned}$$

form a basis for this space.

4.2 Inner product spaces

Definition. An inner product space is a vector space S in which we define a mapping $S \times S \rightarrow C$ associating every ordered pair of elements ξ, η of S with a complex number $< \xi|\eta >$ satisfying the following conditions:

(i) $< \xi|\eta > = < \eta|\xi >^*$ for all $\xi, \eta \in S$,

(ii) $< \xi|\eta + \zeta > = < \xi|\eta > + < \xi|\zeta >$ for all $\xi, \eta, \zeta \in S$,

(iii) $< \xi|a\eta > = a < \xi|\eta >$ for all $\xi, \eta \in S$ and $a \in C$, and

(iv) $< \xi|\xi > > 0$ unless $\xi = 0$.

Condition (i) ensures that $< \xi|\xi >$ is real; (ii) and (iii) show that $< \xi|0 > = 0$ for all ξ and, in particular that $< 0|0 > = 0$. The zero vector is the only one whose inner product with itself is zero. We have

$$< a\xi|\eta > = a^* < \xi|\eta > \ .$$

Two vectors ξ and η are said to be orthogonal if $<\xi|\eta>=0$. Thus the zero vector is orthogonal to all elements of S and is the only one orthogonal to itself (see Sec. 6 for another form of inner product.)

Examples of inner product spaces are:

1. The set of n-tuples of complex numbers $\xi=(x_1,x_2,\dots x_n)$, $\eta=(y_1,y_2,\dots \dots y_n)$ with
$$<\xi|\eta>=\sum_{i=1}^{n} x_i^* y_i \ .$$
2. The set ℓ^2 of vectors of the form $\xi=(x_1,x_2\dots)$ where
$$\sum_{i=1}^{\infty}|x_i|^2$$
is a convergent series.
3. The set L^2 of square integrable functions $\psi(x_1,x_2,\dots x_N)$ of N variables in some appropriately defined domain with
$$<\phi|\psi>=\int dx_1 dx_2\dots d_{x_N}\phi^*(x_1,x_2,\dots x_N)\psi(x_1,x_2,\dots x_N)$$

The norm of a vector is defined by
$$\| \xi \| = <\xi|\xi>^{\frac{1}{2}} \ ,$$
and a "distance" between ξ and η by
$$d(\xi,\eta)=\|\xi-\eta\| \ .$$
With this definition an inner product space is a metric space.

4.3 Hilbert spaces

Definition. A Hilbert space is a complete inner product space. A space is complete if every Cauchy sequence of vectors belonging to it possesses a limit which is, itself, an element of the space. We remind the reader that a Cauchy sequence $\{\phi_n\}$ is one such that for every $\epsilon>0$ we can find a positive number $N(\epsilon)$ depending on ϵ such that for all m and n larger than $N(\epsilon)$,
$$\|\phi_m-\phi_n\|<\epsilon \ .$$
It is clear the set of complex-valued functions of a real variable, continuous in an interval $a\le x\le b$ form an inner product space with
$$<\phi|\psi>=\int_a^b \phi^*(x)\psi(x)dx \ ,$$

but it is not a Hilbert space because there are Cauchy sequences of continuous functions whose limits are discontinuous. All examples, 1, 2, and 3 in Sec. 2 above are Hilbert spaces provided in 3 the concept of integral employed is that of Lebesgue.

We now state a few theorems about vectors in Hilbert spaces. We omit the proofs but discuss their intuitive meanings.

Theorem 1. Let S be a Hilbert space and M one of its non-empty, complete, proper subspaces (thus M is itself a Hilbert space). Let ψ be a vector of S not in M. Then there is a unique element ϕ of M such that $\| \psi - \phi \|$ is a minimum.

An example will clarify the content of the theorem. Let S be the 3-dimensional Euclidean space and M a plane in this space. Take the origin O in the plane and identify ψ with a vector $\overrightarrow{OP}$ to a point P not on M. The vector ϕ is identified with $\overrightarrow{OQ}$ where Q is the foot of the perpendicular dropped from P to the plane M.

Theorem 2. With the assumptions stated in theorem 1, the vector ψ can always be expressed in a unique way as

$$\psi = \phi + \eta$$

where ϕ is the vector in M determined in theorem 1 and η is orthogonal to every vector in M.

The intuitive meaning of this theorem is again clarified by the example used in theorem 1. The decomposition of $\overrightarrow{OP}$ in

$$\overrightarrow{OP} = \overrightarrow{OQ} + \overrightarrow{QP}$$

where $\overrightarrow{QP}$, identified here with η is perpendicular to all vectors in the plane M.

The set of all vectors in S orthogonal to all vectors in M is called the orthogonal complement of M and is labeled $M_\perp$. This theorem states that every vector in S can be uniquely expressed as the sum of a vector in M and a vector in $M_\perp$.

A mapping Ω of a Hilbert space S into a second Hilbert space S' is linear if, when $\phi \to \Omega\phi$ and $\psi \to \Omega\psi$, then

$$\Omega(\phi + \psi) = \Omega\phi + \Omega\psi$$

and

$$\Omega(a\psi) = a\Omega\psi \ ,$$

for all ϕ and ψ in S and all complex numbers a. For the most part we will limit our considerations to the case in which $S' = S$ or a subset of S. Such mappings are often called linear operators.

We define the adjoint or Hermitian conjugate of Ω as an operator $\Omega^\dagger$ such that

$$< \phi|\Omega\psi > = < \Omega^\dagger\phi|\psi >$$

for any pair ϕ, ψ of elements of S; $\Omega^\dagger$, like Ω, is linear. A unitary operator U is one satisfying

$$UU^\dagger = U^\dagger U = E \ .$$

Thus, unitary operators have inverses and

$$U^{-1} = U^{\dagger} .$$

They preserve inner products, *i.e.*,

$$< U\phi|U\psi >=< U^{\dagger}U\phi|\psi >=< \phi|\psi > .$$

If the adjoint of Ω, $\Omega^{\dagger}$, is equal to Ω we say that the operator is Hermitian or self-adjoint.

If A and B are linear operators

$$(AB)^{\dagger} = B^{\dagger}A^{\dagger} .$$

If Ω is Hermitian, then

$$U = e^{i\Omega}$$

is unitary. In fact it can be shown that every unitary operator can be expressed in this form.

Let M be a complete subspace of a Hilbert space S. We say that M is invariant under an operator Ω if

$$\phi \in M \Rightarrow \Omega\phi \in M .$$

We now prove a theorem which we will need in the theory of group representations.

Theorem 3. If M, a complete, proper, subspace of a Hilbert space S is invariant under a Hermitian operator H, then the orthogonal complement $M_{\perp}$ of M is, likewise, invariant under H.

The proof of this theorem is so simple that we give it here. We need to show that if $\eta \in M_{\perp}$ so does $H\eta \in M_{\perp}$. Let ξ be an arbitrary element of M. Every vector in $M_{\perp}$ is orthogonal to every element of M. Now

$$< H\eta|\xi >=< \eta|H\xi >= 0$$

Thus $H\eta$ is orthogonal to every vector in M and, hence, belongs to $M_{\perp}$.

The theorem is also valid for unitary operators. Since M is invariant under a unitary operator U and U is non-singular, every element of M can be written as $U\xi$ with $\xi \in M$. Then, for $\eta \in M_{\perp}$

$$< U\eta|U\xi >=< \eta|\xi >= 0$$

so that $U\eta \in M_{\perp}$. This result also follows from the fact that U can always be expressed as e^{iH} with H Hermitian.

4.4 The linear groups

The purpose of this section is to provide the nomenclature for groups of linear transformations acting on a vector space; linear groups for short.

Let V be a vector space and $\{\epsilon_i\}$ one of its bases; every element ξ of V is a linear combination of the ϵ_i:

$$\xi = \sum_i x_i \epsilon_i \ . \tag{4.1}$$

Let A be a linear mapping of V into V. The action of A on ξ is determined by that on each member of the basis$\{\epsilon_i\}$. Thus, if $\kappa_i = A\epsilon_i$, then

$$\xi \to \xi' = A\xi = \sum_j x_j \kappa_j \ . \tag{4.2}$$

Since the vectors κ_j are elements of V, they can be written as linear combinations of elements of the basis $\{\epsilon_i\}$; thus,

$$\kappa_j = \sum_i \epsilon_i A_{ij} \tag{4.3}$$

and

$$\xi' = A\xi = \sum_{i,j} \epsilon_i A_{ij} x_j = \sum_i \epsilon_i x'_i \tag{4.4}$$

with

$$x'_i = \sum_j A_{ij} x_j \ . \tag{4.5}$$

This shows that, with respect to the particular basis $\{\epsilon_i\}$, the operator A can be represented by a matrix whose elements are the numbers A_{ij} in Eq. (4.5). If the null space of A consists of the zero vector only, then

$$\sum_i c_i \kappa_i = A(\sum_i c_i \epsilon_i) = 0$$

if, and only if, $\sum_i c_i \epsilon_i = 0$ which is only possible if the c_i are all zero. We say that A is non-singular and we can obtain the ϵ_i's as functions of the κ_i's. Then the determinant of the matrix whose elements are A_{ij} cannot vanish.

The set of all non-singular transformation in a n-dimensional inner product space V_n is called the general linear group $GL(n)$. Some times it may be necessary to specify if the space is real (R) or complex (C). The elements of $GL(n)$ are represented by the totality of $n \times n$ matrices with non-vanishing determinants.

The elements of $GL(n)$, themselves, form a vector space if we define the sum $A+B$ of two linear transformations A and B and the product of A by a number a by the requirements

$$(A+B)\xi = A\xi + B\xi \ , \quad A(a\xi) = aA\xi \ ,$$

for all ξ in V_n. Clearly it is enough to verify these conditions on the elements of a basis.

It is immediately found that

$$(A+B)_{ij} = A_{ij} + B_{ij} \ , \quad (aA)_{ij} = aA_{ij} \ .$$

The product of two linear transformations B and A is defined as the successive application of A and B. Thus, if $\xi \to \xi' = A\xi$ and $\xi' \to \xi'' = B\xi'$, then $\xi \to \xi'' = BA\xi$. The matrix representation of BA in the basis $\{\epsilon_i\}$ is

$$(BA)_{ij} = \sum_k B_{ik} A_{kj} \ . \tag{4.6}$$

Having endowed V_n with an inner product, by means of the Gram-Schmidt orthogonalization procedure, we can construct a basis $\{\epsilon_1, \epsilon_2, \dots \epsilon_n\}$ whose vectors satisfy

$$< \epsilon_i | \epsilon_j >= \delta_{ij} \ , \tag{4.7}$$

i.e., any two vectors in the set are orthogonal and each has norm equal to unity (*i.e.*, is normalized).

A Hermitian operator A is represented by a $n \times n$ matrix whose elements are

$$A_{ij} =< \epsilon_i | A\epsilon_j >=< A\epsilon_i | \epsilon_j >=< \epsilon_j | A\epsilon_i >^* = A^*_{ji} \ . \tag{4.8}$$

The matrix elements of an operator B are likewise $B_{ij} =< \epsilon_i | B\epsilon_j >$. Now

$$B^*_{ij} =< B\epsilon_j | \epsilon_i >=< \epsilon_j | B^\dagger \epsilon_i >= (B^\dagger)_{ji} \ . \tag{4.9}$$

The trace of an operator A is defined by

$$\chi(A) = \sum_i A_{ii} \ . \tag{4.10}$$

$\chi(A)$ is invariant under change of basis. Now

$$\chi(AB) = \chi(BA) \tag{4.11}$$

since

$$\sum_i (AB)_{ii} = \sum_{i,j} A_{ij} B_{ji} = \sum_j (BA)_{jj} \ .$$

The set of square matrices $\{A, B, \dots\}$ of n-dimensions can also be viewed as an inner product space with the definition

$$< A | B >= Tr(A^\dagger B) \ . \tag{4.12}$$

The proof of this statement is left to the reader.

The special linear group in n-space is the set of all $n \times n$ matrices whose determinant is unity. It is denoted by $SL(n)$ ($SL(n, C)$ if the vector space is complex).

The unitary group $U(n)$ is the set of all unitary transformations in V_n. Since $U^\dagger U = E$, $\| U \|$ is a complex number of modulus equal to 1. If we restrict ourselves to the elements U such that $\| U \| = 1$, the group is called the special unitary group $SU(n)$. Clearly $SU(n)$ is a subgroup of $U(n)$ which is, in turn, a subgroup of $GL(n)$. When the space is real, the orthogonal transformations in n-space form a group called $O(n)$; $SO(n)$ is the subgroup of $O(n)$ consisting of those elements with determinant equal to unity.

4.5 Homomorphism between SO(3) and SU(2)

The rotation group SO(3) in 3-space has been studied in some detail in chapter 1. We recall that each one of its elements requires three real numbers for its specification ($\hat{n}$ and φ or the three Euler angles α, β, γ, for example). We will now establish that SO(3) is homomorphic to SU(2). A typical element of SU(2) is

$$U = \begin{pmatrix} a & b \\ -b^* & a^* \end{pmatrix} \tag{5.1}$$

where

$$|a|^2 + |b|^2 = 1 \ . \tag{5.2}$$

Calling the real and imaginary parts of a and b by a_1 an a_2 and b_1 and b_2, respectively, Eq. (5.14) reads

$$a_1^2 + a_2^2 + b_1^2 + b_2^2 = 1 \tag{5.3}$$

describing the points on the surface of a four-dimensional sphere of unit radius. Three real numbers are enough to specify a point on this surface and hence, also, to specify an element u of SU(2).

We will now establish the homomorphism of SO(3) to SU(2).

A point $\boldsymbol{r} = (x_1, x_2, x_3)$ in three space can be described by a 2×2 Hermitian matrix of zero trace. There exists a one-to-one correspondence between $\boldsymbol{r} = (x_1, x_2, x_3)$ and

$$X(\boldsymbol{r}) = \begin{pmatrix} x_3 & x_1 - ix_2 \\ x_1 + ix_2 & -x_3 \end{pmatrix} . \tag{5.4}$$

This matrix can be rewritten in the form

$$X(\boldsymbol{r}) = x_1\sigma_1 + x_2\sigma_2 + x_3\sigma_3 = \boldsymbol{r} \cdot \boldsymbol{\sigma} \tag{5.5}$$

where

$$\sigma_1 = \begin{pmatrix} 0 & 1 \\ 1 & 0 \end{pmatrix} , \ \sigma_2 = \begin{pmatrix} 0 & -i \\ i & 0 \end{pmatrix} , \ \sigma_3 = \begin{pmatrix} 1 & 0 \\ 0 & -1 \end{pmatrix} \tag{5.6}$$

are the well-known Pauli matrices.

For every element of u we make the transformation

$$X(\boldsymbol{r}) \to X'(\boldsymbol{r}) = uX(\boldsymbol{r})u^{-1} \ . \tag{5.7}$$

Now $X'(\boldsymbol{r})$ is like $X(\boldsymbol{r})$, a Hermitian matrix of zero trace so that three real numbers x_1', x_2', x_3' exist such that

$$X'(\boldsymbol{r}) = \begin{pmatrix} x_3' & x_1' - ix_2' \\ x_1' + ix_2' & -x_3' \end{pmatrix} = \boldsymbol{\sigma} \cdot \boldsymbol{r}' \ . \tag{5.8}$$

The relation between x_1', x_2', x_3' and x_1, x_2, x_3 is linear. Furthermore

$$\| X' \| = \| uXu^{-1} \| = \| X \|$$

or

$$-x_1'^2 - x_2'^2 - x_3'^2 = -x_1^2 - x_2^2 - x_3^2 \, .$$

Thus the transformation $\boldsymbol{r} \to \boldsymbol{r}'$ is orthogonal. Explicitly it is given by

$$\begin{aligned} x_1' &= \frac{1}{2}(a^2 + a^{*2} - b^2 - b^{*2})x_1 + \frac{i}{2}(-a^2 + a^{*2} - b^2 + b^{*2})x_2 - (ab + a^*b^*)x_3, \\ x_2' &= \frac{i}{2}(a^2 - a^{*2} - b^2 + b^{*2})x_1 + \frac{1}{2}(a^2 + a^{*2} + b^2 + b^{*2})x_2 - i(ab - a^*b^*)x_3, \\ x_3' &= (ab^* + a^*b)x_1 - i(ab^* - a^*b)x_2 + (aa^* - bb^*)x_3 \, . \end{aligned} \tag{5.9}$$

For each $u \in SU(2)$ we have an element of O(3). Now, the determinant of the transformation $\boldsymbol{R}(u)$ corresponding to u, being orthogonal can only be $+1$ or -1. For $a = 1, b = 0$, $\boldsymbol{R}(u)$ is the unit transformation. But $\| \boldsymbol{R}(u) \|$ for arbitrary u is a continuous function of u, and, we can say a continuous function of the points on the surface of the unit sphere (5.3). This being the case, if a value of u existed for which $\| \boldsymbol{R}(u) \| = -1$, this function must take all intermediate points between -1 and 1. But this is impossible because $\| \boldsymbol{R}(u) \|$ is either 1 or -1. Thus, for all u, $\boldsymbol{R}(u) \in SO(3)$.

If u_1 and u_2 are any two elements of SU(2)

$$u_1 u_2 X(\boldsymbol{r})(u_1 u_2)^{-1} = u_1[u_2 X(\boldsymbol{r}) u_2^{-1}]u_1^{-1}$$

so that

$$\boldsymbol{R}(u_1 u_2) = \boldsymbol{R}(u_1)\boldsymbol{R}(u_2) \, . \tag{5.10}$$

We note that

$$\boldsymbol{R}(-u) = \boldsymbol{R}(u) \, , \tag{5.11}$$

so that there are two elements of SU(2) which yield the same rotation. It is required that we show that there are no others. If u and u' give the same rotation

$$uX(\boldsymbol{r})u^{-1} = u'X(\boldsymbol{r})u'^{-1}$$

for all $\boldsymbol{r}$. This implies that $u'^{\dagger}u$ commutes with all three Pauli matrices. The matrix $u'^{\dagger}u$ is unitary. The only unitary matrices which commute with all three Pauli matrices are σ_0 and $-\sigma_0$ where σ_0 is

$$\sigma_0 = \begin{pmatrix} 1 & 0 \\ 0 & 1 \end{pmatrix} , \tag{5.12}$$

the unit 2×2 matrix. Thus $u' = \pm u$.

We have not yet completed our discussion; we have shown that for every u there is a rotation $\boldsymbol{R}(u)$. It is necessary to show that the mapping $\boldsymbol{R}(u)$ of SU(2) to SO(3) is onto. In other words we need to show that for every rotation there is an element $u \in SU(2)$ such that $\boldsymbol{R}(u)$ is that particular rotation. We do this by actually constructing the element u. Let the rotation be determined by the Euler angles α, β, γ. Then $\boldsymbol{R}(\alpha, \beta, \gamma)$ is given by Eq. (2.27) in chapter 1. Now

$$u^{(3)}(\alpha) = \begin{pmatrix} e^{-i\alpha/2} & 0 \\ 0 & e^{i\alpha/2} \end{pmatrix} \tag{5.13}$$

generates

$$\boldsymbol{R}^{(3)}(\alpha) = \begin{pmatrix} \cos\alpha & -\sin\alpha & 0 \\ \sin\alpha & \cos\alpha & 0 \\ 0 & 0 & 1 \end{pmatrix} ,$$

a rotation by α about the 3-axis. Consider now a real element of SU(2), namely

$$u^{(2)}(\beta) = \begin{pmatrix} \cos\frac{\beta}{2} & -\sin\frac{\beta}{2} \\ \sin\frac{\beta}{2} & \cos\frac{\beta}{2} \end{pmatrix} . \tag{5.14}$$

This gives

$$\boldsymbol{R}^{(2)}(\beta) = \begin{pmatrix} \cos\beta & 0 & \sin\beta \\ 0 & 1 & 0 \\ -\sin\beta & 0 & \cos\beta \end{pmatrix} ,$$

a rotation by β about the 2-axis. Thus, the element

$$u(\alpha,\beta,\gamma) = u^{(3)}(\alpha)u^{(2)}(\beta)u^{(3)}(\gamma) = \begin{pmatrix} a & b \\ -b^* & a^* \end{pmatrix} \tag{5.15}$$

where

$$a = e^{-\frac{i}{2}(\alpha+\gamma)}\cos\frac{\beta}{2} \quad , \quad b = -e^{-\frac{i}{2}(\alpha-\gamma)}\sin\frac{\beta}{2} \tag{5.16}$$

generates $\boldsymbol{R}(\alpha,\beta,\gamma)$. This completes the proof: SO(3) is homomorphic to SU(2) with the two-to-one homomorphism $\boldsymbol{R}(u)$.

4.6 The Lorentz group

This is not the place to give a systematic development of the special theory of relativity. We limit ourselves to a few remarks of a general nature and concentrate on the relation of the principle of relativity and symmetry. We assume the reader to be familiar with both the epistemological analysis of the concepts of space and time and with the empirical basis of the theory.

The special theory of relativity generalizes the principle of Galilean relativity to all physical phenomena, not just to those of mechanics. It is assumed that there exists an infinite set of reference frames, called inertial, moving relative to one another with constant velocities, in which all the laws of physics have the same form and in particular in which **free** particles move with uniform rectilinear velocities. It is assumed that the speed of light in the vacuum is the same in all inertial frames and independent of the state of motion of the source.

Observers in different inertial frames compare their experiences according to rules consistent with the principle of relativity. In an inertial frame S consider two events characterized by (x_1, y_1, z_1, t_1) and (x_2, y_2, z_2, t_2). We define the interval between these events by

$$(\Delta s_{12})^2 = c^2\Delta t^2 - \Delta x^2 - \Delta y^2 - \Delta z^2$$

where $\Delta t = t_2 - t_1$, $\Delta x = x_2 - x_1$, $\Delta y = y_2 - y_1$, and $\Delta z = z_2 - z_1$. Intervals are classified as time-like if $(\Delta s_{12})^2 > 0$ and space-like if $(\Delta s_{12})^2 < 0$. If $(\Delta s_{12})^2 > 0$ and $t_2 > t_1$ event "one" can be related to event two as its cause since a signal from "one" to "two" would not need to travel with a speed greater than that of light. For space like events no such relation is possible. If (x_1, y_1, z_1, t_1) corresponds to a light signal being emitted at (x_1, y_1, z_1) and time t_1 and (x_2, y_2, z_2, t_2) describes its arrival at (x_2, y_2, z_2) at time t_2, then $(\Delta s_{12})^2 = 0$. Let us consider now observers at S and S', two inertial frames. Since a free particle in S moves with uniform rectilinear motion, the only way for such a description in S' to be likewise is that the law of transformation of (x, y, z, t) in S to (x', y', z', t) in S' be linear. Furthermore, because the speed of light in empty space is the same for all observers, if $(\Delta s_{12})^2 = 0$ in S, $(\Delta s'_{12})^2$ must be zero in S'. Thus we must have

$$c^2 \Delta t^2 - \Delta x^2 - \Delta y^2 - \Delta z^2 = c^2(\Delta t')^2 - (\Delta x')^2 - (\Delta y')^2 - (\Delta z')^2 \,.$$

Selecting the first event in such a manner that (0,0,0,0) in S coincides with (0,0,0,0) in S' we conclude that

$$(\Delta s)^2 = c^2 t^2 - x^2 - y^2 - z^2$$

must be an invariant with respect to the transformation from S to S'. The set of linear transformations leaving this quantity invariant is called the Lorentz group.

The Lorentz group can be generated by generators similar to I_1, I_2 and I_3 used in our discussion of rotations in 3-space.

In order to simplify the analysis we introduce a new type of vector space. We describe the events by vectors $\boldsymbol{x}$ in this space with the notation

$$x^0 = ct, x^1 = x, x^2 = y, x^3 = z. \tag{6.1}$$

Each $\boldsymbol{x} = (x^0, x^1, x^2, x^3)$; $(x^0 = ct, \boldsymbol{r})$ is called an event. We define a set of basis vectors in the space: $\boldsymbol{e}_0, \boldsymbol{e}_1, \boldsymbol{e}_2, \boldsymbol{e}_3$ given by columns with entries (1,0,0,0), (0,1,0,0), (0,0,1,0) and (0,0,0,1), respectively so that we can write

$$\boldsymbol{x} = x^\mu \boldsymbol{e}_\mu \quad , \quad \mu = 0, 1, 2, 3 \tag{6.2}$$

where a sum over μ is implied. We shall use Greek indices when the range of the index is 0,1,2,3 and Latin indices when it is 1,2,3.

This space of special relativity is called Minkowski space, its points being the events $\boldsymbol{x}$. It differs from that to which we have become accustomed in that intervals are not positive definite. As a consequence the interval between two events may be zero but the components of the interval need not vanish. Accordingly we define a real inner product space with the following postulates:

(i) The space of events is an Abelian group with respect to addition: $\boldsymbol{x} = \{x^\mu\}$, $\boldsymbol{y} = \{y^\mu\}$, $\boldsymbol{x} + \boldsymbol{y} = \{x^\mu + y^\mu\}$,

(ii) for all events $\boldsymbol{x}$ and $\boldsymbol{y}$, $< \boldsymbol{x}|\boldsymbol{y} > = < \boldsymbol{y}|\boldsymbol{x} >$,

(iii) for all events $\boldsymbol{x}, \boldsymbol{y}, \boldsymbol{z}$, $< \boldsymbol{x}|\boldsymbol{y}+\boldsymbol{z} >==< \boldsymbol{x}|\boldsymbol{y} > + < \boldsymbol{x}|\boldsymbol{z} >$

(iv) For all $\boldsymbol{x}$ and $\boldsymbol{y}$ in the space and any real number α

$$< \boldsymbol{x}|\alpha\boldsymbol{y} >= \alpha < \boldsymbol{x}|\boldsymbol{y} > \ .$$

(v) $< \boldsymbol{x}|\boldsymbol{y} >= 0$ for all $\boldsymbol{y}$ implies that $\boldsymbol{x} = 0$.

We note that this definition differs from that given in Sec. 2 in the last postulate only.

The inner product of $\boldsymbol{x} = (x^0, x^1, x^2, x^3)$ and $\boldsymbol{y} = (y^0, y^1, y^2, y^3)$ is defined by

$$< \boldsymbol{x}|\boldsymbol{y} >= x^0y^0 - x^1y^1 - x^2y^2 - x^3y^3 = x^0y^0 - \boldsymbol{x}\cdot\boldsymbol{y} \ . \tag{6.3}$$

The products of the basis vectors $\boldsymbol{e}_0, \boldsymbol{e}_1, \boldsymbol{e}_2, \boldsymbol{e}_3$ among themselves are

$$\begin{aligned} < \boldsymbol{e}_\mu|\boldsymbol{e}_\nu >= 1 \quad & if \quad \mu = \nu = 0 \\ -1 \quad & if \quad \mu = \nu = 1,2,3 \\ 0 \quad & if \quad \mu \neq \nu \end{aligned}$$

We call this set

$$g_{\mu\nu} =< \boldsymbol{e}_\mu|\boldsymbol{e}_\nu >= \begin{pmatrix} 1 & 0 & 0 & 0 \\ 0 & -1 & 0 & 0 \\ 0 & 0 & -1 & 0 \\ 0 & 0 & 0 & -1 \end{pmatrix} \ . \tag{6.4}$$

The tensor $g_{\mu\nu}$ is called the metric tensor.

The components x^μ of the event $\boldsymbol{x} = x^\mu e_\mu$ are called its contravariant components. The covariant components are defined by

$$x_\mu =< \boldsymbol{e}_\mu|\boldsymbol{x} >= g_{\mu\nu}x^\nu \ . \tag{6.5}$$

The most general linear transformation between events in inertial frame S to those in the inertial frame S' is

$$x'^\mu = a^\mu{}_\nu x^\nu \tag{6.6}$$

assuming the zero events coincide in both frames. The requirement that

$$x'^\mu x'_\mu = x^\lambda x_\lambda \tag{6.7}$$

imposes the restriction that

$$g_{\mu\nu}x'^\mu x'^\nu = g_{\kappa\lambda}x^\kappa x^\lambda \tag{6.8}$$

be satisfied for all $\boldsymbol{x}$, so that

$$g_{\mu\nu}a^\mu{}_\kappa a^\kappa{}_\lambda = g_{\kappa\lambda} \tag{6.9}$$

An object $\{V^\mu\}$ with four components which transform in the same way as $\{x^\mu\}$ is called a four-vector. Thus, in S', the components of $\{\boldsymbol{V}\}$ are

$$V'^\mu = a^\mu{}_\nu V^\nu \ , \tag{6.10}$$

If $\phi(\boldsymbol{x})$ is a scalar quantity, *i.e.*, a field invariant under Lorentz transformations, its gradient

$$\frac{\partial \phi}{\partial x^\mu} = \partial_\mu \phi \tag{6.11}$$

are the components of a covariant vector.

The operation

$$g_{\mu\nu} V^\nu = V_\mu \tag{6.12}$$

is called an "index lowering" operation. Thus if

$$\{V^\mu\} = (V^0, \boldsymbol{V}) \ , \tag{6.13}$$

then

$$\{V_\mu\} = (V^0, -\boldsymbol{V}) \ . \tag{6.14}$$

Thus to invert the operation we write

$$V^\mu = g^{\mu\nu} V_\nu \tag{6.15}$$

where $g^{\mu\nu} = g_{\mu\nu}$ for all μ and ν. The operation (6.15) is called an "index raising" operation. The tensors $g^{\mu\nu}$ and $g_{\mu\nu}$ are related by

$$g^{\mu\lambda} g_{\lambda\nu} = g^\mu_\nu = \delta^\mu_\nu = \begin{cases} 1 & if \quad \mu = \nu \\ 0 & if \quad \mu \neq \nu \end{cases} \tag{6.16}$$

These definitions allow us to write the condition (6.9) for $a^\mu{}_\nu$ to be a member of the Lorentz group in the equivalent forms

$$a_{\nu\kappa} a^\nu{}_\lambda = g_{\kappa\lambda} \tag{6.17}$$

and

$$a_{\nu\kappa} a^{\nu\lambda} = g^\lambda_\kappa \ . \tag{6.18}$$

Equation (6.18) shows that the determinant of $a^\mu{}_\nu$ is

$$\| a \| = \pm 1 \tag{6.19}$$

To solve Eq. (6.6) for $\boldsymbol{x}$ in terms of $\boldsymbol{x}'$ we multiply Eq. (6.6) by $a_{\mu\lambda}$ to obtain

$$a_{\mu\lambda} x'^\mu = a_{\mu\lambda} a^\mu{}_\nu x^\nu = g_{\lambda\nu} x^\nu = x_\lambda \ . \tag{6.20}$$

Raising λ we have

$$x^\lambda = a_\mu{}^\lambda x'^\mu \tag{6.21}$$

The four-gradient in Eq. (6.11) transforms as

$$\partial'_\mu \phi = \frac{\partial x^\lambda}{\partial x'^\mu} \partial_\lambda \phi = a_\mu{}^\lambda \partial_\lambda \phi \ , \tag{6.22}$$

i.e., in the same way as x_ν. The index raising operation on the differential operator ∂_λ is defined by

$$\partial^\mu = g^{\mu\nu}\partial_\nu \ . \tag{6.23}$$

Now

$$\partial'^\mu = g^{\mu\nu}\partial'_\nu = g^{\mu\nu}a_\nu{}^\lambda\partial_\lambda = a^\mu{}_\lambda\partial^\lambda \ . \tag{6.24}$$

Before constructing the generators of the Lorentz group we establish a homomorphism between it and SL(2,C). We will denote by L a subgroup of the Lorentz group which we will call the set of all proper Lorentz transformations. A typical element of SL(2,C) is

$$v = \begin{pmatrix} a & b \\ c & d \end{pmatrix} \tag{6.25}$$

where a, b, c and d are complex numbers satisfying

$$ad - bc = 1 \tag{6.26}$$

The inverse of v is

$$v^{-1} = \begin{pmatrix} d & -b \\ -c & a \end{pmatrix} \tag{6.27}$$

As in the case of the homomorphism of SO(3) to SU(2) we define a Hermitian matrix $X(\boldsymbol{x})$ for every event $\boldsymbol{x}$ in an inertial frame S:

$$X(\boldsymbol{x}) = \begin{pmatrix} x^0 + x^3 & x^1 - ix^2 \\ x^1 + ix^2 & x^0 - x^3 \end{pmatrix} = x^\mu\sigma_\mu \tag{6.28}$$

where σ_0 is the unit 2×2 matrix and $\sigma_1, \sigma_2, \sigma_3$ the Pauli matrices. We note that

$$\| X(\boldsymbol{x}) \| = (x^0)^2 - (x^1)^2 - (x^2)^2 - (x^3)^2 = x_\mu x^\mu \ . \tag{6.29}$$

We now form

$$X'(\boldsymbol{x}) = vX(\boldsymbol{x})v^\dagger \ . \tag{6.30}$$

The matrix $X'(\boldsymbol{x})$ is Hermitian so that it can be written in the form

$$X'(\boldsymbol{x}) = \begin{pmatrix} x'^0 + x'^3 & x'^1 - ix'^2 \\ x'^1 + ix'^2 & x'^0 - x'^3 \end{pmatrix} = x'^\mu\sigma_\mu \ . \tag{6.31}$$

The quantities x'^μ are linear in the x^μ and, since

$$\| X'(\boldsymbol{x}) \| = x'^\mu x'_\mu = \| vX(x)v^\dagger \| = \| X(\boldsymbol{x}) \| = x^\mu x_\mu$$

the transformation generated is a Lorentz transformation. We denote this transformation by $a(v)$ where a stands for the 4×4 matrix $a^\mu{}_\nu(v)$. If $v = \sigma_0$, a member of SL(2,C), $a(\sigma_0)$ is the identity transformation $x'^\mu = x^\mu$ whose determinant is 1. The transformation $a(v)$ is a continuous function of v so that, by an argument similar to that used in Sec. 5, $\| a(v) \| = 1$. Thus, this procedure does not generate all Lorentz transformation. But this is not all, *e.g.*, the transformation $x'^\mu = -x^\mu$ has determinant $+1$ but is not attained by any transformation of the type (6.30). In general, every Lorentz transformation preserves time-like vectors (*i.e.*, those for which $x^\mu x_\mu > 0$. But time-like vectors are classified into two sets depending on whether x^0 is positive or negative. No element of one class can be joined continuously to one of the other starting from the identity. Thus we have generated only those Lorentz transformations whose determinants are +1 and have the property of preserving the events in the forward light cone. This set contains the so-called proper Lorentz transformations.

Now, $a(v)$ is a homomorphism between SL(2,C) and the group of proper Lorentz transformations. In fact, from

$$v_1(v_2 X v_2^\dagger)v_1^\dagger = (v_1 v_2)X(v_1 v_2)^\dagger$$

it follows that

$$a(v_1)a(v_2) = a(v_1 v_2) \ . \tag{6.32}$$

We recall that $\boldsymbol{e}_0$ is represented by the 2×2 matrix σ_0 and that $v\sigma_0 v^\dagger = vv^\dagger$. Thus v yields $\boldsymbol{e}_0$ back only if it is unitary, *i.e.*, an element of SU(2), say u. Thus the proper subgroup SU(2) of SL(2,C) generates transformations $a(u)$ with the property

$$a(u)\boldsymbol{e}_0 = \boldsymbol{e}_0 \ .$$

Hence, since $\boldsymbol{e}_0$ is invariant under $a(u)$ so is its orthogonal complement $\boldsymbol{e}_{0\perp} = x^i \boldsymbol{e}_i$. These are just the rotations in 3-space.

To generate all proper Lorentz transformations we start from the identity

$$a^\mu{}_\nu = \delta^\mu_\nu \ .$$

A transformation that differs from the identity by an infinitesimal amount is of the form

$$a^\mu{}_\nu = \delta^\mu_\nu + \Delta\omega^\mu{}_\nu \ . \tag{6.33}$$

The invariance of intervals between events requires that

$$\Delta\omega_{\kappa\lambda} + \Delta\omega_{\lambda\kappa} = 0 \tag{6.34}$$

Thus, the $\Delta\omega_{\mu\nu}$ form a 4×4 antisymmetric matrix only six of whose elements are independent corresponding to "rotations" in the six coordinate planes $(\boldsymbol{e}_\mu, \boldsymbol{e}_\nu)$, $\mu \neq \nu$. Thus, to generate the proper Lorentz transformations we require six generators. For example, an infinitesimal rotation in the $(\boldsymbol{e}_1, \boldsymbol{e}_2)$ plane is of the form

$$x'^0 = x^0 \ , \ x'^1 = x^1 - \Delta\omega_{12}x^2 \ , \ x'^2 = \Delta\omega_{12}x^1 + x^2 \ , \ x'^3 = x^3 \tag{6.35}$$

For a transformation to a frame S' moving with respect to S with infinitesimal velocity along the 3-axis we write

$$x'^0 = x^0 - \Delta\omega_{30}x^3 \ , \ x'^1 = x^1, x'^2 = x^2 \ , \ x'^3 = -\Delta\omega_{30}x^0 + x^3 \ . \tag{6.36}$$

The most general infinitesimal Lorentz transformation of the form (6.33) is of the form

$$\begin{aligned}\Delta \boldsymbol{a} &= E + \frac{1}{2}\Delta\omega_{\mu\nu} I^{\mu\nu} \\ &= E + \Delta\omega_{01} I^{01} + \Delta\omega_{02} I^{02} + \Delta\omega_{03} I^{03} + \Delta\omega_{23} I^{23} + \Delta\omega_{31} I^{31} \\ &\quad + \Delta\omega_{12} I^{12} ,\end{aligned} \tag{6.37}$$

where the quantities $I^{\mu\nu}$ are four by four matrices with the property

$$I^{\mu\nu} = -I^{\nu\mu} \tag{6.38}$$

Such matrices can be read off from the transformations (6.35), (6.36) and similar ones for other rotations. For example

$$I^{03} = \begin{pmatrix} 0 & 0 & 0 & 1 \\ 0 & 0 & 0 & 0 \\ 0 & 0 & 0 & 0 \\ 1 & 0 & 0 & 0 \end{pmatrix} , \quad I^{12} = \begin{pmatrix} 0 & 0 & 0 & 0 \\ 0 & 0 & -1 & 0 \\ 0 & 1 & 0 & 0 \\ 0 & 0 & 0 & 0 \end{pmatrix} .$$

We introduce six matrices K_1, K_2, K_3 and S_1, S_2, S_3 by setting

$$I^{0i} = K_i \tag{6.39}$$

and

$$I^{ij} = -i\epsilon^{ij\ell} S_\ell . \tag{6.40}$$

These matrices are

$$K_1 = \begin{pmatrix} 0 & 1 & 0 & 0 \\ 1 & 0 & 0 & 0 \\ 0 & 0 & 0 & 0 \\ 0 & 0 & 0 & 0 \end{pmatrix} , \quad K_2 = \begin{pmatrix} 0 & 0 & 1 & 0 \\ 0 & 0 & 0 & 0 \\ 1 & 0 & 0 & 0 \\ 0 & 0 & 0 & 0 \end{pmatrix} , \quad K_3 = \begin{pmatrix} 0 & 0 & 0 & 1 \\ 0 & 0 & 0 & 0 \\ 0 & 0 & 0 & 0 \\ 1 & 0 & 0 & 0 \end{pmatrix} , \tag{6.41}$$

$$S_1 = \begin{pmatrix} 0 & 0 & 0 & 0 \\ 0 & 0 & 0 & 0 \\ 0 & 0 & 0 & -i \\ 0 & 0 & i & 0 \end{pmatrix} , \quad S_2 = \begin{pmatrix} 0 & 0 & 0 & 0 \\ 0 & 0 & 0 & i \\ 0 & 0 & 0 & 0 \\ 0 & -i & 0 & 0 \end{pmatrix} , \quad S_3 = \begin{pmatrix} 0 & 0 & 0 & 0 \\ 0 & 0 & -i & 0 \\ 0 & i & 0 & 0 \\ 0 & 0 & 0 & 0 \end{pmatrix} , \tag{6.42}$$

They have the following properties

(i) $$K_i^3 = K_i , \tag{6.43}$$

(ii) $$S_i^3 = S_i , \tag{6.44}$$

(iii) $$[K_i, K_j] = i\epsilon_{ij\ell} S_\ell , \tag{6.45}$$

(iv) $$[S_i, S_j] = i\epsilon_{ij\ell}S_\ell \ , \tag{6.46}$$

and

(v) $$[S_i, K_j] = i\epsilon_{ij\ell}K_\ell \ . \tag{6.47}$$

A finite Lorentz transformation can be regarded as the limit, as N tends to infinity, of N identical infinitesimal Lorentz transformations taken in succession. Let

$$\Delta\omega_{0i} = -\frac{\omega_i}{N} \tag{6.48}$$

and

$$\Delta\omega_{ij} = \frac{1}{N}\epsilon_{ij\ell}\Omega_\ell \ . \tag{6.49}$$

Then Eq. (6.37) becomes

$$\Delta \boldsymbol{a} = \underset{\sim}{E} - \frac{1}{N}\boldsymbol{\omega}\cdot\boldsymbol{K} - \frac{i}{N}\boldsymbol{\Omega}\cdot\boldsymbol{S} \tag{6.50}$$

and

$$\boldsymbol{a} \ = \ \lim_{N\to\infty}(\Delta\boldsymbol{a})^N = e^{-\omega\cdot\mathbf{K} - i\boldsymbol{\Omega}\cdot\mathbf{S}} \ . \tag{6.51}$$

Consider the case $\boldsymbol{\omega} = (0, 0, \omega)$, $\boldsymbol{\Omega} = 0$. Then

$$\begin{aligned} \boldsymbol{a} \ &= \ e^{-\omega K_3} = \sum_{n=0}^{\infty}\frac{(-\omega)^n}{n!}K_3^n = E - K_3\sinh\omega + K_3^2(\cosh\omega - 1) \\ &= \begin{pmatrix} \cosh\omega & 0 & 0 & -\sinh\omega \\ 0 & 1 & 0 & 0 \\ 0 & 0 & 1 & 0 \\ -\sinh\omega & 0 & 0 & \cosh\omega \end{pmatrix} . \end{aligned} \tag{6.52}$$

Thus, this choice of $\boldsymbol{\omega}$ and $\boldsymbol{\Omega}$ corresponds to the transformation

$$\begin{aligned} x'^0 \ &= \ x^0\cosh\omega - x^3\sinh\omega \quad , \quad x'^1 = x^1 \ , \\ x'^2 \ &= \ x^2 \quad , \quad x'^3 = -x^0\sinh\omega + x^3\cosh\omega , \end{aligned} \tag{6.53}$$

ordinarily called a boost. The origin in the primed coordinate system moves with respect to the unprimed one with velocity V ($x'^3 = 0$) given by

$$V = \frac{cx^3}{x^0} = c\tanh\omega = c\beta \tag{6.54}$$

where the last equality defines β. Let

$$\gamma = (1 - \beta^2)^{-\frac{1}{2}} = \cosh\omega \tag{6.55}$$

Then Eqs. (6.53) give the Lorentz transformation in the familiar form

$$t' = \gamma(t - \frac{\beta z}{c}) \quad , \quad x' = x \quad , \quad y' = y \quad , \quad z' = \gamma(z - \beta ct) \ . \tag{6.56}$$

If we define

$$W_i^{\pm} = \frac{1}{2}(S_i \pm K_i) \ , \tag{6.57}$$

use of Eqs. (6.45) – (6.47) gives

$$[W_i^{\pm}, W_j^{\pm}] = i\epsilon_{ij\ell} W_\ell^{\pm} \tag{6.58}$$

and

$$[W_i^{\pm}, W_j^{\mp}] = 0 \ . \tag{6.59}$$

Then

$$\boldsymbol{a} = e^{-\mathbf{W}^{+}\cdot(\boldsymbol{\omega}+\mathbf{i}\boldsymbol{\Omega})} e^{\mathbf{W}^{-}\cdot(\boldsymbol{\omega}-\mathbf{i}\boldsymbol{\Omega})} \ . \tag{6.60}$$

Chapter 5

The Finite subgroups of O(3)

5.1 Finite subgroups of SO(3)

Let H be a finite subgroup of SO(3) of order h and R be one of its elements distinct from the identity E. Consider a point $\boldsymbol{\xi} = (x_1, x_2, x_3)$ on the surface of the unit sphere Σ $(x_1^2 + x_2^2 + x_3^2 = 1)$. The set $H\boldsymbol{\xi}$ containing the vectors $R\boldsymbol{\xi}$ for all $R \in H$ is called the orbit of $\boldsymbol{\xi}$ under H. In SO(3) the orbit of any point of Σ is the whole surface Σ.

Given a point $\boldsymbol{\xi}$ on Σ we consider the subset of all elements R of H such that $R\boldsymbol{\xi} = \boldsymbol{\xi}$. Since $R^{-1}\boldsymbol{\xi} = \boldsymbol{\xi}$ and $R\boldsymbol{\xi} = \boldsymbol{\xi}$ and $S\boldsymbol{\xi} = \boldsymbol{\xi}$ imply that $RS\boldsymbol{\xi} = \boldsymbol{\xi}$, this subset is a subgroup of H. We name it the isotropy group of $\boldsymbol{\xi}$ under H and label it H_ξ. Let H_ξ be of order h_ξ. The number of distinct points in the orbit of $\boldsymbol{\xi}$ is equal to the number of distinct cosets of H_ξ. In fact, if two elements of H, not in H_ξ, are such that $A\boldsymbol{\xi} = \boldsymbol{\eta}$ and $B\boldsymbol{\xi} = \boldsymbol{\eta}$, then $A^{-1}B\boldsymbol{\xi} = \boldsymbol{\xi}$ so that A and B belong in the same coset of H_ξ. Thus, the number of distinct points in the orbit of $\boldsymbol{\xi}$ is equal to n_ξ the number of distinct cosets of H_ξ and

$$n_\xi h_\xi = h \; . \tag{1.1}$$

Consider now the set X of all points of Σ which are left invariant by at least one element of H, excluding the identity. We know that each element of H other than E leaves two diametrically opposite points of Σ unchanged. There are, therefore, $2(h-1)$ such points some of which may coincide. Let us imagine that we have the orbits of all points in X and assume that there are r distinct orbits in X. Each of these orbits corresponds to one or another isotropy group H_i of order h_i. Then

$$2(h-1) = \sum_{i=1}^{r} \frac{h}{h_i}(h_i - 1) \; . \tag{1.2}$$

Dividing by h

$$2 - \frac{2}{h} = r - \sum_{i=1}^{r} \frac{1}{h_i} \, . \tag{1.3}$$

This condition imposes severe restrictions on the possible finite subgroups of SO(3). Since X contains the points left invariant by at least one element of $H - E$, $h_i \geq 2$. Thus the right-hand side of Eq. (1.3) must satisfy

$$r - \sum_{i=1}^{r} \frac{1}{h_i} \geq \frac{r}{2}$$

so that

$$r < 4 \, .$$

But r cannot be unity because $h_i \leq h$ implies that $r \geq 2$. Thus, there are only two possible values of r, namely 2 or 3.

If $r = 2$, then

$$\frac{2}{h} = \frac{1}{h_1} + \frac{1}{h_2} \, . \tag{1.4}$$

This condition can only be verified if

$$h_1 = h_2 = h \, . \tag{1.5}$$

This means that H consists of the rotations by integral multiples of $2\pi/h$ about a fixed axis. The invariant vectors being the north and south poles of Σ with respect to the axis of rotation. Each of these points is, by itself, an orbit.

If $r = 3$ we can take $h_1 \leq h_2 \leq h_3$ without loss of generality and write

$$\frac{1}{h_1} + \frac{1}{h_2} + \frac{1}{h_3} = 1 + \frac{2}{h} \, . \tag{1.6}$$

Clearly $h_1 = 2$ because if $h_1 \geq 3$ the left-hand side of Eq. (1.6) would be ≤ 1 so that the equation cannot be verified by any value of h. Similarly $h_2 < 4$. The only solutions for (h_1, h_2, h_3) are

$$\left(2, 2, \frac{h}{2}\right) \quad , \quad h = 4, 6, 8 \ldots$$

$$(2, 3, 3) \quad , \quad h = 12$$

$$(2, 3, 4) \quad , \quad h = 24$$

and

$$(2, 3, 5) \quad , \quad h = 60$$

The groups with two orbits are the cyclic groups of rotations by $2\pi/h$ designated by $C_h \, : \, C_1, C_2, C_3 \ldots.$

We now turn to the four cases in which there are three distinct orbits. We consider them one by one denoting the three orbits by Q_1, Q_2 and Q_3.

(i) $(2, 2, \frac{1}{2}h)$ where the order of the group can be $h = 4, 6, 8, \ldots$. Since $h_3 = \frac{1}{2}h = 2, 3, 4, \ldots$ the orbit Q_3 consists of two elements and, hence, the isotropy group for any element of Q_3 consists of rotations by $4\pi/h$ $(h = 4, 6, 8, \ldots)$, *i.e.*, $C_{h/2}$. For any point ξ belonging to either Q_1 or Q_2 the isotropy group contains the identity and a rotation by π. Thus, since the elements of $C_{h/2}$ rotates ξ in a plane normal to its axis, the orbits Q_1 and Q_2 consist of points which are vertices of a regular polygon of $h/2$ sides. These groups are denoted by $D_{h/2}$ $(D_2, D_3, D_4, \ldots)$. The group D_2 consists of the identity and three rotations by π about orthogonal axes. It contains these four elements (any two rotations by π about orthogonal axes commute; but see later) and is a realization of the Viergruppe. The groups D_3 and D_4 are the symmetry elements of an equilateral triangle and of a square, respectively (we assume both sides of these figures are not distinguished one from the other; see Fig. 5.1).

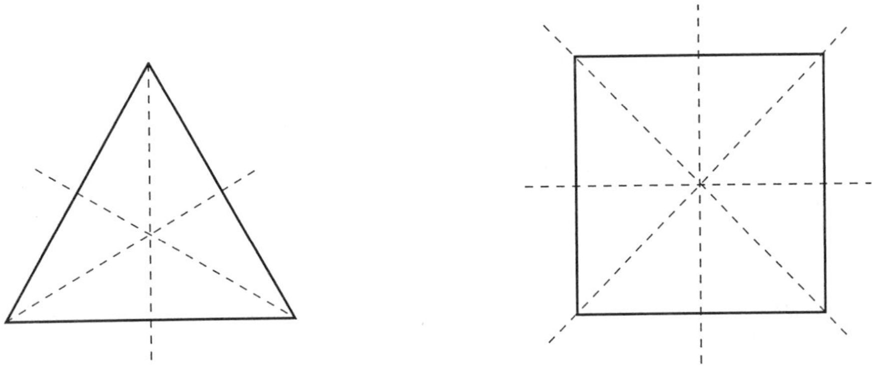

Figure 5.1

(ii) $(2, 3, 3,), h = 12$

The numbers of points in the orbits Q_1, Q_2 and Q_3 are $n_1 = 6$, $n_2 = n_3 = 4$. Thus, there are six poles corresponding to rotations by π about three different axes and two sets of rotational axes. The symmetry elements of a regular tetrahedron are of this type. They form, in fact, the only group of rotations of this form. To show this, let ξ_1, ξ_2, ξ_3 and ξ_4 be the poles of Q_2. The isotropy group of, say, ξ_4 contains a rotation by $2\pi/3$. The action of the isotropy group of ξ_4 on ξ_1, ξ_2 and ξ_3 transforms them among themselves so that the segments joining them to ξ_4 are of the same length. Since this can be stated of ξ_1, ξ_2 and ξ_3 as well, these four poles form the vertices of a regular tetrahedron. This group is denoted by T. Figure 5.2 shows a regular tetrahedron inscribed in a cube.

(iii) $(2, 3, 4), h = 24$

Here $n_1 = 12, n_2 = 8, n_3 = 6$. The isotropy group of an element of Q_3 has four elements and is thus the group C_4 of rotations by $\pi/2$. It follows that the six points in the orbit Q_3 are equidistant and that this is the group

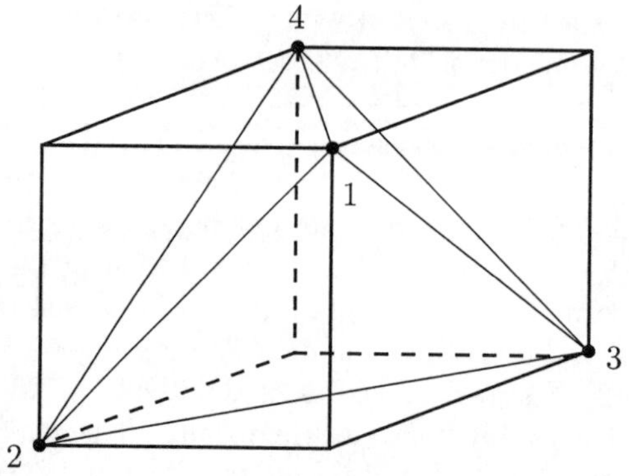

Figure 5.2

of rotations which leave a regular octahedron invariant. It is also the group of rotations leaving a cube in an invariant position. This group of 24 elements is labeled by the symbol O.

(iv) (2,3,5) $h = 60$.

Here $n_1 = 30, n_2 = 20, n_3 = 12$. Let ξ_1 and ξ_2 be two diametrically opposite elements of Q_3. The remaining ten poles cannot all lie in the equator with respect to ξ_1 and ξ_2 as north and south poles because if they were there would exist a five-fold axis in the equatorial plane which would rotate some of the remaining poles out of the plane. Therefore five of these poles must be in the upper hemisphere while the other five are in the lower hemisphere. The first five are equidistant from ξ_1 while the second five are at the same distance from ξ_2. One concludes that each element of Q_3 has five points of Q_3 as nearest neighbors at equal distances. Thus, the rotations of this group are those leaving a regular icosahedron invariant. This figure has 20 faces each one of which is an equilateral triangle. From each vertex there emanates five edges each of which is shared by another vertex and two faces. Thus, if we denote by v the number of vertices, by f that of faces and by e the number of edges we have $f = 20$, $e = 30$, and $v = 12$ since $v = 60/5$. We note that these numbers verify the Euler condition

$$f - e + v = 2$$

valid for any simply connected polyhedron.

We summarize these results in the following table which exhausts all possible finite subgroups of SO(3).

Table of finite rotation groups.

Orbits	Order	Schoenflies notation
(h, h)	h	C_h
$(2, 2, k)$	$h = 2k$	$D_k \quad (k \geq 2)$
(2,3,3)	$h = 12$	T
(2,3,4)	$h = 24$	O
(2,3,5)	$h = 60$	I

5.2 The crystallographic restriction

Not all the finite groups listed in Sec. 1 can be rotation groups of crystals. To discuss the restriction imposed by the translational symmetry of crystals we recall that the atomic arrangement of a crystal is invariant with respect to translations by

$$\boldsymbol{n} = n_1\boldsymbol{a}_1 + n_2\boldsymbol{a}_2 + n_3\boldsymbol{a}_3 \tag{2.1}$$

where $\boldsymbol{a}_1, \boldsymbol{a}_2$ and $\boldsymbol{a}_3$ are the primitive translations and n_1, n_2 and n_3 are integers. We define the reciprocal vectors $\boldsymbol{\tau}_1, \boldsymbol{\tau}_2$ and $\boldsymbol{\tau}_3$ by the nine conditions

$$\boldsymbol{\tau}_i \cdot \boldsymbol{a}_j = \delta_{ij} \ . \tag{2.2}$$

These yield

$$\boldsymbol{\tau}_1 = \frac{\boldsymbol{a}_2 \times \boldsymbol{a}_3}{v_0} \ , \quad \boldsymbol{\tau}_2 = \frac{\boldsymbol{a}_3 \times \boldsymbol{a}_1}{v_0} \ , \quad \boldsymbol{\tau}_3 = \frac{\boldsymbol{a}_1 \times \boldsymbol{a}_2}{v_0} \tag{2.3}$$

where

$$v_0 = \boldsymbol{a}_1 \cdot (\boldsymbol{a}_2 \times \boldsymbol{a}_3) = [\boldsymbol{a}_1, \boldsymbol{a}_2, \boldsymbol{a}_3] \tag{2.4}$$

is the volume of the primitive cell.

It is easy to show that the identity dyadic $\boldsymbol{E}$ is

$$\boldsymbol{E} = \sum_i \boldsymbol{\tau}_i \boldsymbol{a}_i = \sum_i \boldsymbol{a}_i \boldsymbol{\tau}_i \tag{2.5}$$

and that

$$\sum_i \boldsymbol{a}_i \times \boldsymbol{\tau}_i = 0 \ . \tag{2.6}$$

In fact, if $\boldsymbol{u}$ is an arbitrary vector and

$$\boldsymbol{u} = \sum_i u_i \boldsymbol{a}_i \ ,$$

then

$$u_i = \boldsymbol{u} \cdot \boldsymbol{\tau}_i$$

so that

$$\boldsymbol{u} = \sum_i \boldsymbol{a}_i \boldsymbol{\tau}_i \cdot \boldsymbol{u} \ .$$

The other equality follows from the fact $\boldsymbol{a}_1, \boldsymbol{a}_2, \boldsymbol{a}_3$ are the reciprocal vectors of $\boldsymbol{\tau}_1, \boldsymbol{\tau}_2$ and $\boldsymbol{\tau}_3$.

To prove Eq. (2.6) we show that $\sum_i \boldsymbol{\tau}_i \times \boldsymbol{a}_i$ is orthogonal to all three vectors $\boldsymbol{a}_1, \boldsymbol{a}_2$ and $\boldsymbol{a}_3$. In fact,

$$\sum_i (\boldsymbol{\tau}_i \times \boldsymbol{a}_i) \cdot \boldsymbol{a}_j = \sum_i \boldsymbol{\tau}_i \cdot (\boldsymbol{a}_i \times \boldsymbol{a}_j)$$

but $\boldsymbol{a}_i \times \boldsymbol{a}_j = v_0 \sum_k \epsilon_{ijk} \boldsymbol{\tau}_k$ so that

$$\sum_i (\boldsymbol{\tau}_i \times \boldsymbol{a}_i) \cdot \boldsymbol{a}_j = v_0 \sum_{i,k} \epsilon_{ijk} \boldsymbol{\tau}_i \cdot \boldsymbol{\tau}_k \equiv 0 \ .$$

We recall that a rotation by φ about an axis $\boldsymbol{n}$ transforms $\boldsymbol{r}$ into

$$\boldsymbol{r}' = \hat{n}\hat{n} \cdot \boldsymbol{r} + (\boldsymbol{E} - \hat{n}\hat{n}) \cdot \boldsymbol{r} \cos\varphi + \hat{n} \times \boldsymbol{r} \sin\varphi \ .$$

In order for such a rotation to be consistent with the translational symmetry it must transform lattice vectors into lattice vectors. This is satisfied if, and only if,

$$\hat{n}\hat{n} \cdot \boldsymbol{a}_i + (\boldsymbol{E} - \hat{n}\hat{n}) \cdot \boldsymbol{a}_i \cos\varphi + \hat{n} \times \boldsymbol{a}_i \sin\varphi = \sum_j \ell_{ij} \boldsymbol{a}_j \tag{2.7}$$

where the ℓ_{ij} are integers. We obtain

$$\ell_{ij} = \hat{n} \cdot \boldsymbol{\tau}_j \boldsymbol{a}_i \cdot \hat{n} + (\boldsymbol{\tau}_j \cdot \boldsymbol{a}_i - \hat{n} \cdot \boldsymbol{\tau}_j \boldsymbol{a}_i \cdot \hat{n}) \cos\varphi + \hat{n} \cdot (\boldsymbol{a}_i \times \boldsymbol{\tau}_j) \sin\varphi \ .$$

The trace of the matrix $[\ell_{ij}]$ is

$$\sum_i \ell_{ii} = 1 + 2\cos\varphi = p \tag{2.8}$$

Thus, $1 + 2\cos\varphi$ must be an integer and must lie between -1 and 3:

$$\cos\varphi = \frac{p-1}{2}$$

p	$\cos\varphi$	φ	$n = 2\pi/\varphi$
-1	-1	π	2
0	$-\frac{1}{2}$	$2\pi/3$	3
1	0	$\pi/2$	4
2	$\frac{1}{2}$	$\pi/3$	6
3	1	2π	1

We conclude that the only finite subgroups of SO(3) consistent with the translational symmetry of crystals are the 11 groups

$$C_1, C_2, C_3, C_4, C_6, D_2, D_3, D_4, D_6, T, O \tag{2.9}$$

Let C_φ be a rotation by φ. It transforms the vector $\boldsymbol{r}$ into

$$\boldsymbol{r}' = C_\varphi \boldsymbol{r} \ .$$

Now if S is an element of O(3)

$$S\boldsymbol{r}' = SC_\varphi S^{-1} S\boldsymbol{r}$$

so that $SC_\varphi S^{-1}$ transforms $S\boldsymbol{r}$ into $S\boldsymbol{r}'$, *i.e.*, if the image of $\boldsymbol{r}$ is $\boldsymbol{r}' = C_\varphi \boldsymbol{r}$ under C_φ, then the image of $S\boldsymbol{r}$ is $SC_\varphi \boldsymbol{r}$ under $SC_\varphi S^{-1}$. Thus $SC_\varphi S^{-1}$ performs the same transformation as C_φ once all vectors have been transformed by S. $SC_\varphi S^{-1}$ and C_φ are conjugate, *i.e.*, they belong in the same conjugate class. Since C_φ leaves the axis of rotation $\hat{n}$ invariant, $SC_\varphi S^{-1}$ leaves the new axis of rotation $S\hat{n}$ also invariant. In fact

$$C_\varphi \hat{n} = \hat{n}$$

implies that

$$SC_\varphi S^{-1} S\hat{n} = S\hat{n}\ .$$

It is important to recognize that all rotations by φ are in the same conjugate class in SO(3) but that this need not be true for all the finite subgroups of SO(3) (or of O(3) for that matter) because the operation S transforming $\hat{n}$ into $S\hat{n}$ need not be an element of the group. This is the case, for example, for the groups T and D_4.

We give below the class structure of the 11 groups listed in (2.9), where we give the Schoenflies and Hermann-Mauguin notations.

C_1	1	E
C_2	2	E, C_2
C_3	3	E, C_3, C_3^{-1}
C_4	4	E, C_4, C_2, C_4^{-1}
C_6	6	$E, C_6, C_3, C_2, C_3^{-1}, C_6^{-1}$
D_2	222	E, C_2, C_2', C_2''
D_3	32	$E, 2C_3, 3C_2'$
D_4	422	$E, 2C_4, C_2, 2C_2', 2C_2''$
D_6	62	$E, C_2, 2C_3, 2C_6, 3C_2', 3C_2''$
T	23	$E, 3C_2, 4C_3, 4C_3^{-1}$
O	432	$E, 8C_3, 3C_2, 6C_4, 6C_2'$

The transformation C_φ in dyadic form is

$$\begin{aligned}
\boldsymbol{C}_\varphi &= \hat{n}\hat{n} + (\boldsymbol{E} - \hat{n}\hat{n})\cos\varphi + \hat{n} \times \boldsymbol{E}\sin\varphi\ ,\\
\boldsymbol{C}_1 &= \boldsymbol{E}\ ,\\
\boldsymbol{C}_2 &= 2\hat{n}\hat{n} - \boldsymbol{E}\ ,\\
\boldsymbol{C}_3 &= \frac{3}{2}\hat{n}\hat{n} - \frac{1}{2}\boldsymbol{E} + \frac{1}{2}\sqrt{3}\hat{n} \times \boldsymbol{E}\ ,\\
\boldsymbol{C}_4 &= \hat{n}\hat{n} + \hat{n} \times \boldsymbol{E}\ ,\\
\boldsymbol{C}_6 &= \frac{1}{2}\hat{n}\hat{n} + \frac{1}{2}\boldsymbol{E} + \frac{1}{2}\sqrt{3}\hat{n} \times \boldsymbol{E}\ .
\end{aligned}$$

5.3 Finite subgroups of O(3)

The eleven subgroups of SO(3) obtained in Sec. 2 are, of course, subgroups of O(3) as well. We consider now subgroups of O(3) containing elements not included in SO(3). Let H be such a finite subgroup of O(3). The mapping

$$H \to \{1, -1\}$$

defined by the homomorphism

$$R \in H \;,\quad h(R) = \parallel R \parallel$$

defines the kernel K of this homomorphism as a finite subgroup of SO(3). If S_1 and S_2 are any two elements of H not in K, $S_1^{-1}S_2K = K$ so that $S_2K = S_1K$. Thus, there are only two cosets of K and H contains, therefore, twice as many elements as K. One can immediately construct eleven finite subgroups of O(3) by forming the direct products of the eleven groups listed in Sec. 2 by the group

$$C_i = \{E, i\}$$

where i is the inversion. These are

C_i	$\bar{1}$	E, i
C_{2h}	$2/m$	E, C_2, i, σ_h
S_6	$\bar{3}$	$E, C_3, C_3^{-1}, i, S_6^{-1}, S_6$
C_{4h}	$4/m$	$E, C_4, C_2, C_4^{-1}, i, S_4^{-1}, \sigma_h, S_4$
C_{6h}	$6/m$	$E, C_6, C_3, C_2, C_3^{-1}, C_6^{-1}, i, S_3^{-1}, S_6^{-1}, \sigma_h, S_6, S_3$
D_{2h}	mmm	$E, C_2, C_2', C_2'', i, \sigma_v, \sigma_v', \sigma_v''$
D_{3d}	$\bar{3}m$	$E, 2C_3, 3C_2', i, 2S_6, 3\sigma_d$
D_{4h}	$4/mmm$	$E, 2C_4, C_2, 2C_2', 2C_2'', i, 2S_4, \sigma_h, 2\sigma_v, 2\sigma_d$
D_{6h}	$6/mmm$	$E, C_2, 2C_3, 2C_6, 3C_2', 3C_2'', i, \sigma_h, 2S_6, 2S_3, 3\sigma_d, 3\sigma_v$
T_h	$m3$	$E, 3C_2, 4C_3, 4C_3^{-1}, i, 3\sigma_h, 4S_6^{-1}, 4S_6$
O_h	$m3m$	$E, 8C_3, 3C_2, 6C_4, 6C_2', i, 8S_6, 3\sigma_h, 6S_4, 6\sigma_d$

In addition to these there are finite subgroups of O(3) containing improper operations while the inversion i is not present. They are obtained from the eleven rotation groups in (2.9).

For example, in addition to T, a regular tetrahedron has six mirror reflection planes and six improper rotations by $\pi/2$ about the three two-fold axes. These operations, together with the rotations in T form a group called T_d. They permute the vertices of the tetrahedron so that it is a subgroup of the symmetric group S_4. Since S_4 also contains 24 elements it follows that T_d and S_4 are isomorphic (later we will define a group also called S_4 which should not be confused with the group of permutations labeled by the same symbol). We recall that the group O also has 24 elements and is isomorphic with S_4. Hence T_d and O are isomorphic but they are not equivalent representations of S_4 because no similarity transformation $S^{-1}RS$ exists with $S \in O(3)$ taking one into the other because while O is a subgroup of SO(3), T_d is not because it contains improper operations.

The construction of the remaining finite subgroups of O(3) is now a simple matter. It is enough to start with the groups in (2.9) and include reflection mirror planes and improper rotations which are consistent with the finite nature of the subgroups.

The group C_S is the simplest. It contains the identity E and a mirror plane. Associated with C_2 we have C_{2v} and S_4; with C_6, C_{6v} with D_2, D_{2d}; and with

T, T_d. The notation v means a vertical mirror plane where by vertical we mean the direction of the axis of the highest rotational symmetry. By h we mean a horizontal mirror plane, *i.e.*, one normal to the axis of highest rotational symmetry. By d we mean a dihedral mirror plane, *i.e.*, one that bisects the angle between axes of 2-fold rotational symmetry if these are present.

We list below the remaining ten finite subgroups of O(3).

(E)	C_S	$2/m$	E, σ
(C_2)	C_{2v}	$mm2$	$E, C_2, \sigma_v, \sigma_v'$
	S_4	$\bar{4}$	E, S_4^{-1}, C_2, S_4
(C_3)	C_{3h}	$3/m$	$E, S_3^{-1}, C_3, \sigma_h, C_3^{-1}, S_3$
	C_{3v}	$3m$	$E, 2C_3, 3\sigma_v$
	D_{3h}	$\bar{6}2m$	$E, \sigma_h, 2C_3, 2S_3, 3C_2', 3\sigma_v$
(C_4)	C_{4v}	$4mm$	$E, 2C_4, C_2, 2\sigma_v, 2\sigma_d$
(C_6)	C_{6v}	$6mm$	$E, C_2, 2C_3, 2C_6, 3\sigma_d, 3\sigma_v$
(D_2)	D_{2d}	$\bar{4}2m$	$E, 2S_4, C_2, 2C_2', 2\sigma_d$
(T)	T_d	$\bar{4}3m$	$E, 8C_3, 3C_2, 6\sigma_d, 6S_4$

The dyadic associated with a reflection in a plane normal to $\hat{n}$ is

$$\boldsymbol{\sigma} = \boldsymbol{E} - 2\hat{n}\hat{n} \ . \tag{3.1}$$

Thus

$$\boldsymbol{S}_\varphi = \boldsymbol{\sigma} \cdot \boldsymbol{C}_\varphi = -\hat{n}\hat{n} + (\boldsymbol{E} - \hat{n}\hat{n}) \cos\varphi + \hat{n} \times \boldsymbol{E} \sin\varphi \ . \tag{3.2}$$

We remark that the dyadics $\boldsymbol{C}_\varphi - \boldsymbol{E}$ and $\boldsymbol{S}_\varphi + \boldsymbol{E}$ are singular, *i.e.*, they transform all vectors into vectors normal to $\hat{n}$ since $(\boldsymbol{C}_\varphi - \boldsymbol{E}) \cdot \hat{n} = (\boldsymbol{S}_\varphi + \boldsymbol{E}) \cdot \hat{n} = 0$. If $\boldsymbol{C}$ represents a rotation consistent with crystal symmetry

$$\boldsymbol{C} = \sum_{i,j} \ell_{ij} \boldsymbol{a}_i \boldsymbol{\tau}_j \ ,$$

with ℓ_{ij} being integers. Then

$$\boldsymbol{C} - \boldsymbol{E} = \sum_{i,j} (\ell_{ij} - \delta_{ij}) \boldsymbol{a}_i \boldsymbol{\tau}_j \ .$$

Since the ℓ_{ij} are integers $\boldsymbol{C} - \boldsymbol{E}$ can be written in either of the two ways $\sum_j \boldsymbol{u}_j \boldsymbol{\tau}_j$ or $\sum \boldsymbol{a}_i \boldsymbol{v}_i$ where $\boldsymbol{u}_j$ are vectors of the direct lattice and $\boldsymbol{v}_i$ vectors of the reciprocal lattice. Now $\boldsymbol{C} - \boldsymbol{E}$ is planar so that the vectors $\boldsymbol{u}_j$ and $\boldsymbol{v}_i$ are normal to $\hat{n}$. Hence, $\hat{n}$ is parallel to a vector of the reciprocal lattice or to a vector of the direct lattice.

Chapter 6

Theory of Group Representations

6.1 Orthogonality theorems

We consider a group H and a group G of linear transformations in a vector space $\mathcal{S}$. We say that G is a representation of H if G is homomorphic to H. The linear transformations of G may be expressed as matrices once a choice of basis vectors in $\mathcal{S}$ has been made. In applications to quantum mechanics the space $\mathcal{S}$ is often a subspace of a particular Hilbert space. It is convenient in such cases to describe the states of $\mathcal{S}$ in terms of an orthonormal basis of vectors.

The element of G associated with $a \in H$ will be denoted here by $\Gamma(a)$. Since G is homomorphic to H

$$\Gamma(a)\Gamma(b) = \Gamma(ab) \tag{1.1}$$

for an two elements a, b of H.

We have already encountered examples of group representations. The 3×3 orthogonal matrices $\Gamma(R)$ constitute a representation of the group O(3) of orthogonal transformations in 3-space. $\| \Gamma(R) \| = \pm 1$ is a representation of O(3). The totally symmetric or identity representation of a group $H = \{e, a, b, \ldots\}$ is $\Gamma(a) = 1$, the identity operation, for every element of H.

The dimension of a representation G of a group H is the dimension of the space or subspace on which the elements of G act.

Definition. The presentations Γ and Γ' of a group $H = \{e, a, b, \ldots\}$ acting on the same vector space $\mathcal{S}$ are said to be equivalent if a non-singular transformation S on $\mathcal{S}$ exists such that

$$\Gamma'(a) = S^{-1}\Gamma(a)S \tag{1.2}$$

for all $a \in H$.

Theorem 1.1: Maschke's theorem.

Every representation of a group $H = \{e, a, b, \ldots\}$ of finite order h is equivalent to a unitary representation, *i.e.*, to one whose elements are unitary transformations.

Proof: Let $a \in H$ be an arbitrary element of the group and ψ and ϕ any two vectors in the space of the representation Γ. We define an inner product of the form

$$\{\psi|\phi\} = \frac{1}{h}\sum_{a\in H} < \Gamma(a)\psi|\Gamma(a)\phi > \ . \tag{1.3}$$

The verification that this is indeed an inner product is left to the reader. For any $b \in H$ we have

$$\begin{aligned}
\{\Gamma(b)\psi|\Gamma(b)\phi\} &= \frac{1}{h}\sum_{a\in H} < \Gamma(a)\Gamma(b)\psi|\Gamma(a)\Gamma(b)\phi > \\
&= \frac{1}{h}\sum_{a\in H} < \Gamma(ab)\psi|\Gamma(ab)\phi > \\
&= \frac{1}{h}\sum_{a\in H} < \Gamma(a)\psi|\Gamma(a)\phi > = \{\psi|\phi\}
\end{aligned}$$

where we used Cayley's theorem. Thus, the operations $\Gamma(b)$ for all $b \in H$ are unitary with respect to the inner product $\{\cdots|\cdots\}$. Let $\{\epsilon_i\}$ be an orthonormal basis of the space of the representation Γ with respect of the initial inner product, *i.e.*,

$$< \epsilon_i|\epsilon_j > = \delta_{ij} \ .$$

In general $\{\epsilon_i|\epsilon_j\} \neq \delta_{ij}$. However, linear dependence has nothing to do with the definition of the inner product. Thus, the $\{\epsilon_i\}$ are linearly independent regardless of the definition of the inner product. By means of the Gram-Schmidt orthogonalization procedure we can find a set of linear combinations of ϵ_i which are orthogonal and normalized with respect to the new inner product. Let these be ϕ_i:

$$\phi_i = \sum_j \epsilon_j S_{ji} \ .$$

Since the ϕ_i are linearly independent, the matrix $[S_{ji}]$ is non-singular. Let

$$\xi = x_i\epsilon_i \ , \quad \eta = y_i\epsilon_i \ .$$

Then

$$\{S\xi|S\eta\} = x_i^* y_j\{S\epsilon_i|S\epsilon_j\} = x_i^* y_j\{\phi_i|\phi_j\} = x_i^* y_i = < \xi|\eta > \ .$$

This means that

$$\begin{aligned}
< S^{-1}\Gamma(a)S\psi|S^{-1}\Gamma(a)S\phi > &= \{\Gamma(a)S\psi|\Gamma(a)S\phi\} \\
&= \{S\psi|S\phi\} = < \psi|\phi > \ .
\end{aligned}$$

Thus the operators $\Gamma'(a) = S^{-1}\Gamma(a)S$ are unitary.

A representation Γ of a group $H = \{e, a, b, \ldots\}$ can be regarded as a set of linear transformations in the space C^ℓ of ℓ-dimensional vectors with complex components. Let $\{\phi_i\}$ be an orthonormal basis in the space of Γ; a vector

$$\xi = \sum_i x_i \phi_i$$

can be represented by the point $(x_1, x_2, \ldots x_\ell)$ of C^ℓ, ℓ being the dimension of the space of Γ. The action of $\Gamma(a)$ on a vector ϕ_i of the basis is a linear combination of these same vectors, *i.e.*,

$$\Gamma(a)\phi_i = \sum_{j=1}^{\ell} \phi_j \Gamma_{ji}(a) \ . \tag{1.4}$$

Thus, associated with the operation $\Gamma(a)$ there is a matrix, also denoted by $\Gamma(a)$, whose components are $\Gamma_{ji}(a)$. Needless to say, the matrix elements $\Gamma_{ji}(a)$ depend on the choice of basis. The transformed of

$$\xi = \sum_i x_i \phi_i$$

under $\Gamma(a)$ is

$$\Gamma(a)\xi = \sum_i x_i \Gamma(a)\phi_i = \sum_{i,j} \phi_j \Gamma_{ji}(a) x_i = \xi' \tag{1.5}$$

whose components are

$$x'_j = \sum_i \Gamma_{ji}(a) x_i \ . \tag{1.6}$$

Given a set of representations $\Gamma^{(1)}, \Gamma^{(2)}, \ldots \Gamma^{(n)}$ of dimensions $\ell_1, \ell_2, \ldots \ell_n$ of a group, we can construct a representation of dimension $\ell_1 + \ell_2 + \ldots + \ell_n$ in the space formed by the union of the spaces of $\Gamma^{(1)}, \Gamma^{(2)}, \ldots \Gamma^{(n)}$. A representation obtained in this manner is denoted by

$$\begin{aligned} \Gamma &= \Gamma_1 + \Gamma_2 + \cdots + \Gamma_n \\ &\text{or} \\ \Gamma &= \Gamma_1 \oplus \Gamma_2 \oplus \cdots \oplus \Gamma_n \ . \end{aligned} \tag{1.7}$$

We remark that some of the $\Gamma^{(i)}$ may be equal and that two different subspaces may give representations described by identical matrices. A matrix is said to be in block form if it can be written in the form

$$\begin{pmatrix} \cdot & \cdot & \cdot & 0 & 0 & \\ \cdot & \cdot & \cdot & 0 & 0 & \\ \cdot & \cdot & \cdot & 0 & 0 & \\ 0 & 0 & 0 & \cdot & \cdot & \\ 0 & 0 & 0 & \cdot & \cdot & \\ & & & & & \ddots \end{pmatrix}$$

where the dots $(\cdot)$ indicate non-zero elements. The representation Γ of a group $H = \{e, a, b, \ldots\}$ is said to be in block form if all matrices $\Gamma(a)$ are in this form with the same block structure.

Let $\{\phi_i\}$ be an orthonormal basis for the space of an ℓ-dimensional representation Γ of a group $H = \{e, a, b, \ldots\}$ and ξ a vector in this space:

$$\xi = \sum_{i=1}^{\ell} x_i \phi_i \ .$$

If Γ is in block form with an upper block of dimension ℓ_1, and if $x_i = 0$ for $i \geq \ell_1 + 1$, then $\Gamma(a)\xi$ is also a vector of the same form, *i.e.*, its components beyond the ℓ_1th vanish. If $\phi_1, \phi_2, \ldots \phi_{\ell_1}$ form a set of linearly independent vectors whose components beyond the ℓ_1th are all zero, we say that they span a vector space invariant under the operations $\Gamma(a)$ of the group representation.

Let us imagine now a representation $\{\Gamma(a)\}$ of $H = \{e, a, b, \ldots\}$ of dimension n and assume that there exist $\ell < n$ linearly independent vectors $\phi_1, \phi_2, \ldots \phi_\ell$ spanning a subspace of Γ invariant under all the operations of the group. This means that

$$\Gamma(a)\phi_i = \sum_{j=1}^{\ell} \Gamma_{ji}(a)\phi_j \tag{1.8}$$

for all $a \in H$. We can construct a basis for the n-dimensional space of Γ consisting of $\phi_1, \phi_2, \ldots \phi_\ell$ and $n - \ell$ vectors $\phi_{\ell+1}, \phi_{\ell+2}, \ldots \phi_n$ which are linearly independent and are not in the space spanned by $\phi_1, \phi_2, \ldots \phi_\ell$. Then, for $i = \ell + 1, \ell + 2, \ldots n$

$$\Gamma(a)\phi_i = \sum_{j=1}^{\ell} \Gamma_{ji}(a)\phi_j + \sum_{j=\ell+1}^{n} \Gamma_{ji}(a)\phi_j \quad i = \ell + 1, \ell + 2, \ldots n \ . \tag{1.9}$$

This shows that the matrix whose elements are $\Gamma_{ji}(a)$ is of the form

$$\Gamma(a) = \begin{pmatrix} P & Q \\ O & S \end{pmatrix} \tag{1.10}$$

where P and S are square matrices of dimensions ℓ and $n - \ell$, respectively; Q is a rectangular matrix with ℓ rows and $n - \ell$ columns and O is a zero matrix with $n - \ell$ rows and ℓ columns. When this occurs we say that the representation is reducible. In other words, a representation Γ of a group H, acting on a vector space V_n of dimension n, is reducible if there exists a non-empty proper subspace of V_n which is invariant under all the operations of the group. If such a subspace does not exist, then we say that the representation is irreducible.

A representation Γ of a group H in the form of a set of linear transformations in a space V_n is said to be completely reducible if for every non-empty subspace M of V_n invariant under the operations of the group, the orthogonal complement $M_\perp$ of M is also invariant. We immediately conclude that a reducible unitary representation of a group is completely reducible. In such a case the matrix

Q in (1.10) is null. From Maschke's theorem we deduce that every reducible representation of a finite group is completely reducible.

Two lemmas due to Schur provide criteria for the reducibility of group representations. We now discuss them in some detail.

Lemma 1. A linear transformation that commutes with all the transformations of an irreducible representation Γ of a group is a constant, *i.e.*, a multiple of the unit operator.

Proof: Let H be the group, a one of its elements and $\Gamma(a)$ the transformation associated with a. We need to prove that, if

$$M\Gamma(a) = \Gamma(a)M \tag{1.11}$$

for all $a \in H$, then M is a multiple of the unit operator. If $M = 0$ there is nothing to prove so that we assume that $M \neq 0$. Let $\mathcal{S}$ be the vector space on which Γ acts. The set of vectors $\{M\xi\}$ for all $\xi \in \mathcal{S}$ is a vector space which we denote by $M(\mathcal{S})$. In general $M(\mathcal{S})$ is a subspace of $\mathcal{S}$. From Eq. (1.11)

$$M\Gamma(a)\xi = \Gamma(a)M\xi$$

for all $a \in H$. Thus $\Gamma(a)$ acting on $M\xi$ gives an element of $M(\mathcal{S})$. But this is just the statement that $M(\mathcal{S})$ is invariant under Γ. Since Γ is irreducible $M(\mathcal{S})$ cannot be a proper subspace of $\mathcal{S}$ and, since $M \neq 0$ it cannot be just the zero vector. Thus $M(\mathcal{S}) = \mathcal{S}$. Thus, the null space of M, *i.e.*, the totality of vectors ξ such that $M\xi = 0$, contains the single element $\xi = 0$. Therefore, M is nonsingular. We suppose that M is bounded and, hence, continuous. It must have, at least, one invariant direction, *i.e.*, a vector $\phi_m \neq 0$ such that $M\phi_m = m\phi_m$. We now define

$$M' = M - mE$$

where E is the unit operator. The null space of M' contains all vectors parallel to ϕ_m. But M' satisfies Eq (1.11) and hence, the same requirements as M. If M' were non-zero we have a contradiction. Thus $M' = 0$ and, thus,

$$M = mE\ .$$

This completes the proof of the lemma.

Lemma 2. Consider a group $H = \{e, a, b, \ldots\}$ and two of its non-equivalent irreducible representations $\Gamma^{(1)}$ and $\Gamma^{(2)}$ acting on vector spaces $\mathcal{S}_1$ and $\mathcal{S}_2$ of dimensions ℓ_1 and ℓ_2, respectively. Let M be a linear mapping from $\mathcal{S}_1$ to $\mathcal{S}_2$ satisfying

$$M\Gamma^{(1)}(a) = \Gamma^{(2)}(a)M \tag{1.12}$$

for all $a \in H$. Then $M \equiv 0$.

Proof: Suppose $M \neq 0$. Then M maps $\mathcal{S}_1$ into a subspace $M(\mathcal{S}_1)$ of $\mathcal{S}_2$. If ξ is a vector of $\mathcal{S}_1$, then, by virtue of Eq. (1.12)

$$M\Gamma^{(1)}(a)\xi = \Gamma^{(2)}(a)M\xi$$

so that $M(\mathcal{S}_1)$ is an invariant subspace of $\mathcal{S}_2$ under $\Gamma^{(2)}$. Since $\Gamma^{(2)}$ is irreducible and $M \neq 0$,

$$M(\mathcal{S}_1) = \mathcal{S}_2 \ .$$

Since the dimension of $M(\mathcal{S}_1)$ is less than or equal to ℓ_1, we conclude that

$$\ell_2 \leq \ell_1 \ .$$

For the time being we assume that the operations in $\Gamma^{(1)}$ and $\Gamma^{(2)}$ are unitary. Then, from Eq. (1.12)

$$\Gamma^{(1)}(a)M^\dagger = M^\dagger\Gamma^{(2)}(a)$$

for all $a \in H$. Now $M^\dagger$ maps $\mathcal{S}_2$ into a subspace $M^\dagger(\mathcal{S}_2)$ of $\mathcal{S}_1$. As before, since $\Gamma^{(1)}$ is irreducible we conclude that

$$M^\dagger(\mathcal{S}_2) = \mathcal{S}_1$$

and that

$$\ell_1 \leq \ell_2 \ .$$

This means that $\ell_1 = \ell_2$ and that M is non-singular so that the representations $\Gamma^{(1)}$ and $\Gamma^{(2)}$ are equivalent. Since this in contradiction with the hypothesis,

$$M = 0 \ .$$

To remove the restriction to unitary representations we show that if $\Gamma(a)$ is a representation of H, then

$$D(a) = \Gamma^\dagger(a^{-1})$$

is also a representation. In fact, for any two elements a and b of the group

$$\begin{aligned} D(a)D(b) &= \Gamma^\dagger(a^{-1})\Gamma^\dagger(b^{-1}) = (\Gamma(b^{-1})\Gamma(a^{-1}))^\dagger \\ &= \Gamma^\dagger((ab)^{-1}) = D(ab) \ . \end{aligned}$$

Furthermore, if Γ is irreducible, then so is D because if there existed a proper subspace of $\mathcal{S}^\dagger$ invariant under all elements of D, then, a corresponding proper subspace of $\mathcal{S}$ would not be invariant under all operations of Γ.

Under these conditions we deduce that

$$D^{(1)}(a)M^\dagger = M^\dagger D^{(2)}(a)$$

for all a and the proof that $\ell_1 \leq \ell_2$ follows the same lines as before.

Corollary. A necessary and sufficient condition for a unitary representation Γ of a group $H = \{e, a, b, \ldots\}$ to be irreducible is that the totality of the linear transformations M in the space of Γ such that, for all $a \in H$,

$$M\Gamma(a) = \Gamma(a)M \ , \tag{1.13}$$

be multiples of the unit operator.

Proof: By virtue of lemma 1, the condition is necessary. To prove that it is sufficient we suppose that all M satisfying Eq. (1.13) are multiples of the unit transformation but that Γ is reducible. In this case there exists of proper subspace V of the space $\mathcal{S}$ of Γ which is invariant under all $\Gamma(a)$ of the group. The orthogonal complement $V_\perp$ of V is also invariant under all $\Gamma(a)$. Then, it is possible to find a transformation M such that $M\xi_1 = m_1\xi_1$ if $\xi_1 \in V$ and $M\xi_2 = m_2\xi_2$ if $\xi_2 \in V_\perp$ with $m_1 \neq m_2$. This contradicts the assumption and proves the corollary.

Theorem 1.2: Given two irreducible, inequivalent, unitary representations $\Gamma^{(1)}$ and $\Gamma^{(2)}$ of a finite group $H = \{e, a, b, \ldots\}$ of order h, we have

$$\sum_{a\in H} \Gamma^{(1)*}_{ij}(a)\Gamma^{(2)}_{k\ell}(a) = 0 \tag{1.14}$$

and

$$\sum_{a\in H} \Gamma^{(1)*}_{ij}(a)\Gamma^{(1)}_{k\ell}(a) = \frac{h}{\ell_1}\delta_{ik}\delta_{j\ell} \tag{1.15}$$

where $\ell_1(\ell_2)$ is the dimension of $\Gamma^{(1)}(\Gamma^{(2)})$.

Proof: (i) Consider a linear transformation X mapping the space of $\Gamma^{(2)}$ into that of $\Gamma^{(1)}$. In matrix form it is a rectangular matrix with ℓ_1 rows and ℓ_2 columns. We form

$$M = \sum_{a\in H} \Gamma^{(1)}(a^{-1})X\Gamma^{(2)}(a) \ . \tag{1.16}$$

For every element b of H

$$\begin{aligned} M\Gamma^{(2)}(b) &= \sum_{a\in H} \Gamma^{(1)}(b)\Gamma^{(1)}(b^{-1})\Gamma^{(1)}(a^{-1})X\Gamma^{(2)}(a)\Gamma^{(2)}(b) \\ &= \Gamma^{(1)}(b)\sum_{a\in H} \Gamma^{(1)}((ab)^{-1})X\Gamma^{(2)}(ab) \\ &= \Gamma^{(1)}(b)M \ , \end{aligned}$$

where in the last equality we made use of Cayley's theorem. From lemma 2, $M = 0$ independently of the choice of X. Thus

$$\sum_{a\in H}\sum_{ik} \Gamma^{(1)*}_{ij}(a)X_{ik}\Gamma^{(2)}_{k\ell}(a) = 0$$

Taking X so that all its elements vanish except X_{ik} which is equal to unity we obtain Eq. (1.14).

(ii) If Γ is an irreducible representation of H and we define

$$M = \sum_{a\in H} \Gamma(a^{-1})X\Gamma(a) \ , \tag{1.17}$$

then we conclude that

$$M\Gamma(b) = \Gamma(b)M$$

for every $b \in H$. By virtue of lemma 1, $M = c(X)E$, where $c(X)$ is a number which depends on the choice of X.

In matrix form Eq. (1.17) reads

$$\sum_{ik}\sum_{a\in H}\Gamma_{ji}(a^{-1})X_{ik}\Gamma_{k\ell}(a) = c(X)\delta_{j\ell} .$$

Setting $\ell = j$ and summing over j we obtain

$$\sum_{ik}X_{ik}\sum_{a\in H}\sum_{j}\Gamma_{ji}(a^{-1})\Gamma_{kj}(a) = \ell c(X)$$

where ℓ is the dimension of Γ. But

$$\sum_{j}\Gamma_{ji}(a^{-1})\Gamma_{kj}(a) = \Gamma_{ki}(aa^{-1}) = \delta_{ik}$$

so that

$$c(X) = \frac{h}{\ell}TrX . \tag{1.18}$$

We now chose X so that all elements of X vanish except X_{ik} which is equal to unity; we obtain

$$\sum_{a\in H}\Gamma^*_{ij}(a)\Gamma_{k\ell}(a) = \frac{h}{\ell}\delta_{ik}\delta_{j\ell} \tag{1.19}$$

where we used the fact that Γ is in unitary form. We can combine Eqs. (1.14) and (1.15) into

$$\sum_{a\in H}\Gamma^{(i)*}_{k\ell}(a)\Gamma^{(j)}_{mn}(a) = \frac{h}{\ell_i}\delta_{ij}\delta_{km}\delta_{\ell n} \tag{1.20}$$

where $\Gamma^{(i)}$ and $\Gamma^{(j)}$ are irreducible representations of H; here $\delta_{ij} = 1$ only if $\Gamma^{(i)}$ and $\Gamma^{(j)}$ are **identical** unitary, irreducible representations of H. This restriction will be circumvented later.

This theorem is called the great orthogonality theorem. It may be written in the form

$$\sum_{a\in H}\left(\frac{\ell_i}{h}\right)^{\frac{1}{2}}\Gamma^{(i)*}_{k\ell}(a)\left(\frac{\ell_j}{h}\right)^{\frac{1}{2}}\Gamma^{(j)}_{mn}(a) = \delta_{ij}\delta_{km}\delta_{\ell n} . \tag{1.21}$$

We can view this result as expressing the orthogonality of vectors whose components, labeled by the elements of the group, are

$$\left(\frac{\ell_i}{h}\right)^{\frac{1}{2}}\Gamma^{(i)}_{k\ell}(a) . \tag{1.22}$$

The appropriate space is C^h. Since no more than h such vectors can be mutually orthogonal we conclude that the number of inequivalent irreducible representations of a finite group is, necessarily, finite. Let this number be N_Γ. The representation $\Gamma^{(i)}$ contributes ℓ_i^2 such vectors. The number of orthogonal vectors is, then

$$\ell_1^2 + \ell_2^2 + \cdots + \ell_{N_\Gamma}^2 .$$

We conclude that

$$\ell_1^2 + \ell_2^2 + \cdots + \ell_{N_\Gamma}^2 \leq h \, . \tag{1.23}$$

Definition. The character of a representation Γ of a group $H = \{e, a, b, \ldots\}$ is the set $\{\chi(a)\}$ of the traces of the matrices $\Gamma(a)$ of the representation, *i.e.*,

$$\chi(a) = \sum_i \Gamma_{ii}(a) \, . \tag{1.24}$$

Two equivalent representations have identical characters for, if

$$\Gamma'(a) = S^{-1}\Gamma(a)S \, ,$$

then

$$Tr\Gamma'(a) = Tr\Gamma(a) \, .$$

Thus, for the purpose of determining the character of a representation it is not necessary for it to be in unitary form.

If b and a are conjugate elements of H, then $u \in H$ exists for that $b = u^{-1}au$ and

$$\begin{aligned} \chi(b) &= Tr\Gamma(b) = Tr\Gamma(u^{-1}au) = Tr[\Gamma(u^{-1})\Gamma(a)\Gamma(u)] \\ &= Tr[\Gamma(a)\Gamma(u)\Gamma(u^{-1})] = Tr\Gamma(a) = \chi(a) \, . \end{aligned}$$

Thus, while determining the character of a representation, it is enough to obtain the trace of one element in each class. The trace of $\Gamma(e)$ is the dimension of the representation.

Theorem 1.3: If $\{\chi^{(i)}(a)\}$ and $\{\chi^{(j)}(a)\}$ are the characters of $\Gamma^{(i)}$ and $\Gamma^{(j)}$, two irreducible representations of a finite group $H = \{e, a, b, \ldots\}$, then

$$\sum_{a \in H} \chi^{(i)*}(a)\chi^{(j)}(a) = h\delta_{ij} \tag{1.25}$$

where $\delta_{ij} = 1$ if the representations are equivalent and zero otherwise.

Proof: Set $\ell = k$ and $n = m$ in Eq. (1.20) and sum over k and m. Then Eq. (1.25) follows immediately. Labeling the conjugate classes by $C_\mu(\mu = 1, 2, \ldots N_c)$ where N_c is their number, Eq. (1.25) implies

$$\sum_{\mu=1}^{N_c} n_\mu \chi^{(i)*}(C_\mu)\chi^{(j)}(C_\mu) = h\delta_{ij} \tag{1.26}$$

where n_μ is the number of elements in class C_μ and $\chi(C_\mu)$ the trace of an element in that class. Equation (1.26) can be interpreted as an orthogonality relation in a complex vector space of N_c dimensions. There are as many vectors whose components are

$$\left(\frac{n_\mu}{h}\right)^{\frac{1}{2}} \chi^{(i)}(C_\mu)$$

as there are inequivalent, irreducible representations of H. Since the number of mutually orthogonal vectors cannot exceed the dimension of the space

$$N_\Gamma \leq N_c \ . \tag{1.27}$$

The advantage of theorem 1.3 over the great orthogonality theorem 1.2 is that it does not depend on the particular set of basis vectors used in the description of the space of Γ. The importance of the character is a consequence of the next few theorems.

Theorem 1.4: A necessary and sufficient condition for two irreducible representations of a finite group to be equivalent is that their characters be the same.

Proof: (i) the necessity has already been proved

(ii) the condition is sufficient for, if $\chi^{(i)}(a) = \chi^{(j)}(a)$ but $\Gamma^{(i)}$ and $\Gamma^{(j)}$ are inequivalent, then

$$\sum_{a\in H} |\chi^{(i)}(a)|^2 = 0$$

which contradicts

$$\sum_{a\in H} |\chi^{(i)}(a)|^2 = h \ .$$

We are now in a position to establish how an arbitrary representation of a finite group can be reduced into its irreducible components. Let Γ be such a representation of $H = \{e, a, b, \ldots\}$ and suppose its irreducible components are $\Gamma^{(i)}$ with multiplicity b_i, $(i = 1, 2, \ldots N_\Gamma)$ *i.e.*,

$$\Gamma = b_1\Gamma^{(1)} + b_2\Gamma^{(2)} + \ldots + b_{N_\Gamma}\Gamma^{(N_\Gamma} \tag{1.28}$$

where $b_i = 0, 1, 2, \ldots$. Then for $a \in H$

$$\chi(a) = \sum_{i=1}^{N_\Gamma} b_i\chi^{(i)}(a) \ , \tag{1.29}$$

or

$$\chi(C_\mu) = \sum_{j=1}^{N_\Gamma} b_j\chi^{(j)}(C_\mu) \tag{1.30}$$

Multiplying both sides of Eq. (1.30) by $n_\mu\chi^{(i)*}(C_\mu)$ and summing over μ we obtain

$$b_i = \frac{1}{h}\sum_{\mu=1}^{N_c} n_\mu\chi^{(i)*}(C_\mu)\chi(C_\mu) \ . \tag{1.31}$$

Equation (1.31) is the basic formula for the reduction of an arbitrary representation. In practice we often carry out reductions by inspection.

From Eq. (1.30) we obtain

$$\begin{aligned}\sum_{\mu=1}^{N_c} n_\mu|\chi(C_\mu)|^2 &= \sum_{i,j} b_ib_j\sum_{\mu=1}^{N_c} n_\mu\chi^{(i)*}(C_\mu)\chi^{(j)}(C_\mu) \\ &= h\sum_{i=1}^{N_\Gamma} b_i^2\end{aligned} \tag{1.32}$$

where we used Eq. (1.26).

Theorem 1.5: A necessary and sufficient condition for a representation whose character is $\{\chi(C_\mu)\}$ to be irreducible is

$$\sum_{\mu=1}^{N_c} n_\mu |\chi(C_\mu)|^2 = h \tag{1.33}$$

Proof: If the representation is irreducible only one of the coefficients b_i in Eq. (1.32) is different from zero and is itself unity so that (1.33) follows. If (1.33) is valid, then from Eq. (1.32)

$$\sum_{i=1}^{N_\Gamma} b_i^2 = 1$$

so that all the b_i are zero except one which is unity. Thus the reduction of the representation consists of a single irreducible component. Equation (1.33) can also be written as

$$\sum_{a\in H} |\chi(a)|^2 = h \ .$$

Theorem 1.6: The sum of the linear transformations of an irreducible representation of a group belonging to a conjugate class is a multiple of the unit map.

Proof: Let $a_1, a_2, \dots a_n$ be the elements of a conjugate class of a group H and Γ one of its irreducible representations. Form

$$M = \sum_{i=1}^{n} \Gamma(a_i) \ .$$

For any $b \in H$

$$\Gamma(b^{-1})M\Gamma(b) = \sum_{i=1}^{n} (b^{-1}a_i b) = M$$

so that

$$M\Gamma(b) = \Gamma(b)M \ .$$

By virtue of lemma 1, M is a multiple of the unit map. We remark that this result is valid for infinite as well as for finite groups.

Consider now a finite group and, for each set of equivalent irreducible representations and for each conjugate class, $\Gamma^{(i)}$ and C_μ, respectively, we form

$$M_\mu^{(i)} = \sum_{k=1}^{n_\mu} \Gamma^{(i)}(a_k^\mu) = m_\mu^{(i)} E^{(i)} \tag{1.34}$$

where $E^{(i)}$ is the unit operation (dimension ℓ_i) in the space of $\Gamma^{(i)}$, a_k^μ is the kth element in the class C_μ; and $m_\mu^{(i)}$ is a number appropriate to class μ and to $\Gamma^{(i)}$. Taking the trace of both sides of Eq. (1.34) we find

$$n_\mu \chi^{(i)}(C_\mu) = \ell_i m_\mu^{(i)} \ . \tag{1.35}$$

The product of $M_\mu^{(i)}$ and $M_\nu^{(i)}$ is

$$M_\mu^{(i)} M_\nu^{(i)} = \sum_{k=1}^{n_\mu} \sum_{\ell=1}^{n_\nu} \Gamma^{(i)}(a_k^\mu a_\ell^\nu) \ .$$

The sum on the right hand side extends over the products in the set

$$[C_\mu C_\nu] \ .$$

In Sec. 3.8. we saw that

$$[C_\mu C_\nu] = \sum_{\lambda=1}^{N_c} n_{\mu\nu\lambda} C_\lambda \ .$$

Thus

$$M_\mu^{(i)} M_\nu^{(i)} = \sum_{\lambda=1}^{N_c} n_{\mu\nu\lambda} M_\lambda^{(i)} \ . \tag{1.36}$$

Combining Eqs. (1.34) and (1.36) we obtain

$$m_\mu^{(i)} m_\nu^{(i)} = \sum_{\lambda=1}^{N_c} n_{\mu\nu\lambda} m_\lambda^{(i)} \ .$$

From (1.35) we obtain

$$n_\mu n_\nu \chi^{(i)}(C_\mu) \chi^{(i)}(C_\nu) = \ell_i \sum_{\lambda}^{N_c} n_{\mu\nu\lambda} n_\lambda \chi^{(i)}(C_\lambda) \ . \tag{1.37}$$

The Regular representation

In Sec 3.4, we discussed the multiplication tables of groups. In those tables the entries were arranged in the same way for columns and rows. We now consider arranging the rows in a certain order but the columns are arranged in the order of the reciprocals of the elements in the rows. This is shown in the table for a group $H = \{e, a_2, a_3 \ldots a_h\}$. The main diagonal of the table contains at each position the identity element e. We remark that the inverses of the elements form a rearrangement of the elements.

H	e	a_2^{-1}	a_3^{-1}	$\ldots$	a_h^{-1}
e	e	a_2^{-1}	a_3^{-1}	$\ldots$	a_h^{-1}
a_2	a_2	e	$a_2 a_3^{-1}$	$\ldots$	$a_2 a_h^{-1}$
a_3	a_3	$a_3 a_2^{-1}$	e	$\ldots$	$a_3 a_h^{-1}$
$\vdots$	$\vdots$				$\vdots$
a_h	a_h	$a_h a_2^{-1}$	$a_h a_2^{-1}$	$\ldots$	e

We now associate with the element a of H an $h \times h$ matrix in which unity is placed in every position on the table above in which a appears and a zero everywhere else. We claim that the set of matrices formed in this way constitute a representation of H which we call the regular representation, $\Gamma^{(R)}$. Formally

$$\Gamma^{(R)}_{ij}(a) = \begin{cases} 1 & \text{if } a_i a_j^{-1} = a \\ 0 & \text{otherwise} . \end{cases}$$

Clearly $\Gamma^{(R)}(e)$ is the unit $h \times h$ matrix. To prove that $\Gamma^{(R)}$ is indeed a representation of H we consider

$$\sum_k \Gamma^{(R)}_{ik}(a)\Gamma^{(R)}_{kj}(b)$$

where a and b are any two elements of H. For given i and j the term k in the sum above is different from zero only if

$$a_i a_k^{-1} = a \ , \ \ a_k a_j^{-1} = b$$

are simultaneously verified. But each of these conditions is enough to define a_k uniquely. This means that the terms in the sum over k vanish unless

$$a_i a_j^{-1} = ab \ ,$$

in which case the sum is equal to unity. Thus

$$\sum_k \Gamma^{(R)}_{ik}(a)\Gamma^{(R)}_{kj}(b) = \Gamma^{(R)}_{ij}(ab)$$

or

$$\Gamma^{(R)}(a)\Gamma^{(R)}(b) = \Gamma^{(R)}(ab) \ .$$

This establishes our statement. In addition it also shows that the matrices $\Gamma^{(R)}$ are non-singular since

$$\Gamma^{(R)}(a)\Gamma^{(R)}(a^{-1}) = \Gamma^{(R)}(e) = E_h \ .$$

Theorem 1.7. The Celebrated Theorem. The reduction of the regular representation of a finite group contains each of its inequivalent irreducible representations as many times as its dimension, *i.e.*,

$$\Gamma^{(R)} = \sum_{i=1}^{N_\Gamma} \ell_i \Gamma^{(i)} \ . \tag{1.38}$$

Proof: Let

$$\Gamma^{(R)} = \sum_{i=1}^{N_\Gamma} b_i \Gamma^{(i)} \ .$$

Use of Eq. (1.31) gives

$$b_i = \frac{1}{h}\sum_{\mu=1}^{N_c} n_\mu \chi^{(i)*}(C_\mu)\chi^{(R)}(C_\mu) = \frac{1}{h}\chi^{(i)}(e)h = \ell_i$$

since all $\chi^{(R)}(C_\mu) = 0$ except $\chi^{(R)}(e)$ which equals h. Taking the trace of

$$\Gamma^{(R)}(e) = \sum_{i=1}^{N_\Gamma} \ell_i \Gamma^{(i)}(e)$$

we obtain

$$\sum_{i=1}^{N_\Gamma} \ell_i^2 = h \ . \tag{1.39}$$

Thus, in the inequality (1.23) the strict equality always holds.

As a consequence of theorem 1.3 we concluded that the number of inequivalent irreducible representations of a finite group cannot exceed the number of conjugate classes. We now prove that $N_\Gamma = N_c$ for which we first make a few preliminary developments. We consider first the product $[C_\mu C_\nu]$ of two classes in which we keep repeated elements. This product contains the identity e, if and only if, an element a of C_μ exists such that a^{-1} is in C_ν. If this is the case for any element b of C_μ we can find $u \in H$ such that $b = u^{-1}au$. Then $b^{-1} = u^{-1}a^{-1}u \in C_\nu$. Thus $[C_\mu C_\nu]$ contains e only if the classes C_μ and C_ν consist entirely of each other's inverses. The number of times e appears in this product is, therefore

$$n_{\mu\nu 1} = n_\mu \delta_{\mu'\nu} \tag{1.40}$$

where we denote by $C_{\mu'}$ the conjugate class consisting of the inverses of the elements in C_μ. It may, of course, happen that $C_{\mu'} = C_\mu$ (*e.g.*, in T the classes $4C_3$ and $4C_3^{-1}$ contain each other's inverses but in T_d, $8C_3$ contains all C_3's as well as their inverses).

Since representations of finite groups are always equivalent to unitary representations

$$\chi(C_{\mu'}) = Tr\Gamma(a^{-1}) = Tr\Gamma^\dagger(a) = \chi^*(C_\mu) \ . \tag{1.41}$$

Furthermore

$$n_{\mu'} = n_\mu \ . \tag{1.42}$$

Theorem 1.8. The N_c vectors associated with the conjugate classes of a finite group H of order h, whose i-component are

$$\left(\frac{n_\mu}{h}\right)^{\frac{1}{2}} \chi^{(i)}(C_\mu) \quad ; \quad i = 1, 2, \dots N_\Gamma \ ; \ \mu = 1, 2, \dots N_c$$

are orthonormal, *i.e.*,

$$\sum_{i=1}^{N_\Gamma} \chi^{(i)*}(C_\mu)\chi^{(i)}(C_\nu) = \frac{h}{n_\mu}\delta_{\mu\nu} \tag{1.43}$$

where $\delta_{\mu\nu} = 1$ if the classes C_μ and C_ν are the same and zero otherwise.
Proof: From Eq. (1.37)

$$\sum_{i=1}^{N_\Gamma} n_{\mu'} n_\nu \chi^{(i)}(C_{\mu'})\chi^{(i)}(C_\nu) = \sum_{\lambda=1}^{N_c} n_{\mu'\nu\lambda} n_\lambda \sum_{i=1}^{N_\Gamma} \ell_i \chi^{(i)}(C_\lambda)$$

where the sum over i extends over all inequivalent irreducible representations of H. But

$$\sum_{i=1}^{N_\Gamma} \ell_i \chi^{(i)}(C_\lambda)$$

is nothing other than the trace of an element in the class C_λ in the regular representation. Thus, this sum vanishes except when $C_\lambda = C_1 = \{e\}$ in which case the sum is $\ell_1^2 + \ell_2^2 + \ldots + \ell_{N_\Gamma}^2 = h$. Thus,

$$\sum_{i=1}^{N_\Gamma} n_{\mu'} n_\nu \chi^{(i)}(C_{\mu'}) \chi^{(i)}(C_\nu) = h n_{\mu'\nu 1} \ .$$

The validity of Eq. (1.43) follows immediately from this result and Eqs. (1.40), (1.41) and (1.42).

The orthogonality of the N_c vectors in Eq. (1.43) shows that

$$N_c \leq N_\Gamma \ . \tag{1.44}$$

The inequalities (1.27) and (1.44) can only be consistent if

$$N_c = N_\Gamma \ ; \tag{1.45}$$

We conclude that the number of inequivalent irreducible representations of a finite group is equal to the number of its conjugate classes.

6.2 Character tables of finite groups

We have seen that the irreducible representations of a finite group are uniquely determined by their characters. In many applications of group theory a knowledge of the characters and basis vectors of the irreducible representations of particular symmetry groups is all that is required. For this reason we devote this section to an explanation of how character tables are constructed.

We consider finite groups of orthogonal transformations in three-dimensional space, *i.e.*, finite subgroups of O(3). Let $\psi(\boldsymbol{r}_1, \boldsymbol{r}_2, \ldots \boldsymbol{r}_N)$ be a function of the positions, $\boldsymbol{r}_i (i = 1, 2, \ldots N)$, of N points in space. This function may represent the potential energy of a mechanical system or the wave function of a quantum mechanical system composed of N point particles. If the object is transformed by performing an operation R of O(3), each position vector $\boldsymbol{r}_i$ is transformed into

$$\boldsymbol{r}'_i = R\boldsymbol{r}_i \ , \quad i = 1, 2, \ldots N \ . \tag{2.1}$$

The physical object described by $\psi(\boldsymbol{r}_1, \boldsymbol{r}_2, \ldots \boldsymbol{r}_N)$, having been transformed, is characterized by a new function

$$P_R \psi(\boldsymbol{r}'_1, \boldsymbol{r}'_2, \ldots \boldsymbol{r}'_N)$$

such that it has at $\boldsymbol{r}'_1, \boldsymbol{r}'_2, \ldots \boldsymbol{r}'_N$ the same numerical value that ψ had at $\boldsymbol{r}_1, \boldsymbol{r}_2, \ldots$ $\ldots \boldsymbol{r}_N$. If R were a rotation $P_R\psi$ represents the same object as ψ rigidly rotated to its new position. Thus

$$P_R\psi(\boldsymbol{r}'_1, \boldsymbol{r}'_2, \ldots \boldsymbol{r}'_N) = \psi(\boldsymbol{r}_1, \boldsymbol{r}_2, \ldots \boldsymbol{r}_N) = \psi(R^{-1}\boldsymbol{r}'_1, R^{-1}\boldsymbol{r}'_2, \ldots R^{-1}\boldsymbol{r}'_N) \ . \quad (2.2)$$

This is the, so-called, active point of view. We suppose that the object of interest has been physically transformed, *e.g.*, rotated. There is also a passive view in which we assume the physical object is fixed in space but the reference frame has been transformed. In the latter case the axes of the reference frame have been transformed according to R^{-1}.

To simplify the writing we use functions of the position $\boldsymbol{r} = (x, y, z)$ of a single point. Our results are, clearly, generally valid.

For any subgroup $\{R\}$ of O(3), there is a group $\{O_R\}$ of operations on functions. If $\boldsymbol{r} \to \boldsymbol{r}' = R\boldsymbol{r}$ and $\boldsymbol{r}' \to \boldsymbol{r}'' = S\boldsymbol{r}' = SR\boldsymbol{r}$, then

$$P_S P_R \psi(\boldsymbol{r}'') = P_R\psi(\boldsymbol{r}') = \psi(\boldsymbol{r}) = \psi((SR)^{-1}r'') = P_{SR}\psi(\boldsymbol{r}'') \ .$$

Thus

$$P_S P_R = P_{SR} \ , \quad (2.3)$$

and the operations $\{P_R\}$ form a group.

In the L^2 space of functions $\psi(\boldsymbol{r})$, with inner product

$$< \psi|\phi >= \int \psi^*(\boldsymbol{r})\phi(\boldsymbol{r})d\boldsymbol{r}$$

we have

$$\begin{aligned} < P_R\psi|P_R\phi > &= \int (P_R\psi(\boldsymbol{r}'))^*(P_R\phi(\boldsymbol{r}'))d\boldsymbol{r}' \\ &= \int \psi^*(\boldsymbol{r})\phi(\boldsymbol{r}) \left| \frac{\partial(x', y', z')}{\partial(x, y, z)} \right| d\boldsymbol{r} \\ &= < \psi|\phi > \end{aligned}$$

since the Jacobian of the transformation $\boldsymbol{r} \to \boldsymbol{r}'$ has absolute value equal to unity. This shows that P_R is a unitary operator in L^2.

A rotation by φ about the z-axis of a Cartesian coordinate system transforms $(x, y, z) \to (x', y', z')$ according to

$$\begin{aligned} x' &= x\cos\varphi - y\sin\varphi \\ y' &= x\sin\varphi + y\cos\varphi \\ z' &= z \ . \end{aligned}$$

Thus

$$\begin{aligned} P_{C_\varphi}x' &= x = x'\cos\varphi + y'\sin\varphi \ , \\ P_{C_\varphi}y' &= y = -x'\sin\varphi + y'\cos\varphi \ , \\ P_{C_\varphi}z' &= z = z' \ . \end{aligned}$$

The components V_x, V_y, V_z of a vector field transform as

$$\begin{aligned} P_{C\varphi}V_1 &= V_1\cos\varphi + V_2\sin\varphi \\ P_{C\varphi}V_2 &= -V_1\sin\varphi + V_2\cos\varphi \\ P_{C\varphi}V_3 &= V_3 \end{aligned} \tag{2.4}$$

where we use V_1, V_2, V_3 instead of V_x, V_y, V_z. Thus $P_{C\phi}$ is represented by the matrix

$$\begin{pmatrix} \cos\varphi & \sin\varphi & 0 \\ -\sin\varphi & \cos\varphi & 0 \\ 0 & 0 & 1 \end{pmatrix}$$

whose trace is

$$\chi_V(C_\varphi) = 1 + 2\cos\varphi \ . \tag{2.5}$$

This result is clearly valid for all rotations by φ because we can transform a rotation by φ about an arbitrary axis into one about the z-axis by means of a similarity transformation involving a rotation.

The transformation

$$S_\varphi = \sigma_h C_\varphi \tag{2.6}$$

where σ_h is a reflection in the (x, y) plane leads to

$$\begin{aligned} P_{S\varphi}V_1 &= V_1\cos\varphi + V_2\sin\varphi \\ P_{S\varphi}V_2 &= -V_1\sin\varphi + V_2\cos\varphi \\ P_{S\varphi}V_3 &= -V_3 \end{aligned} \tag{2.7}$$

so that

$$\chi_V(S_\varphi) = -1 + 2\cos\varphi \ . \tag{2.8}$$

For a mirror plane

$$\chi_V(\sigma) = 1 \tag{2.9}$$

while the inversion i gives

$$\chi_V(i) = -3 \ . \tag{2.10}$$

The components of a pseudovector transform as those of a vector under rotations but are multiplied by -1 under improper operations. Thus

$$\chi_A(C_\varphi) = 1 + 2\cos\varphi \ , \tag{2.11}$$

$$\chi_A(S_\varphi) = 1 - 2\cos\varphi \ , \tag{2.12}$$

$$\chi_A(\sigma) = -1, \tag{2.13}$$

and

$$\chi_A(i) = 3 \ . \tag{2.14}$$

Here we denote the pseudovector by A (antisymmetric second rank tensor).

Depending on the particular group in question, representations generated by a vector or by a pseudovector vector need not be irreducible. As an example we consider D_3, the group of rotations leaving an equilateral triangle invariant. The rotations are C_3 and C_3^{-1} about an axis through the center of gravity of the triangle normal to its plane and C_2, C_2', C_2'' three rotations by π about the altitudes of the triangle.

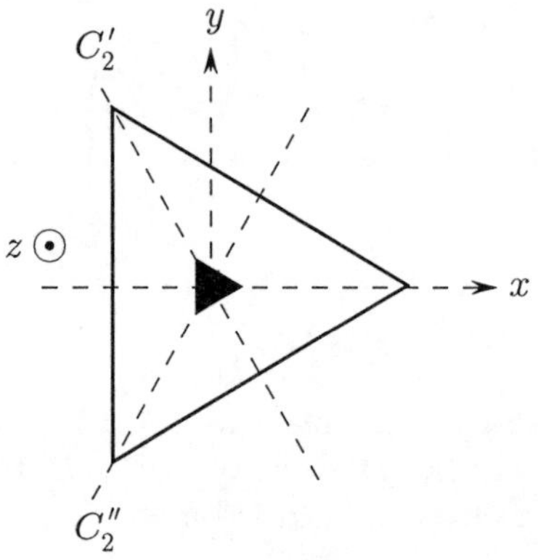

Figure 6.1

The conjugate classes are

$$E, \{C_3, C_3^{-1}\} \ , \ \{C_2, C_2', C_2''\}$$

We write

$$D_3 = \{E, 2C_3, 3C_2\} \tag{2.15}$$

This group has three conjugate classes and, hence, three non-equivalent irreducible representations. We see immediately that the representation generated by the vector (x, y, z) is reducible because the subspaces of vectors (x, y) in the plane of the triangle and of vectors z, normal to its plane are invariant under all operations of the group.

According to the celebrated theorem, the dimensions ℓ_1, ℓ_2, and ℓ_3 of the irreducible representations must satisfy

$$\ell_1^2 + \ell_2^2 + \ell_3^2 = 6 \ .$$

This restriction has only one solution, namely $\ell_1 = \ell_2 = 1, \ell_3 = 2$. We denote these representations by Γ_1, Γ_2 and Γ_3. The first is that in which every element of the group is represented by the unit operator or the number 1. The functions $x^2 + y^2$ and z^2 are invariant with respect to all operations of the group: Γ_1 is the totally symmetric or identity representation of D_3. The functions which remain invariant under all operations of the group are said to belong to Γ_1. A function transforming as the z-component of a vector remains invariant under

E, C_3 and C_3^{-1} but changes sign under C_2, C_2' and C_2''. Such a function generates a representation $\Gamma(a) = 1$ for $a = E, C_3, C_3^{-1}$ and $\Gamma(a) = -1$ for $a = C_2, C_2'$ or C_2''. We call this Γ_2. Application of theorem 1.5 verifies its irreducibility. It is , of course, evident since reduction is not possible. Finally the components x, y of a vector in the plane of the triangle transform as

$$\begin{aligned} P_{C_3} x &= x \cos\frac{2\pi}{3} + y \sin\frac{2\pi}{3} = -\frac{1}{2}x + \frac{1}{2}\sqrt{3}y \\ P_{C_3} y &= -x \sin\frac{2\pi}{3} + y \cos\frac{2\pi}{3} = -\frac{1}{2}\sqrt{3}x - \frac{1}{2}y \end{aligned}$$

Thus, the matrix associated with C_3 is

$$[P_{C_3}] = \begin{pmatrix} -\frac{1}{2} & \frac{1}{2}\sqrt{3} \\ -\frac{1}{2}\sqrt{3} & -\frac{1}{2} \end{pmatrix} .$$

Similarly

$$[P_{C_3^{-1}}] = \begin{pmatrix} -\frac{1}{2} & -\frac{1}{2}\sqrt{3} \\ \frac{1}{2}\sqrt{3} & -\frac{1}{2} \end{pmatrix} .$$

The action of P_{C_2} on the components x and y of a vector is $P_{c_2}x = x$, $P_{c_2}y = -y$ so that

$$[P_{C_2}] = \begin{pmatrix} 1 & 0 \\ 0 & -1 \end{pmatrix} .$$

The character table of D_3 is, therefore

D_3	E	$2C_3$	$3C_2$	Basis Functions
Γ_1	1	1	1	$x^2 + y^2, z^2$
Γ_2	1	1	-1	z
Γ_3	2	-1	0	(x, y) ; $(yz, -zx)$; $(x^2 - y^2, -2xy)$

Since D_3 contains only proper rotations, the components of a pseudovector transform, under its symmetry operations, in the same manner as the components of a vector. Instead of x and y as basis functions for Γ_3 we could have selected, for example

$$u_1 = -\frac{i}{\sqrt{2}}(x + iy) \ , \ u_{-1} = \frac{i}{\sqrt{2}}(x - iy) .$$

Then

$$P_{C_3} u_1 = e^{-i2\pi/3} u_1 = \omega^2 u_1 \ ; \ P_c u_{-1} = \omega u_{-1} \ , \ \omega = e^{\frac{2\pi i}{3}} .$$

With respect to these functions

$$[P_{C_3}] = \begin{pmatrix} \omega^2 & 0 \\ 0 & \omega \end{pmatrix}$$

whose trace is $\omega^2 + \omega = -1$.

Consider a symmetric second rank tensor $\boldsymbol{\alpha}$ whose components α_{ij} verify

$$\alpha_{ij} = \alpha_{ji} .$$

The components of this tensor transform as $V_i V_j$ where V_i's are components of a vector. Then

$$\alpha_{11} + \alpha_{22} \text{ and } \alpha_{33}$$

behaving as $x^2 + y^2$ and as z^2, respectively belong to Γ_1. The remaining components are

$$\alpha_{23}, -\alpha_{31} \quad \in \Gamma_3$$

and

$$\alpha_{11} - \alpha_{22}, -2\alpha_{12} \quad \in \Gamma_3 .$$

The components of an antisymmetric second rank tensor A_{ij} are

$$A_{23} = S_1,\ A_{31} = S_2,\ A_{12} = S_3$$

or

$$\boldsymbol{A} = \begin{pmatrix} 0 & S_3 & -S_2 \\ -S_3 & 0 & S_1 \\ S_2 & -S_1 & 0 \end{pmatrix} .$$

Since under rotations S_1, S_2, S_3 behave as the components of a vector

$$\begin{aligned} A_{12} &= S_3 \in \Gamma_2 \\ A_{23} &= S_1 \ , \ A_{31} = S_2 \in \Gamma_3 . \end{aligned}$$

A simple example of a group containing improper operations is C_{3v}. This group contains the orthogonal operations that carry the equilateral triangle, already discussed in relation to D_3, onto itself without exposing the opposite face of the figure. The new operations are reflections in "vertical" planes containing the z-axis (vertical axis) and the altitudes of the triangle. These reflections are denoted by σ_v, σ_v' and σ_v''. The conjugate class structure of this group is

$$C_{3v} = \{E, 2C_3, 3\sigma_v\} . \tag{2.16}$$

C_{3v} is isomorphic to D_3 since they are both isomorphic to the symmetric group S_3 (the group of permutations of three objects; in this case the permutations of the vertices of the triangle). Thus, the character table of C_{3v} is identical to that of D_3 but the basis functions are as shown in the table below.

C_{3v}	E	$2C_3$	$3\sigma_v$	Basis Functions
Γ_1	1	1	1	z , z^2 , x^2+y^2 , $(2z^2-x^2-y^2)$
Γ_2	1	1	-1	S_z , $y^3 - 3x^2y$
Γ_3	2	-1	0	(x,y) ; $(S_y - S_x)$; (zx, zy) ; $(x^2-y^2, -2xy)$

The function $y^3 - 3x^2y$, belonging to Γ_2 is obtained forming the vector product of (x, y) and $(x^2 - y^2, -2xy)$.

The representations $\Gamma_1, \Gamma_2, \Gamma_3$ are often denoted by A_1, A_2 and E, particularly in the chemical literature.

A simple Abelian group is

$$C_3 = \{E, C_3, C_3^{-1}\}$$

whose character table is

	C_3	E	C_3	C_3^{-1}	Basis Functions ($\omega = e^{i2\pi/3}$)
A_1	Γ_1	1	1	1	z , x^2+y^2 , z^2
E	Γ_2	1	ω^2	ω	$u_1 = -\frac{i}{\sqrt{2}}(x+iy)$, u_{-1}^2
	Γ_3	1	ω	ω^2	$u_{-1} = \frac{i}{\sqrt{2}}(x-iy)$, u_1^2

Since C_3 is Abelian every element is in a class by itself. Thus, there are as many inequivalent irreducible representations as elements in the group, each being one dimensional as is required by the celebrated theorem. We note that Γ_2 and Γ_3 are complex conjugate representations. In chemistry they are denoted together by E for reasons to be discussed later.

We now turn our attention to the cubic groups T, T_d and O. The groups T_h and O_h will be studied later after we discuss direct products of groups.

We consider a cube whose edges are two units of length and inscribe in it a regular tetrahedron $P_1P_2P_3P_4$. We take the origin at the center of the cube and the x, y and z axes parallel to its edges. The vertices of the tetrahedron are (see Fig. 6.2)

$$\begin{array}{ll} P_1 & (1,1,1) \\ P_2 & (1,-1,-1) \\ P_3 & (-1,1,-1) \\ P_4 & (-1,-1,1)\ . \end{array}$$

The operations leaving the tetrahedron invariant are

(i) E, the identity

(ii) $4C_3$, four rotations by $2\pi/3$ about the axes $[111]$, $[1\bar{1}\bar{1}]$, $[\bar{1}1\bar{1}]$, and $[\bar{1}\bar{1}1]$

(iii) $4C_3^{-1}$, four rotations by $-2\pi/3$ about the same axes as in (ii)

(iv) $3C_2$, three rotations by π about $\hat{x}, \hat{y}$ and $\hat{z}$

(v) $6\sigma_d$, six mirror planes bisecting the (y, z), (z, x) and (x, y) planes

(vi) $6S_4$, six improper rotations by $\pm\pi/2$ about the axes $\hat{x}, \hat{y}, \hat{z}$ ($3S_4$ and $3S_4^{-1}$).

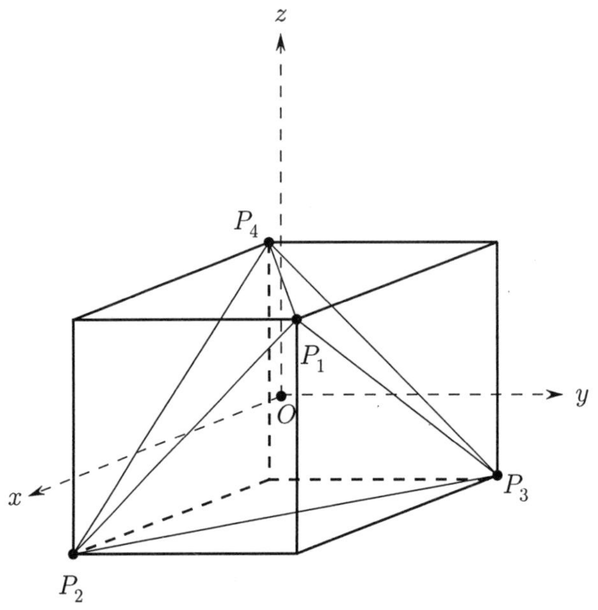

Figure 6.2

We study first the group T containing the rotations in (i)–(iv). Since a rotation by π about $\hat{z}$ transforms [111] into $[\bar{1}\bar{1}1]$ the rotations by $2\pi/3$ about these axes are in the same class. However, there is no operation in the group that changes the polarity of [111], *i.e.*, into $[\bar{1}\bar{1}\bar{1}]$. This means that C_3 and C_3^{-1} are not conjugate operations. The conjugate classes of T are, thus

$$T = \{E, 3C_2, 4C_3, 4C_3^{-1}\} .$$

We now analyze the action of the operations of T on the components (x, y, z) of a vector and on the functions

$$u_1 = z^2 + \omega^2 x^2 + \omega y^2 = \frac{1}{2}(2z^2 - x^2 - y^2) - \frac{i}{2}\sqrt{3}(x^2 - y^2)$$

and

$$u_2 = z^2 + \omega^2 x^2 + \omega^2 y^2 = \frac{1}{2}(2z^2 - x^2 - y^2) + \frac{i}{2}\sqrt{3}(x^2 - y^2)$$

with

$$\omega = e^{2\pi i/3} .$$

Operation	Axis of rotation	Transformed functions
E		$xyz \quad u_1 \quad u_2$
C_2	$\hat{x}$	$x\bar{y}\bar{z} \quad u_1 \quad u_2$
C_2	$\hat{y}$	$\bar{x}y\bar{z} \quad u_1 \quad u_2$
C_2	$\hat{z}$	$\bar{x}\bar{y}z \quad u_1 \quad u_2$
C_3	[111]	$yzx \quad \omega u_1 \quad \omega^2 u_2$
C_3	$[1\bar{1}\bar{1}]$	$\bar{y}z\bar{x} \quad \omega u_1 \quad \omega^2 u_2$
C_3	$[\bar{1}1\bar{1}]$	$\bar{y}\bar{z}x \quad \omega u_1 \quad \omega^2 u_2$
C_3	$[\bar{1}\bar{1}1]$	$y\bar{z}\bar{x} \quad \omega u_1 \quad \omega^2 u_2$
C_3^{-1}	[111]	$zxy \quad \omega^2 u_1 \quad \omega u_2$
C_3^{-1}	$[1\bar{1}\bar{1}]$	$\bar{z}\bar{x}y \quad \omega^2 u_1 \quad \omega u_2$
C_3^{-1}	$[\bar{1}1\bar{1}]$	$z\bar{x}\bar{y} \quad \omega^2 u_1 \quad \omega u_2$
C_3^{-1}	$[\bar{1}\bar{1}1]$	$\bar{z}x\bar{y} \quad \omega^2 u_1 \quad \omega u_2$

In this table $\bar{x}, \bar{y}$ and $\bar{z}$ stand for $-x, -y$ and $-z$, respectively. We note that $x^2+y^2+z^2$ and xyz are invariant under all operations of T. Thus, these functions belong to the totally symmetric representation.

According to the theorems proved in Sec. 1, T has four inequivalent irreducible representations of dimensions ℓ_1, ℓ_2, ℓ_3 and ℓ_4 such that

$$\ell_1^2 + \ell_2^2 + \ell_3^2 + \ell_4^2 = 12 \; .$$

The only solution is $\ell_1 = \ell_2 = \ell_3 = 1$, $\ell_4 = 3$. The components of a vector generate a representation whose character is

$$\begin{array}{cccc} E & 3C_2 & 4C_3 & 4C_3^{-1} \\ 3 & -1 & 0 & 0 \end{array}$$

By virtue of theorem 1.5 of Sec. 1 this representation is irreducible. u_1 generates the representation $(1, 1, \omega, \omega^2)$ while u_2 gives $(1, 1, \omega^2, \omega)$. We also note that yz, zx and xy transform exactly as x, y and z. These considerations are sufficient to obtain the character table of T, namely

T	E	$3C_2$	$4C_3$	$4C_3^{-1}$	Basis functions
Γ_1	1	1	1	1	$x^2+y^2+z^2;\; xyz$
Γ_2	1	1	ω	ω^2	u_1
Γ_3	1	1	ω^2	ω	u_2
Γ_4	3	-1	0	0	$(x, y, z);\; (yz, zx, xy)$

Considering the improper operations σ_d and S_4 in (v) and (vi) above we obtain the group T_d. The class structure of T_d is

$$T_d = \{E, 8C_3, 3C_2, 6\sigma_d, 6S_4\} \; .$$

C_3 and C_3^{-1} are in the same class because σ_d transforms rotations by $2\pi/3$ into rotations by $-2\pi/3$. T_d has, therefore, five inequivalent irreducible representations of dimensions $\ell_1, \ell_2, \ell_3, \ell_4$ and ℓ_5 obeying the condition

$$\ell_1^2 + \ell_2^2 + \ell_3^2 + \ell_4^2 + \ell_5^2 = 24 \ .$$

The only solution of this equation in integers ≥ 1 is

$$\ell_1 = \ell_2 = 1, \ell_3 = 2, \ell_4 = \ell_5 = 3 \ .$$

The effect of $6\sigma_d$ and $6S_4$ on a vector (x, y, z) and on u_1 and u_2 is given in the table below

Operation		Tranformed Functions		
E		xyz	u_1	u_2
σ	$\perp \hat{e}$			
σ_d	$[\bar{1}10]$	yxz	u_2	u_1
σ_d	$[110]$	$\bar{y}\bar{x}z$	u_2	u_1
σ_d	$[0\bar{1}1]$	xzy	ωu_2	$\omega^2 u_1$
σ_d	$[011]$	$x\bar{z}\bar{y}$	ωu_2	$\omega^2 u_1$
σ_d	$[10\bar{1}]$	zyx	$\omega^2 u_2$	ωu_1
σ_d	$[101]$	$\bar{z}y\bar{x}$	$\omega^2 u_2$	ωu_1
S	axis			
S_4	$\hat{x}$	$\bar{x}z\bar{y}$	ωu_2	$\omega^2 u_1$
S_4^{-1}	$\hat{x}$	$\bar{x}\bar{z}y$	ωu_2	$\omega^2 u_1$
S_4	$\hat{y}$	$\bar{z}\bar{y}x$	$\omega^2 u_2$	ωu_1
S_4^{-1}	$\hat{y}$	$z\bar{y}\bar{x}$	$\omega^2 u_2$	ωu_1
S_4	$\hat{z}$	$y\bar{x}\bar{z}$	u_2	u_1
S_4^{-1}	$\hat{z}$	$\bar{y}x\bar{z}$	u_2	u_1

$\hat{e}$ is a unit vector normal to the mirror plane.

The group T_d has two one-dimensional representations which are clearly the totally symmetric one and that associating to each element its determinant as an orthogonal transformation. These are

E	$8C_3$	$3C_2$	$6\sigma_d$	$6S_4$
1	1	1	1	1
1	1	1	−1	−1

A vector generates

E	$8C_3$	$3C_2$	$6\sigma_d$	$6S_4$
3	0	−1	1	−1

while a pseudovector gives

3	0	−1	−1	1

From theorem 1.5 we deduce that these representations are irreducible. The remaining two-dimensional representation is generated by u_1 and u_2 as can be seen from the tables above. The character of the representation generated is

$$2 \quad -1 \quad 2 \quad 0 \quad 0$$

Theorem 1.5 shows that this representation is irreducible. The table of characters of T_d is, therefore,

T_d	E	$8C_3$	$3C_2$	$6\sigma_d$	$6S_4$	Basis Functions
Γ_1	1	1	1	1	1	$x^2+y^2+z^2$; xyz
Γ_2	1	1	1	-1	-1	
Γ_3	2	-1	2	0	0	(u_1, u_2)
Γ_4	3	0	-1	-1	1	(S_x, S_y, S_z) ; $(x(y^2-z^2), y(z^2-x^2), z(x^2-y^2))$
Γ_5	3	0	-1	1	-1	(x,y,z) ; (yz, zx, xy) ; (x^3, y^3, z^3)

Here S_x, S_y and S_z are the components of a pseudovector. The vector product of (x,y,z) and (yz, zx, xy) gives the pseudovector

$$(x(y^2-z^2) \;,\; y(z^2-x^2) \;,\; z(x^2-y^2)) \;.$$

Another notation which is often used in the literature is

$$\begin{array}{ccccc} \Gamma_1 & \Gamma_2 & \Gamma_3 & \Gamma_4 & \Gamma_5 \\ A_1 & A_2 & E & F_1 & F_2 \end{array}$$

We can separate u_1 and u_2 into their real and imaginary parts as follows

$$u_1 = \frac{1}{2}(2z^2-x^2-y^2) - \frac{i}{2}\sqrt{3}(x^2-y^2) = u_2^*$$

We shall often use the following notation to denote functions belonging to the different irreducible representations of T_d:

$$\begin{array}{ll} \Gamma_1 & \alpha \\ \Gamma_2 & \beta \\ \Gamma_3 & \gamma_1, \gamma_2 \\ \Gamma_4 & \delta_1, \delta_2, \delta_3 \\ \Gamma_5 & \epsilon_1, \epsilon_2, \epsilon_3 \;. \end{array}$$

$(\epsilon_1, \epsilon_2, \epsilon_3)$ is a vector and $(\delta_1, \delta_2, \delta_3)$ a pseudovector. γ_1 and γ_2 transform as $2z^2-x^2-y^2$ and $\sqrt{3}(x^2-y^2)$, respectively. Because xyz is an invariant ϵ_1, ϵ_2 and ϵ_3 behave also as yz, zx and xy.

The octahedral group O contains the proper rotations leaving a regular octahedron or a cube invariant. The operations of T are clearly some of its elements. In addition the following 12 rotations leave these regular polyhedra invariant:

(i) six rotations by π about lines bisecting the $(y,z), (z,x)$ and (x,y) planes where x, y and z are the cubic axes $(6C_2)$.

(ii) six rotations by $\pm\frac{\pi}{2}$ about the x, y and z-axes $(6C_4)$

The action of these operations on a vector (x, y, z) and on u_1 and u_2 is given below. We note that xyz is not an invariant in O. It generates, however, a one-dimensional representation:

$$\begin{array}{ccccc} E & 8C_3 & 3C_2(C_4^2) & 6C_2 & 6C_4 \\ 1 & 1 & 1 & -1 & -1 \end{array}$$

The representation generated by a vector is

$$3 \; 0 \; -1 \; -1 \; 1$$

Again, theorem 1.5 shows this representation is irreducible.

Operation	Axis	Transformed Functions		
E		xyz	u_1	u_2
C_2	[011]	$\bar{x}zy$	ωu_2	$\omega^2 u_1$
C_2	[0$\bar{1}$1]	$\bar{x}\bar{z}\bar{y}$	ωu_2	$\omega^2 u_1$
C_2	[101]	$z\bar{y}x$	$\omega^2 u_2$	ωu_1
C_2	[10$\bar{1}$]	$\bar{z}\bar{y}\bar{x}$	$\omega^2 u_2$	ωu_1
C_2	[110]	$yx\bar{z}$	u_2	u_1
C_2	[$\bar{1}$10]	$\bar{y}\bar{x}\bar{z}$	u_2	u_1
C_4	[100]	$xz\bar{y}$	ωu_2	$\omega^2 u_1$
C_4^{-1}	[100]	$x\bar{z}y$	ωu_2	$\omega^2 u_1$
C_4	[010]	$\bar{z}yx$	$\omega^2 u_2$	ωu_1
C_4^{-1}	[010]	$zy\bar{x}$	$\omega^2 u_2$	ωu_1
C_4	[001]	$y\bar{x}z$	u_2	u_1
C_4^{-1}	[001]	$\bar{y}xz$	u_2	u_1

We denote by ϵ_1, ϵ_2 and ϵ_3 functions that transform as yz, zx and xy, respectively. The manner in which such functions transform is deduced from the previous table and is given below. The character of the representation generated by $\epsilon_1, \epsilon_2, \epsilon_3$ is

$$\begin{array}{ccccc} E & 8C_3 & 3C_2 & 6C_2 & 6C_4 \\ 3 & 0 & -1 & 1 & -1 \end{array}$$

By virtue of theorem 1.5 this representation is irreducible. The components of a vector (x, y, z) generate the representation

$$3 \; 0 \; -1 \; -1 \; 1$$

which is clearly irreducible.

Operation	Axis	Transformed Functions		
E		ϵ_1	ϵ_2	ϵ_3
C_2	[100]	ϵ_1	$-\epsilon_2$	$-\epsilon_3$
C_2	[010]	$-\epsilon_1$	ϵ_2	$-\epsilon_3$
C_2	[001]	$-\epsilon_1$	$-\epsilon_2$	ϵ_3
C_3	[111]	ϵ_2	ϵ_3	ϵ_1
C_3	$[1\bar{1}\bar{1}]$	$-\epsilon_2$	ϵ_3	$-\epsilon_1$
C_3	$[\bar{1}1\bar{1}]$	$-\epsilon_2$	$-\epsilon_3$	ϵ_1
C_3	$[\bar{1}\bar{1}1]$	ϵ_2	$-\epsilon_3$	$-\epsilon_1$
C_3^{-1}	[111]	ϵ_3	ϵ_1	ϵ_2
C_3^{-1}	$[1\bar{1}\bar{1}]$	$-\epsilon_3$	$-\epsilon_1$	ϵ_2
C_3^{-1}	$[\bar{1}1\bar{1}]$	ϵ_3	$-\epsilon_1$	$-\epsilon_2$
C_3^{-1}	$[\bar{1}\bar{1}1]$	$-\epsilon_3$	ϵ_1	$-\epsilon_2$
C_2	[011]	ϵ_1	$-\epsilon_3$	$-\epsilon_2$
C_2	$[0\bar{1}1]$	ϵ_1	ϵ_3	ϵ_2
C_2	[101]	$-\epsilon_3$	ϵ_2	$-\epsilon_1$
C_2	$[10\bar{1}]$	ϵ_3	ϵ_2	ϵ_1
C_2	[110]	$-\epsilon_2$	$-\epsilon_1$	ϵ_3
C_2	$[\bar{1}10]$	ϵ_2	ϵ_1	ϵ_3
C_4	[100]	$-\epsilon_1$	$-\epsilon_3$	ϵ_2
C_4	[010]	ϵ_3	$-\epsilon_2$	$-\epsilon_1$
C_4	[001]	$-\epsilon_2$	ϵ_1	$-\epsilon_3$
C_4^{-1}	[100]	$-\epsilon_1$	ϵ_3	$-\epsilon_2$
C_4^{-1}	[010]	$-\epsilon_3$	$-\epsilon_2$	ϵ_1
C_4^{-1}	[001]	ϵ_2	$-\epsilon_1$	$-\epsilon_3$

The character table of O is

O	E	$8C_3$	$3C_2 = C_4^2$	$6C_2$	$6C_4$	Basis Functions
Γ_1	1	1	1	1	1	$x^2 + y^2 + z^2$
Γ_2	1	1	1	-1	-1	xyz
Γ_3	2	-1	2	0	0	$(u_1, u_2); (2z^2 - x^2 - y^2, \sqrt{3}(x^2 - y^2))$
Γ_4	3	0	-1	-1	1	$(x, y, z); (x^3, y^3, z^3)$
Γ_5	3	0	-1	1	-1	$(yz, zx, xy); (x(y^2 - z^2), y(z^2 - x^2), z(x^2 - y^2))$

Problem. Construct the character tables of C_{6v} and D_6 (see Fig. 6.3).
Solution:

$$C_{6v} = \{E, C_2, 2C_3, 2C_6, 3\sigma_d, 3\sigma_v\}$$

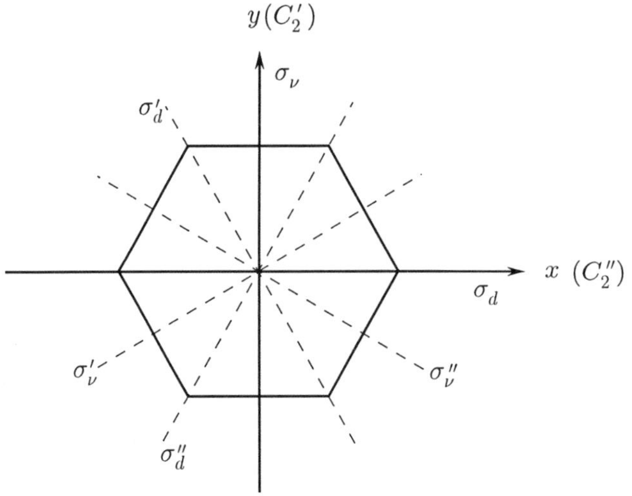

Figure 6.3

C_{6v}	E	C_2	$2C_3$	$2C_6$	$3\sigma_d$	$3\sigma_v$	Basis Functions
Γ_1	1	1	1	1	1	1	$z; x^2+y^2; z^2; z^3; z(x^2+y^2)$
Γ_2	1	1	1	1	−1	−1	S_z
Γ_3	1	−1	1	−1	1	−1	x^3-3xy^2
Γ_4	1	−1	1	−1	−1	1	y^3-3x^2y
Γ_5	2	−2	−1	1	0	0	$(x,y); (zx,zy); (S_y,-S_x)$
Γ_6	2	2	−1	−1	0	0	$(x^2-y^2,-2xy); (z(x^2-y^2),-2xyz)$

$$D_6 = \{E, C_2, 2C_3, 2C_6, 3C_2', 3C_2''\}$$

D_6	E	C_2	$2C_3$	$2C_6$	$3C_2'$	$3C_2''$	Basis Functions
Γ_1	1	1	1	1	1	1	$x^2+y^2; z^2$
Γ_2	1	1	1	1	−1	−1	$z; z^3; z(x^2+y^2)$
Γ_3	1	−1	1	−1	1	−1	y^3-3x^2y
Γ_4	1	−1	1	−1	−1	1	x^3-3xy^2
Γ_5	2	−2	−1	1	0	0	$(x,y); (yz,-zx)$
Γ_6	2	2	−1	−1	0	0	$(x^2-y^2,-2xy)$

6.3 Two important theorems

The purpose of this section is to prove two theorems which are the basis of many applications of the theory of groups in quantum mechanics.

Theorem 3.1. The inner product of two vectors belonging to different irreducible representations of a finite group or to different rows of the same irreducible, unitary representation is equal to zero.

Proof: Let $\{\phi_m^{(i)}\}$ $(m = 1, 2, \ldots \ell_i)$ and $\{\psi_n^{(j)}\}$ $(n = 1, 2, \ldots \ell_j)$ be orthonormal bases generating the irreducible representations $\Gamma^{(i)}$ and $\Gamma^{(j)}$, respectively of a group $H = \{e, a, b, \ldots\}$. For every $a \in H$,

$$P_a \phi_m^{(i)} = \sum_p \phi_p^{(i)} \Gamma_{pm}^{(i)}(a) \ , \ P_a \psi_n^{(j)} = \sum_q \psi_q^{(j)} \Gamma_{qn}^{(j)}(a) \ .$$

Then

$$< P_a \phi_m^{(i)} | P_a \psi_n^{(j)} > = \sum_{p,q} \Gamma_{pm}^{(i)*}(a) \Gamma_{qn}^{(j)}(a) < \phi_p^{(i)} | \psi_q^{(j)} > \ .$$

Recalling that P_a is unitary, summing over all elements of the group and using the great orthogonality theorem we obtain

$$h < \phi_m^{(i)} | \psi_n^{(j)} > = \sum_{p,q} < \phi_p^{(i)} | \psi_q^{(j)} > \frac{h}{\ell_i} \delta_{ij} \delta_{pq} \delta_{mn}$$

or

$$< \phi_m^{(i)} | \psi_n^{(j)} > = \frac{1}{\ell_i} \delta_{ij} \delta_{mn} \sum_{p=1}^{\ell_i} < \phi_p^{(i)} | \psi_p^{(i)} > \ . \qquad \text{Q.E.D.} \qquad (3.1)$$

We note that setting $i = j$ and $m = n$ we find that

$$< \phi_n^{(i)} | \psi_n^{(i)} > = \frac{1}{\ell_i} \sum_{p=1}^{\ell_1} < \phi_p^{(i)} | \psi_p^{(i)} > \qquad (3.2)$$

is independent of n.

The reader will note that in the first part of the statement of the theorem it was not required that the representation be unitary. But the proof was carried out under this assumption. If $\Gamma^{(i)}$ and $\Gamma^{(j)}$ were generated by vectors not necessarily orthonormal any vector belonging to $\Gamma^{(i)}$ can be expressed as a linear combination of vectors such as $\{\phi_m^{(i)}\}$. Similarly every vector belonging to $\Gamma^{(j)}$ is expressible as a linear combination of orthonormal vectors such as the elements of the set $\{\psi_n^{(j)}\}$. Then, since every $\phi_m^{(i)}$ is orthogonal to every $\psi_n^{(j)}$ when i and j correspond to non-equivalent irreducible representations, every vector belonging to $\Gamma^{(i)}$ is orthogonal to every element in the basis set of $\Gamma^{(j)}$. This completes the proof of the theorem.

This theorem can be extended to vectors belonging to the irreducible representations of infinite groups.

Theorem 3.2. Unsöld's theorem

Let $\{\phi_n^{(i)}\}$ and $\{\psi_n^{(i)}\}$ $(n = 1, 2, \ldots \ell_i)$ be orthonormal bases for the **same** unitary representation of a group $H = \{e, a, b, \ldots\}$. Then

$$\sum_{n=1}^{\ell_i} < \xi | \psi_n^{(i)} >< \phi_n^{(i)} | \eta > = \sum_{n=1}^{\ell_i} < \xi | P_a \psi_n^{(i)} >< P_a \phi_n^{(i)} | \eta > \qquad (3.3)$$

for all operations P_a of H and for any two vectors ξ, η.
Proof:

$$\sum_{n=1}^{\ell_i} < \xi | P_a \psi_n^{(i)} >< P_a \phi_n^{(i)} | \eta > = \sum_{n,p,q} < \xi | \psi_p^{(i)} >< \phi_q^{(i)} | \eta > \Gamma_{pn}^{(i)}(a) \Gamma_{qn}^{(i)*}(a)$$
$$= \sum_{p,q} < \xi | \psi_p^{(i)} >< \psi_q^{(i)} | \eta > \sum_n \Gamma_{pn}^{(i)}(a) \Gamma_{nq}^{(i)}(a^{-1}) \, .$$

The sum over n is $\Gamma_{pq}^{(i)}(e) = \delta_{pq}$ so that Eq. (3.3) follows.

We note that in the proof of this theorem no assumption was made regarding the order of the group or whether the representation was irreducible or not. Thus, the theorem is valid for finite as well as for infinite groups and for any representation in unitary form.

We often use this theorem in the form

$$\sum_{n=1}^{\ell_i} | \psi_n^{(i)} >< \phi_n^{(i)} | = E \, , \tag{3.4}$$

i.e., the left hand side of this equation is proportional to the identity element. For application to spinless particles described by wave functions $\psi_n^{(i)}$ and $\phi_n^{(i)}$ we deduce that

$$\sum_{n=1}^{\ell_i} \psi_n^{(i)*} \phi_n^{(i)}$$

is invariant under all operations of the group. This is the way in which Unsöld proved this theorem which he called the principle of spectroscopic stability. The theorem should not come as a surprise since a special case is familiar to all, namely, if two vectors (U_x, U_y, U_z) and (V_x, V_y, V_z) appear in some physical problem, their components transform in identical ways under the action of $O(3)$ or any of its subgroups and the quantity

$$\boldsymbol{U} \cdot \boldsymbol{V} = U_x V_x + U_y V_y + U_z V_z$$

is invariant under all operations of $O(3)$.

6.4 Direct Products

We introduce now some additional concepts essential for the use of group theory in many physical applications. For the time being we can view them as tools for the construction of representations of groups starting from known representations.

Consider two vector spaces $\mathcal{S}_1$ and $\mathcal{S}_2$ whose elements are $\{\xi_1, \eta_1, \zeta_1, \ldots\}$ and $\{\xi_2, \eta_2, \zeta_2 \ldots\}$, respectively. The Cartesian product $\mathcal{S}_1 \times \mathcal{S}_2$ of $\mathcal{S}_1$ and $\mathcal{S}_2$ consists of the set of all ordered pairs of vectors in which the first belongs to $\mathcal{S}_1$ and the second to $\mathcal{S}_2$. A typical element of $\mathcal{S}_1 \times \mathcal{S}_2$, (ξ_1, η_2) will be written as $\xi_1 \eta_2$. The

Cartesian product $\mathcal{S}_1 \times \mathcal{S}_2$ can be regarded as a vector space if we define among its elements and scalars operations of addition and scalar multiplication. For any scalars a_1, a_2, b_1 and b_2 and any vectors $\xi_1, \xi_2, \eta_1, \eta_2$ we postulate

$$\begin{aligned}(a_1\xi_1 + b_1\eta_1)(a_2\xi_2 + b_2\eta_2) &= a_1a_2\xi_1\xi_2 + a_1b_2\xi_1\eta_2 + b_1a_2\eta_1\xi_2 \\ &\quad + b_1b_2\eta_1\eta_2 \ .\end{aligned}$$

Suppose now that $\mathcal{S}_1$ and $\mathcal{S}_2$ are mapped into $\mathcal{S}_1'$ and $\mathcal{S}_2'$, respectively, by linear transformations A and B, *i.e.*,

$$\xi_1 \to \xi_1' = A\xi_1 \in \mathcal{S}_1' \ , \ \xi_2 \to \xi_2' = B\xi_2 \in \mathcal{S}_2' \ .$$

We define the direct product $A \times B$ of A and B as the linear transformation mapping $\mathcal{S}_1 \times \mathcal{S}_2$ into $\mathcal{S}_1' \times \mathcal{S}_2'$ according to the rule

$$\xi_1\xi_2 \to \xi_1'\xi_2' = (A \times B)(\xi_1\xi_2) = (A\xi_1)(B\xi_2) \in \mathcal{S}_1' \times \mathcal{S}_2' \ .$$

Let $\{\phi_i\}$ and $\{\psi_j\}$ be two sets of linearly independent vectors of $\mathcal{S}_1$ and $\mathcal{S}_2$, respectively, spanning the subspaces $\Sigma_1 \subset \mathcal{S}_1$ and $\Sigma_2 \subset \mathcal{S}_2$. The image of Σ_1 under A is a subspace $A(\Sigma_1) = \Sigma_1' \subset \mathcal{S}_1'$ of $\mathcal{S}_1'$. Similarly we define $B(\Sigma_2) = \Sigma_2' \subset \mathcal{S}_2'$. If $\{\phi_m'\}$ and $\{\psi_n'\}$ are sets of linearly independent vectors spanning Σ_1' and Σ_2', respectively, we have

$$A\phi_i = \sum_m \phi_m' A_{mi} \ , \ B\psi_j = \sum_n \psi_n' B_{nj} \ .$$

Then

$$(A \times B)\phi_i\psi_j = (A\phi_i)(B\psi_j) = \sum_{mn} A_{mi}B_{nj}\phi_m'\psi_n' \ . \tag{4.1}$$

Thus, the matrix representation of $A \times B$ in the map $\Sigma_1 \times \Sigma_2 \to \Sigma_1' \times \Sigma_2'$ is

$$(A \times B)_{mn;ij} = A_{mi}B_{nj} \ . \tag{4.2}$$

This is called the direct product of two matrices. It differs from the usual matrix product in that A and B need not be square or even of the same dimensions. $A \times B$ is called the direct product of A and B. In matrix form the rows and columns are labeled by two indices each (mn is a pair of indices, not a product).

Mostly we will be interested in cases in which $\mathcal{S}_1' = \mathcal{S}_1$ and $\mathcal{S}_2' = \mathcal{S}_2$. If $\Sigma_1' = \Sigma_1$ and $\Sigma_2' = \Sigma_2$, then we say that Σ_1 and Σ_2 are invariant under A and B, respectively. If this is the case the matrices representing A and B are square but not necessarily of the same dimensions. In such a case the trace of $A \times B$ is the product of the traces of A and B. In fact,

$$\chi(A \times B) = \sum_{i,j}(A \times B)_{ij,ij} = \sum_i \sum_j A_{ii}B_{jj} = \chi(A)\chi(B) \ . \tag{4.3}$$

If A and A' map Σ_1 into Σ_1 and B and B' map Σ_2 into Σ_2, then

$$(AA') \times (BB') = (A \times B)(A' \times B') \ . \tag{4.4}$$

It is enough to prove this result in a particular representation:

$$\begin{aligned}[(AA')\times(BB')]_{mn;ij} &= (AA')_{mi}(BB')_{nj} \\ &= \sum_{p=1}^{\ell_1} A_{mp}A'_{pi}\sum_{q=1}^{\ell_2} B_{nq}B'_{qj} \\ &= \sum_{p,q}(A\times B)_{mn;pq}(A'\times B')_{pq;ij} \\ &= [(A\times B)(A'\times B')]_{mn;ij}\end{aligned}$$

Here ℓ_1 and ℓ_2 are the dimensions of Σ_1 and Σ_2, respectively.

Let $H = \{e, a, b, \ldots\}$ be a group and Γ and Γ' two of its representations acting on subspaces $\mathcal{S}$ and $\mathcal{S}'$ of dimensions ℓ and ℓ', respectively. Selecting bases $\{\phi_i\}$ $(i = 1, 2, \ldots \ell)$ and $\{\phi'_j\}$ $(j = 1, 2, \ldots \ell')$ in $\mathcal{S}$ and $\mathcal{S}'$ we have

$$\Gamma(a)\phi_i = \sum_{m=1}^{\ell} \phi_m\Gamma_{mi}(a) \ , \ \Gamma'(a)\phi'_j = \sum_{n=1}^{\ell'} \phi'_n\Gamma'_{nj}(a)$$

for every element $a \in H$. Then

$$\begin{aligned}(\Gamma(a)\times\Gamma'(a))\,\phi_i\phi'_j &= \sum_{m=1}^{\ell}\sum_{n=1}^{\ell'} \phi_m\phi'_n\Gamma_{mi}(a)\Gamma'_{nj}(a) \\ &= \sum_{m=1}^{\ell}\sum_{n=1}^{\ell'} \phi_m\phi'_n\left(\Gamma(a)\times\Gamma'(a)\right)_{mn;ij} \ .\end{aligned} \tag{4.5}$$

The direct products $\{\Gamma(a)\times\Gamma'(a)\}$ form a representation of H because, by virtue of Eq. (4.4)

$$\begin{aligned}(\Gamma(a)\times\Gamma'(a))\,(\Gamma(b)\times\Gamma'(b)) &= (\Gamma(a)\Gamma(b))\times(\Gamma'(a)\Gamma'(b)) \\ &= \Gamma(ab)\times\Gamma'(ab) \ ,\end{aligned} \tag{4.6}$$

for any two elements a, b of H. The character of this representation, denoted here by $\Gamma\times\Gamma'$, is given by

$$\chi^{\Gamma\times\Gamma'}(a) = \chi^{\Gamma}(a)\chi^{\Gamma'}(a) \ . \tag{4.7}$$

In general, the direct product of two irreducible representations of a group is reducible. For example $\Gamma_4\times\Gamma_4$ for the octahedral group O is

E	$8C_3$	$3C_2$	$6C_2$	$6C_4$
9	0	1	1	1 .

Its reduction is

$$\Gamma_4\times\Gamma_4 = \Gamma_1+\Gamma_3+\Gamma_4+\Gamma_5 \ .$$

We note that for any two representations Γ and Γ' of a group,

$$\Gamma\times\Gamma' = \Gamma'\times\Gamma \ .$$

The following is a table of direct products for the groups O and T_d.

$\Gamma_i \times \Gamma_j$	Γ_1	Γ_2	Γ_3	Γ_4	Γ_5
Γ_1	Γ_1	Γ_2	Γ_3	Γ_4	Γ_5
Γ_2		Γ_1	Γ_3	Γ_5	Γ_4
Γ_3			$\Gamma_1+\Gamma_2+\Gamma_3$	$\Gamma_4+\Gamma_5$	$\Gamma_4+\Gamma_5$
Γ_4				$\Gamma_1+\Gamma_3+\Gamma_4+\Gamma_5$	$\Gamma_2+\Gamma_3+\Gamma_4+\Gamma_5$
Γ_5					$\Gamma_1+\Gamma_3+\Gamma_4+\Gamma_5$

Consider two sets of ℓ vectors $\{\phi_i\}$ and $\{\psi_i\}$ $(i = 1, 2, \ldots \ell)$ each set generating **identical** representations of a group $H = \{e, a, b, \ldots\}$. For $a \in H$

$$\Gamma(a)\phi_i = \sum_m \phi_m \Gamma_{mi}(a) \;\; , \;\; \Gamma(a)\psi_j = \sum_n \psi_n \Gamma_{nj}(a) \; .$$

The $\ell(\ell+1)/2$ vectors $\phi_i\psi_j + \phi_j\psi_i$ $(i \leq j)$ in the Cartesian product of the space of Γ by itself transform as

$$\Gamma(a)(\phi_i\psi_j + \phi_j\psi_i) = \sum_{m,n} \Gamma_{mi}(a)\Gamma_{nj}(a)(\phi_m\psi_n + \phi_n\psi_m) \; , \tag{4.8}$$

where in the right hand side the sum over m and n is unrestricted. Equation (4.8) defines a matrix with $\frac{1}{2}\ell(\ell+1)$ rows and columns whose elements are

$$\Gamma_{ii}(a)\Gamma_{jj}(a) \qquad \text{if} \quad m = i \quad \text{and} \quad n = j$$

and

$$\Gamma_{mi}(a)\Gamma_{nj}(a) + \Gamma_{ni}(a)\Gamma_{mj}(a) \qquad \text{if} \quad m < n \;\; , \;\; i < j \; .$$

We denote this matrix by $[\Gamma(a) \times \Gamma(a)] = [\Gamma \times \Gamma](a)$. For any two elements a and b of the group

$$\Gamma(b)\Gamma(a)(\phi_i\psi_j + \phi_j\psi_i) = \sum_{p,q} \Gamma_{pi}(ba)\Gamma_{qj}(ba)(\phi_p\psi_q + \phi_q\psi_p)$$

so that

$$[\Gamma \times \Gamma](ba) = ([\Gamma \times \Gamma](b))([\Gamma \times \Gamma](a)) \; .$$

Thus $[\Gamma \times \Gamma]$ is a representation of the group. It is called the symmetric direct product of Γ by itself. Its character is obtained as follow:

$$\begin{aligned} \chi^{[\Gamma\times\Gamma]}(a) &= \sum_i \Gamma_{ii}(a)\Gamma_{ii}(a) + \sum_{i<j}(\Gamma_{ii}(a)\Gamma_{jj}(a) + \Gamma_{ji}(a)\Gamma_{ij}(a)) \\ &= \frac{1}{2}\sum_{i,j}(\Gamma_{ii}(a)\Gamma_{jj}(a) + \Gamma_{ij}(a)\Gamma_{ji}(a)) \\ &= \frac{1}{2}[(\chi^\Gamma(a))^2 + \chi^\Gamma(a^2)] \; . \end{aligned} \tag{4.9}$$

In the same manner the $\frac{1}{2}\ell(\ell-1)$ vectors $\phi_i\psi_j - \phi_j\psi_i$ $(i \neq j)$ form a representation of H called the antisymmetric direct product of Γ by itself. It is denoted by $\{\Gamma \times \Gamma\}$.

We have

$$\begin{aligned}\Gamma(a)(\phi_i\psi_j - \phi_j\psi_i) &= \sum_{m,n}\Gamma_{mi}(a)\Gamma_{nj}(a)(\phi_m\psi_n - \phi_n\psi_m) \\ &= \sum_{m<n}(\Gamma_{mi}(a)\Gamma_{nj}(a) - \Gamma_{ni}(a)\Gamma_{mj}(a))(\phi_n\psi_m - \phi_m\psi_n)\ .\end{aligned}$$

The character of this representation is

$$\begin{aligned}\chi^{\{\Gamma\times\Gamma\}}(a) &= \sum_{i<j}(\Gamma_{ii}(a)\Gamma_{jj}(a) - \Gamma_{ji}(a)\Gamma_{ij}(a)) \\ &= \frac{1}{2}\sum_{i,j}(\Gamma_{ii}(a)\Gamma_{jj}(a) - \Gamma_{ij}(a)\Gamma_{ji}(a)) \\ &= \frac{1}{2}[(\chi^{\Gamma}(a))^2 - \chi^{\Gamma}(a^2)]\ .\end{aligned} \tag{4.10}$$

For the groups T_d and O

$$[\Gamma_3\times\Gamma_3] = \Gamma_1 + \Gamma_3 \quad , \quad \{\Gamma_3\times\Gamma_3\} = \Gamma_2$$

and

$$[\Gamma_4\times\Gamma_4] = [\Gamma_5\times\Gamma_5] = \Gamma_1+\Gamma_3+\Gamma_5 \quad , \quad \{\Gamma_4\times\Gamma_4\} = \{\Gamma_5\times\Gamma_5\} = \Gamma_4\ .$$

The six functions x^2, y^2, z^2, yz, zx, and xy transform as $[\Gamma_5\times\Gamma_5]$ for T_d and as $[\Gamma_4\times\Gamma_4]$ for O. Since $x^2+y^2+z^2$ belongs to Γ_1 we conclude that the five functions $2z^2-x^2-y^2$, $\sqrt{3}(x^2-y^2)$, yz, zx, xy belong to $\Gamma_3+\Gamma_5$. Since the first two generate Γ_3 the last three belong to Γ_5. These functions behave as d-functions for atoms and as the components of the electric-quadrupole moment of a system of charged particles.

In Sec. 3.9 we introduced the idea of direct product of two groups. Consider the direct product of two groups H_a and H_b in accordance with the restrictions discussed in Sec. 3.9. Let $\Gamma(a_i)$ and $\Gamma'(b_j)$ be representations of H_a and H_b, respectively. Then the transformations

$$\Gamma(a_m)\times\Gamma'(b_n) \qquad m=1,2,\ldots h_a \quad ; \quad n=1,2,\ldots h_b$$

form a representation of $H_a\times H_b$. Here h_a and h_b are the dimensions of H_a and H_b, respectively. The proof follows from Eq. (4.4):

$$\begin{aligned}(\Gamma(a_m)\times\Gamma'(b_n))(\Gamma(a_p)\times\Gamma'(b_q)) &= (\Gamma(a_m)\Gamma(a_p))\times(\Gamma'(b_n)\Gamma'(b_q)) \\ &= \Gamma(a_m a_p)\times\Gamma'(b_n b_q)\ .\end{aligned}$$

Further, if $\Gamma^{(i)}(a_m)$ and $\Gamma^{(j)}(b_n)$ are irreducible representations of H_a and H_b, respectively, then, the transformations

$$\Gamma^{(i\times j)}(a_m b_n) = \Gamma^{(i)}(a_m)\times\Gamma^{(j)}(b_n) \tag{4.11}$$

form an irreducible representation of $H_a \times H_b$. In fact, from Eq. (4.3)

$$\chi^{(i\times j)}(a_m b_n) = \chi^{(i)}(a_m)\chi^{(j)}(b_n)$$

and

$$\sum_{a_m\in H_a}\sum_{b_n\in H_b}\left|\chi^{(i\times j)}(a_m b_n)\right|^2 = \left(\sum_{a_m\in H_a}|\chi^{(i)}(a_m)|^2\right)\left(\sum_{b_n\in H_b}|\chi^{(j)}(b_n)|^2\right) = h_a h_b .$$

But $h_a h_b$ is the order of $H_a \times H_b$ so that by virtue of theorem (1.5) the transformations (4.11) constitute an irreducible representation of $H_a \times H_b$.

We can further prove that all inequivalent irreducible representations of $H_a \times H_b$ are of the form (4.11). If $\Gamma^{(i)}$ and $\Gamma^{(i')}$ are inequivalent irreducible representations of H_a and $\Gamma^{(j)}$ and $\Gamma^{(j')}$ are in the same relation to H_b, then

$$\begin{aligned}\sum_{a\in H_a}\sum_{b\in H_b}&\chi^{(i\times j)*}(ab)\chi^{(i'\times j')}(ab)\\ &= \sum_{a\in H_a}\chi^{(i)*}(a)\chi^{(i')}(a)\sum_{b\in H_b}\chi^{(j)*}(b)\chi^{(j')}(b)\\ &= \delta_{i'i}\delta_{j'j}\ .\end{aligned} \tag{4.12}$$

Now, the number of conjugate classes of $H_a \times H_b$ is equal to the product of the numbers of classes in H_a and H_b. The numbers of inequivalent irreducible representations of H_a and H_b are equal to the number of classes in them, respectively. Thus, the number of inequivalent irreducible representations of $H_a \times H_b$ is simply the product of those in H_a and in H_b. Therefore, Eq. (4.12) shows that the representations (4.11) exhaust the irreducible representations of the direct product of the two groups.

As an example we mention the group O_h, the direct product of O and $C_i = \{E, i\}$. We label this group by the symbols $O_h = O \times C_i$. The table of characters of C_i is

	E	i
Γ^+	1	1
Γ^-	1	−1

The elements of O_h are

$$\begin{aligned}O_h &= \{E, 8C_3, 3C_4^2, 6C_2', 6C_4, i, 8iC_3, 3iC_4^2, 6iC_2, 6iC_4\}\\ &= \{E, 8C_3, 3C_2, 6C_4, 6C_2', i, 8S_6, 3\sigma_h, 6S_4, 6\sigma_d\}\end{aligned}$$

For each irreducible representation of O there are two irreducible representations of O_h corresponding to even and odd basis functions. They are denoted by

$$\Gamma_i^{\pm} \qquad i = 1, 2, 3, \ldots 5\ .$$

or by an index g (gerade) for the even representations under inversion and by u (ungerade) for the odd ones. The wave functions are also labeled by $+$ or $-$ depending on whether they are even or odd under inversion.

The last cubic group is $T_h = T \times C_i$.

In some applications, particularly in molecular spectroscopy, one is often faced with the problem of establishing the symmetry properties of multiple overtones of molecular or crystal vibrations. From the mathematical point of view what is required is to obtain completely symmetric direct products of the form $[\Gamma \times \Gamma \times \ldots \times \Gamma]$. In general, this is a complex problem so that we limit ourselves to a few simple cases. To introduce the subject we give an alternative derivation of Eq. (4.9).

Let $H = \{e, a, b, \ldots\}$ be a group and Γ one of its ℓ-dimensional representations. We select a particular element, say a, and consider the unitary operator P_a acting on vectors in the ℓ-dimensional space of Γ. For this particular operator it is always possible to find an orthonormal basis in the space of Γ in which it is diagonal, *i.e.*, a set $\{\phi_i\}$ such that

$$P_a \phi_i = a_i \phi_i \quad , \quad i = 1, 2, \ldots \ell \ .$$

Of course, the basis will be different for other elements of H. Now

$$\chi(a) = \sum_{i=1}^{\ell} a_i \ .$$

The set of $\frac{1}{2}\ell(\ell+1)$ vectors $\phi_i \phi_j$, $i \leq j$ generates the representation $[\Gamma \times \Gamma]$ whose character is

$$\chi_2(a) = \sum_{i \leq j} a_i a_j = a_1^2 + a_2^2 + \ldots + a_\ell^2 + a_1 a_3 + a_1 a_3 + \ldots + a_{\ell-1} a_\ell \ .$$

Using the algebraic identity

$$\sum_{i<j} a_i a_j = \frac{1}{2} \left[\left(\sum_i a_i \right)^2 - \sum_{i-1}^{n} a_i^2 \right]$$

we obtain

$$\chi_2(a) = \frac{1}{2} [\chi^2(a) + \chi(a^2)] \ . \tag{4.13}$$

For direct products of higher order simple formulas can be obtained for low values of ℓ. For $\ell = 2$ the character of $[\Gamma^n]$ is obtained from

$$\chi_n(a) = \sum_{k=0}^{n} a_1^{n-k} a_2^k \ . \tag{4.14}$$

We start from

$$\chi(a) = a_1 + a_2$$

and use the identity

$$\sum_{k=0}^{n} a_1^{n-k} a_2^k - \frac{1}{2}(a_1 + a_2) \sum_{k=0}^{n-1} a_1^{n-k-1} a_2^k \;=\; \frac{1}{2}(a_1^n + a_2^n) \, . \tag{4.15}$$

We obtain the recursion relation

$$\chi_n(a) = \frac{1}{2}[\chi(a)\chi_{n-1}(a) + \chi_n(a^2)] \, . \tag{4.16}$$

For a three-dimensional representation we use the expression

$$2(a_1 + a_2 + a_3) \sum a_1^{p_1} a_2^{p_2} a_3^{p_3}$$

where the sum is restricted to all non-negative integers such that $p_1 + p_2 + p_3 = n - 1$. In this sum each of a_1^n, a_2^n and a_3^n appears twice. The term

$$a_1^{p_1} a_2^{p_2} a_3^{p_3}$$

with $p_1 + p_2 + p_3$ ($p_i = 0, 1, 2, \ldots$) equal to n, while none of the p_i equals n, appears four times. Thus we obtain the identity

$$\begin{aligned}
& 2(a_1 + a_2 + a_3) \sum_{p_1+p_2+p_3=n-1} a_1^{p_1} a_2^{p_2} a_3^{p_3} \\
- & (a_2 a_3 + a_3 a_1 + a_1 a_2) \sum_{p_1+p_2+p_3=n-2} a_1^{p_1} a_2^{p_2} a_3^{p_3} \\
+ & a_1^n + a_2^n + a_3^n = 3 \sum_{p_1+p_2+p_3=n} a_1^{p_1} a_2^{p_2} a_3^{p_3} \, .
\end{aligned}$$

This identity leads to

$$\chi_n(a) = \frac{1}{3}[2\chi(a)\chi_{n-1}(a) + \frac{1}{2}(\chi(a^2) - \chi^2(a))\chi_{n-2}(a) + \chi(a^n)] \, . \tag{4.17}$$

For $n = 3$

$$\chi_3(a) = \frac{1}{6}\chi^3(a) + \frac{1}{2}\chi(a)\chi(a^2) + \frac{1}{3}\chi(a^3) \, . \tag{4.18}$$

while for $n = 4$

$$\chi_4(a) = \frac{1}{36}\chi^4(a) + \frac{1}{3}\chi^2(a)\chi(a^2) + \frac{2}{9}\chi(a)\chi(a^3) + \frac{1}{12}\chi^2(a^2) + \frac{1}{3}\chi(a^4) \tag{4.19}$$

or

$$\chi_4(a) = \frac{1}{6}\chi^4(a) + \frac{4}{3}\chi(a)\chi(a^3) + \frac{1}{2}\chi^2(a^2) - \frac{1}{2}\chi^2(a)\chi(a^2) - \frac{1}{2}\chi(a^4) \, . \tag{4.20}$$

Problems:

1. Prove that the two expressions (4.19) and (4.20) for $\chi_4(a)$ are equivalent.

2. Find $[\Gamma_5^3]$ for the group T_d. Use this result to show that there is only one homogeneous polynomial of degree three in x, y, z invariant under all operations of T_d.

3. Find $[\Gamma_5^4]$ for T_d and show that there are only two homogeneous polynomials of degree four in x, y, z which are invariant under all operations of the group T_d. By only two we mean two linearly independent polynomials.

Solutions:

1. Use the algebraic identity

$$\begin{aligned} &a_1^4 + a_2^4 + a_3^4 + a_1^3(a_2 + a_3) + a_2^3(a_3 + a_1) + a_3^3(a_1 + a_2) \\ &\quad + (a_1 + a_2 + a_3)a_1a_2a_3 = -\frac{1}{2}(a_1^4 + a_2^4 + a_3^4) + (a_1 + a_2 + a_3)(a_1^3 + a_2^3 + a_3^3) \\ &\quad + \frac{1}{2}(a_1^2 + a_2^2 + a_2^2)^2 + (a_1 + a_2 + a_3)a_1a_2a_3 \end{aligned}$$

2.

	E	$8C_3$	$3C_2$	$6\sigma_d$	$6S_4$
Γ_5	3	0	−1	1	−1
$[\Gamma_5^3]$	10	1	−2	2	0

$$[\Gamma_5^3] = \Gamma_1 + \Gamma_4 + 2\Gamma_5$$

Since Γ_1 appears only once in the reduction there is only one polynomial of degree three in x, y, z invariant under all elements of T_d. This is, of course, xyz.

3.

	E	$8C_3$	$3C_2$	$6\sigma_d$	$6S_4$
$[\Gamma_5^4]$	15	0	3	3	1

$$[\Gamma_5^4] = 2\Gamma_1 + 2\Gamma_3 + \Gamma_4 + 2\Gamma_5$$

Polynomials of degree four

$$x^4 + y^4 + z^4 \quad , \quad y^2z^2 + z^2x^2 + x^2y^2 \qquad \in \Gamma_1$$

$$2z^4 - x^4 - y^4 \quad , \quad \sqrt{3}(x^4 - y^4) \qquad \in \Gamma_3$$

$$2x^2y^2 - z^2(x^2 + y^2) \quad , \quad \sqrt{3}z^2(y^2 - x^2) \qquad \in \Gamma_3$$

$$yz(z^2 - y^2) \quad , \quad zx(x^2 - z^2) \quad , \quad xy(y^2 - x^2) \qquad \in \Gamma_4$$

$$x^2yz \quad , \quad xy^2z \quad , \quad xyz^2 \qquad \in \Gamma_5$$

$$yz(y^2 + z^2) \quad , \quad zx(z^2 + x^2) \quad , \quad xy(x^2 + y^2) \qquad \in \Gamma_5$$

Reader prove that

$$x^4(y^2 - z^2) + y^4(z^2 - x^2) + z^4(x^2 - y^2) \qquad \in \Gamma_2 \, .$$

6.5 Complex conjugate representations

Let Γ be a unitary representation of a group $H = \{e, a, b, \ldots\}$. Taking a particular orthonormal basis in the space of Γ we can regard the set of operations $\{\Gamma(a)\}$ as a collection of unitary matrices. The matrix $\Gamma^*(a)$ is defined as the matrix obtained from $\Gamma(a)$ by taking the complex conjugate of all its elements. The set $\{\Gamma^*(a)\}$ like $\{\Gamma(a)\}$ is a representation of the group since

$$\Gamma(a)\Gamma(b) = \Gamma(ab)$$

for any two elements a, b of H implies that

$$\Gamma^*(a)\Gamma^*(b) = \Gamma^*(ab) \ . \tag{5.1}$$

The character of Γ^* is $\{\chi^*(a)\}$ where $\{\chi(a)\}$ is the character of Γ. By virtue of theorem 1.5, Γ and Γ^* are both reducible or irreducible.

Take Γ irreducible. The following three situations occur:

(a) Γ and Γ^* are equivalent to a real representation of H, *i.e.*, to a representation Γ, all whose matrix elements are real.

(b) Γ and Γ^* are inequivalent so that their characters are not identical.

(c) Γ and Γ^* are equivalent but cannot be made equivalent to a real representation by a similarity transformation.

If Γ and Γ^* are equivalent, a non-singular matrix S exists such that

$$\Gamma(a) = S^{-1}\Gamma^*(a)S \tag{5.2}$$

for every $a \in H$. Since $\Gamma(a)$ is unitary, so is $\Gamma^*(a)$ and Eq. (5.2) implies that

$$S\Gamma(a) = \Gamma^*(a)S = \widetilde{\Gamma}(a^{-1})S \ . \tag{5.3}$$

Taking the Hermitian conjugate of both sides of Eq. (5.3) we obtain

$$\Gamma(a^{-1})S^\dagger = S^\dagger\widetilde{\Gamma}(a) \tag{5.4}$$

for all $a \in H$. Thus, replacing a by a^{-1} we have

$$\Gamma(a)S^\dagger = S^\dagger\widetilde{\Gamma}(a^{-1}) = S^\dagger S\Gamma(a)S^{-1}$$

or

$$\Gamma(a)S^\dagger S = S^\dagger S\Gamma(a) \ . \tag{5.5}$$

By virtue of lemma 1, $S^\dagger S$ must be a multiple of the unit matrix, *i.e.*,

$$S^\dagger S = mE \ .$$

But

$$(S^{\dagger}S)_{ii} = \sum_{j=1}^{\ell} |S_{ji}|^2 > 0$$

so that $m > 0$. If m were zero $|S_{ji}| = 0$ for all i and j and S would be singular contrary to our assumption. Replacing S by $S\sqrt{m}$ allows us to assume the similarity transformation is effected by a unitary matrix. Thus, without loss of generality we can take S unitary and, hence, $m = 1$.

Replacing a by a^{-1} in Eq. (5.3) and taking the transposed of both sides we obtain

$$\widetilde{\Gamma}(a^{-1})\widetilde{S} = \widetilde{S}\Gamma(a) \; . \tag{5.6}$$

Then

$$S^{-1}\widetilde{S}\Gamma(a) = S^{-1}\widetilde{\Gamma}(a^{-1})\widetilde{S} = \Gamma(a)S^{-1}\widetilde{S} \tag{5.7}$$

where use was made of Eq. (5.15). Since Γ is irreducible $S^{-1}\widetilde{S}$ is a constant operator, *i.e.*,

$$\widetilde{S} = cS \tag{5.8}$$

where c is a number. Now $S = c\widetilde{S} = c^2 S$ so that

$$c = \pm 1 \; . \tag{5.9}$$

We shall show that $c = 1$ corresponds to case (a) above and $c = -1$ to case (c). To prove this we remark that, since S is unitary we can find a Hermitian matrix H such that

$$S = e^{iH} \; . \tag{5.10}$$

Now

$$\widetilde{S} = (e^{-iH})^* = e^{iH*} \; . \tag{5.11}$$

When $c = 1$, $\widetilde{S} = S$ so that

$$H^* = H \; .$$

This means that H is a real symmetric matrix. Hence,

$$\Gamma(a) = e^{-iH}\Gamma^*(a)e^{iH} \; ,$$

or

$$e^{\frac{i}{2}H}\Gamma(a)e^{-\frac{1}{2}iH} = e^{-\frac{i}{2}H}\Gamma^*(a)e^{\frac{i}{2}H}$$

is a real representation.

When $c = -1$, $\widetilde{S} = -S$ and

$$e^{iH^*} = -e^{iH} \; .$$

In this case Γ cannot be made equivalent to a real representation. To prove this we proceed by *reductio ad absurdum.* Suppose a unitary transformation U exists such that

$$U^{\dagger}\Gamma(a)U$$

is real for all $a \in H$. Then

$$U^\dagger \Gamma(a) U = \widetilde{U}\Gamma^*(a)U^* = \widetilde{U}e^{iH}\Gamma(a)e^{-iH}U^* \ .$$

This is equivalent to

$$\Gamma(a)U\widetilde{U}e^{iH} = U\widetilde{U}e^{iH}\Gamma(a) \ .$$

By virtue of lemma 1, $U\widetilde{U}e^{iH}$ is a multiple of the unit matrix, *i.e.*,

$$U\widetilde{U}e^{iH} = mE \tag{5.12}$$

so that

$$me^{-iH} = U\widetilde{U} \tag{5.13}$$

and

$$e^{iH} = mU^*U^\dagger \ . \tag{5.14}$$

The complex conjugate of (5.13) gives

$$m^*e^{iH^*} = U^*U^\dagger \ . \tag{5.15}$$

But, when $c = -1$, $e^{iH^*} = -e^{iH}$ so that this implies that

$$U^*U^\dagger = 0$$

which makes U singular in contradiction to the hypothesis.

The reader will notice that the proof does not require U to be unitary but merely non-singular. In fact, Eq. (5.12) is obtained as follows:

$$U^{-1}\Gamma(a)U = U^{*-1}\Gamma^*(a)U^* = U^{*-1}e^{iH}\Gamma(a)e^{-iH}U^*$$

implies that

$$\Gamma(a)UU^{*-1}e^{iH} = UU^{*-1}e^{iH}\Gamma(a) \ .$$

By virtue of lemma 1,

$$UU^{*-1}e^{iH} = mE$$

and

$$me^{-iH} = UU^{*-1} \ .$$

Taking the reciprocal

$$e^{iH} = mU^*U^{-1}$$

Furthermore

$$m^*e^{iH^*} = U^*U^{-1} \ .$$

Hence, as before we conclude that

$$U^*U^{-1} = 0 \ .$$

Frobenius and Schur developed a simple test to determine whether a particular irreducible representation of a finite group is of type a, b or c. Let $H = \{e, a, b, \ldots\}$

be a finite group and Γ one of its unitary irreducible representations. We define the matrix

$$M = \sum_{a\in H} \Gamma^*(a^{-1})X\Gamma(a) = \sum_{a\in H} \widetilde{\Gamma}(a)X\Gamma(a) \tag{5.16}$$

where X is an arbitrary square matrix of the same dimension, say ℓ, as Γ. For every $b \in H$

$$\Gamma^*(b^{-1})M\Gamma(b) = \sum_{a\in H} \Gamma^*((ab)^{-1})X\Gamma(ab) = M$$

so that

$$M\Gamma(b) = \Gamma^*(b)M$$

If Γ and Γ^* are inequivalent (case b), by lemma 2, $M = 0$. If Γ and Γ^* are equivalent, M is a multiple of the matrix S of Eq. (5.14). The constant of proportionality between M and S depends on the matrix X but, in any case

$$M = c\widetilde{M}$$

Thus, for cases a, b and c we have $c = 1, 0$ and -1, respectively. Now

$$\sum_{k,m}\sum_{a\in H} \Gamma_{ki}(a)X_{km}\Gamma_{mj}(a) = c\sum_{a\in H}\sum_{k,m} \Gamma_{mi}(a)X_{km}\Gamma_{kj}(a)$$

We now take $X_{km} = 1$ while all other components of X are zero. We have

$$\sum_{a\in H} \Gamma_{ki}(a)\Gamma_{mj}(a) = c\sum_{a\in H} \Gamma_{mi}(a)\Gamma_{kj}(a) \ .$$

Set $m = i$ and $k = j$ and sum over i and j. This gives

$$\sum_{a\in H} \chi(a^2) = c\sum_{a\in H} \chi^2(a) \ . \tag{5.17}$$

In case b, $c = 0$ and the sum of $\chi(a^2)$ over all elements of the group is equal to zero. When Γ and Γ^* are equivalent, regardless of whether they are of type a or c, $\chi(a)$ is real for all $a \in H$. By virtue of theorem 1.5, the sum of $\chi^2(a)$ is equal to h, the order of the group. Thus, we have the following test allowing us to establish whether a particular irreducible representation of a finite group is of type a, b or c:

$$\sum_{a\in H} \chi(a^2) = \begin{cases} h & \text{in case a} \\ 0 & \text{in case b} \\ -h & \text{in case c} \end{cases} \tag{5.18}$$

Wigner calls representations of type a integer representations and those of type c, half integer representations. If ℓ is the dimension of a representation of type c, since

$$\widetilde{S} = -S$$

$$\| \widetilde{S} \| = \| S \| = (-1)^\ell \| S \|$$

so that ℓ must be even. However, the converse is not true, not all even-dimensional representations are of type c. In fact, Γ_3 of T_d is two-dimensional but of type a.

6.6 The irreducible representations of SO(3)

We have already described the special orthogonal group in three-dimensions. We now investigate its irreducible representations. Let $D(R)$ be a representation of $SO(3)$ with R an arbitrary one of its elements. Suppose that $D(R)$ acts on an appropriate subspace of a Hilbert space. If we denote the vectors in this space by Greek letters, $\psi, \phi, \ldots$ the action of R on ψ will be denoted by

$$D(R)\psi\ .$$

We shall take the operations of D to be unitary so that

$$< D(R)\psi|D(R)\phi >=< \psi|\phi > \tag{6.1}$$

for any two elements ψ and ϕ of the space of D. Since D is a representation of $SO(3)$

$$D(R)D(S) = D(RS) \tag{6.2}$$

where R and S are operations of $SO(3)$. Like in any homomorphism we have

$$D(R^{-1}) = D^{-1}(R) \tag{6.3}$$

and

$$D(E) = E\ . \tag{6.4}$$

Since rotations about a common axis commute we have

$$D_z(\alpha_1)D_z(\alpha_2) = D_z(\alpha_1 + \alpha_2) = D_z(\alpha_2)D_z(\alpha_1)$$

for rotations by α_1 and α_2 about the z-axis of a Cartesian coordinate system. Since for $\alpha = 0$, $D_z(0) = E$, for an infinitesimal angle $\delta\alpha$ we have

$$D_z(\delta\alpha) = E - i\delta\alpha J_z + \cdots \tag{6.5}$$

The operator D_z is unitary so that $D_z^\dagger(\delta\alpha) = D_z(-\delta\alpha)$. Hence

$$J_z^\dagger = J_z\ , \tag{6.6}$$

i.e., J_z is Hermitian. In a similar way we define Hermitian operators J_x and J_y to describe infinitesimal rotations about $\hat{x}$ and $\hat{y}$.

A finite rotation about the z-axis is described by

$$D_z(\alpha) = \lim_{N\to\infty}\left(1 - \frac{i\alpha J_z}{N}\right)^N = e^{-i\alpha J_z}\ . \tag{6.7}$$

We now recall that, to second order in the infinitesimal angles, a rotation by $\delta\varphi_x$ about the x-axis followed by one by $\delta\varphi_y$ about the y-axis is equivalent to the operation in which the order of the rotations is reversed followed by a third rotation by $-\delta\varphi_x\delta\varphi_y$ about the z-axis. Thus,

$$\exp(-i\delta\varphi_y J_y)\exp(-i\delta\varphi_x J_x) = \exp(i\delta\varphi_x\delta\varphi_y J_z)\exp(-i\delta\varphi_x J_x)\exp(-i\delta\varphi_y J_y)$$

Expanding the exponentials we have

$$\left(1 - i\delta\varphi_y J_y - \frac{1}{2}\delta\varphi_y^2 J_y^2 + \cdots\right)\left(1 - i\delta\varphi_x J_x - \frac{1}{2}\delta\varphi_x^2 J_x^2 + \cdots\right)$$
$$= (1 + i\delta\varphi_x\delta\varphi_y J_z + \cdots)\left(1 - i\delta\varphi_x J_x - \frac{1}{2}\delta\varphi_x^2 J_x^2 + \cdots\right)$$
$$\left(1 - i\delta\varphi_y J_y - \frac{1}{2}\delta\varphi_y^2 J_y^2 + \cdots\right)$$

or

$$1 - i\delta\varphi_x J_x - i\delta\varphi_y J_y - \frac{1}{2}\delta\varphi_x^2 J_x^2 - \delta\varphi_x\delta\varphi y J_y J_x - \frac{1}{2}\delta\varphi_y^2 J_y^2 + \cdots$$
$$= 1 - i\delta\varphi_x J_x - i\delta\varphi_y J_y - \delta\varphi_x\delta\varphi_y J_x J_y - \frac{1}{2}\delta\varphi_x^2 J_x^2 - \frac{1}{2}\delta\varphi_y^2 J_y^2$$
$$+ i\delta\varphi_x\delta\varphi y J_z + \cdots .$$

Hence

$$J_x J_y - J_y J_x = iJ_z$$

or

$$[J_x, J_y] = iJ_z \ . \tag{6.8}$$

In a similar way we obtain

$$[J_z, J_x] = iJ_y \tag{6.9}$$

and

$$[J_y, J_z] = iJ_x \tag{6.10}$$

If we define

$$J_x^2 + J_y^2 + J_z^2 = J^2 \tag{6.11}$$

we can immediately show that

$$[J^2, \boldsymbol{J}] = 0 \tag{6.12}$$

Equation (6.12) implies that we can find common eigenvectors for J^2 and one of the components of $\boldsymbol{J}$, say J_z.

For any vector ψ of the space

$$< \psi|J^2|\psi > = \sum_{i=1}^{3} < \psi|J_i^2|\psi > = \sum_{i=1}^{3} < J_i\psi|J_i\psi > \geq 0 \tag{6.13}$$

so that the eigenvalues of J^2 are non-negative. Let μ^2 be an eigenvalue of J^2 and define a unique number $j \geq 0$ such that $j(j+1) = \mu^2$. We denote by $|jm>$ the common eigenvectors of J^2 and J_z satisfying

$$J^2|jm> = j(j+1)|jm> \tag{6.14}$$

and

$$J_z|jm> = m|jm> \tag{6.15}$$

We now define
$$J_{\pm} = J_x \pm iJ_y\,. \tag{6.16}$$
Since J^2 and $J_{\pm}$ commute $J_{\pm}|jm>$ are eigenvectors of J^2 with eigenvalue $j(j+1)$. They are also eigenvectors of J_z. In fact
$$\begin{aligned}[J_z, J_{\pm}] &= [J_z, J_x] \pm i[J_z, J_y] \\ &= iJ_y \pm J_x = \pm J_{\pm}\end{aligned}$$
or
$$J_z J_{\pm} = J_{\pm} J_z \pm J_{\pm}\,.$$
Thus
$$\begin{aligned} J_z J_{\pm}|jm> &= (J_{\pm}J_z \pm J_{\pm})|jm> \\ &= (m \pm 1)J_{\pm}|jm>\ . \end{aligned} \tag{6.17}$$
Thus $J_{\pm}|jm>$ are common eigenvectors of J^2 and J_z with eigenvalues $j(j+1)$ and $m \pm 1$, respectively. The square of the norms of $J_{\pm}|jm>$, namely
$$<jm|J_{\mp}J_{\pm}|jm> = <jm|J^2 - J_z^2 \mp J_z|jm> = (j \mp m)(j \pm m + 1) \tag{6.18}$$
cannot be negative. This implies that $-j \leq m \leq j$. Thus, for given j, m has a largest value $\overline{m}$ and a lowest one $\underline{m}$. This requires that
$$J_{+}|j\overline{m}> = 0 \quad , \quad J_{-}|j\underline{m}> = 0$$
which, in turn shows that $\overline{m} = j$ and $\underline{m} = -j$. Thus, the extremes of m in the inequality are actually achieved. Now applying J_- to $|j\overline{m}>$ we can lower the value of m until we reach $\underline{m} = -\overline{m}$. To go from j to $-j$ in steps of unity is only possible if $2j$ is an integer, the number of values of m being $2j+1$. This shows that for a value of j selected from the sequence
$$j = 0, \frac{1}{2}, 1, \frac{3}{2}, \ldots$$
there are $2j+1$ vectors $|jm>$ generating a $(2j+1)$-dimensional representation of $SO(3)$. Choosing the phase in an appropriate manner, we have
$$J_{\pm}|jm> = [(j \mp m)(j \pm m + 1)]^{\frac{1}{2}}|j, m \pm 1>\ . \tag{6.19}$$
A rotation by the Euler angles α, β, γ is equivalent to rotations by γ about $\hat{z}$, followed by one by β about $\hat{y}$ and by a rotation by α about $\hat{z}$ as shown in Sec. 1.2. Thus
$$D(\alpha, \beta, \gamma) = e^{-i\alpha J_z} e^{-i\beta J_y} e^{-i\gamma J_z}\ . \tag{6.20}$$
The state $|jm>$ after a rotation by the Euler angles α, β, γ becomes
$$|jm; \alpha, \beta, \gamma> = D(\alpha, \beta\gamma)|jm> = \sum_{m'=-j}^{j} |jm'> D^{(j)}_{m'm}(\alpha, \beta, \gamma) \tag{6.21}$$

where

$$
\begin{aligned}
D^{(j)}_{m'm}(\alpha,\beta,\gamma) &= <jm'|D(\alpha,\beta,\gamma)|jm> = e^{-i(m'\alpha+m\gamma)} <jm'|e^{i\beta J_y}|jm> \\
&= e^{-i(m'\alpha+m\gamma)} d^{(j)}_{m'm}(\beta) \ .
\end{aligned}
\tag{6.22}
$$

To obtain explicit expressions for $d^{(j)}_{m'm}(\beta)$ we make use of the homomorphism between $SO(3)$ and $SU(2)$. The set of matrices of $SU(2)$ as well as the totally symmetric representation $\Gamma(u) = 1$ form representations of $SU(2)$. Any representation of $SU(2)$ is either a representation of $SO(3)$ or a double valued representation of $SO(3)$. To construct representations of dimension higher than two we consider direct products of the two dimensional representation by itself. Let ξ_1 and ξ_2 be two orthogonal basis vectors generating $SU(2)$. Under

$$
u = \begin{pmatrix} a & b \\ -b^* & a^* \end{pmatrix}
\tag{6.23}
$$

ξ_1 and ξ_2 transform as

$$
\begin{aligned}
\xi_1' &= u_{11}\xi_1 + u_{21}\xi_2 = a\xi_1 - b^*\xi_2 \\
\xi_2' &= u_{12}\xi_1 + u_{22}\xi_2 = b\xi_1 + a^*\xi_2 \ .
\end{aligned}
\tag{6.24}
$$

We form the vectors

$$
\zeta^{(j)}_\mu = \frac{\xi_1^{j+\mu}\xi_2^{j-\mu}}{[(j+\mu)!(j-\mu)!]^{\frac{1}{2}}}
\tag{6.25}
$$

for $j = 0, \frac{1}{2}, 1, \frac{3}{2}, \ldots,\ \mu = -j, -j+1, \ldots j-1, j$. The quantities $\zeta^{(j)}_\mu$ generate a $(2j+1)$-dimensional representation of $SU(2)$. The transformation induced by u is

$$
\begin{aligned}
\zeta^{(j)}_\mu \to \zeta'^{(j)}_\mu &= \frac{\xi_1'^{j+\mu}\xi_2'^{j-\mu}}{[(j+\mu)!(j-\mu)!]^{\frac{1}{2}}} \\
&= \frac{(a\xi_1 - b^*\xi_2)^{j+\mu}(b\xi_1 + a^*\xi_2)^{j-\mu}}{[(j+\mu)!(j-\mu)!]^{\frac{1}{2}}} \\
&= \sum_{\mu'=-j}^{j} D^{(j)}_{\mu'\mu}(u)\zeta^{(j)}_{\mu'} \ .
\end{aligned}
\tag{6.26}
$$

The form of $D^{(j)}_{\mu'\mu}(u)$ is obtained by carrying out the binomial expansions in Eq. (6.26). We find

$$
\begin{aligned}
D^{(j)}_{\mu'\mu}(u) &= \sum_\kappa (-1)^\kappa \frac{[(j+\mu)!(j-\mu)!(j+\mu')!(j-\mu')!]^{\frac{1}{2}}}{\kappa!(j+\mu-\kappa)!(j-\mu'-\kappa)!(\kappa+\mu'-\mu)!} \\
&\quad \times a^{j+\mu-\kappa} a^{*j-\mu'-\kappa} b^{\kappa+\mu'-\mu} b^{*\kappa} \ ,
\end{aligned}
\tag{6.27}
$$

where the sum over κ extends over all values of κ for which the factorials are defined, *i.e.*, $\kappa \geq 0$, $\kappa \leq j+\mu$, $\kappa \leq j-\mu'$, $\kappa \geq \mu-\mu'$.

The matrices $D^{(j)}(u)$ are unitary as a consequence of the definition (6.25) of the vectors $\zeta^{(j)}$. It is, in fact, for this reason that the form (6.25) was selected. Since u is unitary

$$< \xi_1'|\xi_1' > + < \xi_2'|\xi_2' > = < \xi_1|\xi_1 > + < \xi_2|\xi_2 > \ .$$

Then

$$\begin{aligned}\sum_{\mu=-j}^{j} < \varsigma_\mu^{(j)}|\varsigma_\mu^{(j)} > &= \sum_{\mu=-j}^{j} \frac{< \xi_1|\xi_1 >^{j+\mu} < \xi_2|\xi_2 >^{j-\mu}}{(j+\mu)!(j-\mu)!} \\ &= \sum_{\kappa=0}^{2j} \frac{< \xi_1|\xi_1 >^{\kappa} < \xi_2|\xi_2 >^{2j-\kappa}}{(2j-\kappa)!\kappa!} \\ &= \frac{1}{(2j)!}(< \xi_1|\xi_1 > + < \xi_2|\xi_2 >)^{2j} \ .\end{aligned}$$

Since $< \xi_1|\xi_1 > + < \xi_2|\xi_2 >$ is invariant under u, so is

$$\sum_{\mu=-j}^{j} < \varsigma_\mu^{(j)}|\varsigma_\mu^{(j)} > \ .$$

The representations $D^{(j)}(u)$ are irreducible. To establish it we use the corollary to Schur's lemmas by showing that the totality of the matrices M such that

$$MD^{(j)}(u) = D^{(j)}(u)M \tag{6.28}$$

for all $u \in SU(2)$ are multiples of the unit matrix. In fact, from Eq. (6.28)

$$\sum_{\mu''} M_{\mu\mu''} D^{(j)}_{\mu''\mu'}(u) = \sum_{\mu''} D^{(j)}_{\mu\mu''}(u) M_{\mu''\mu'} \tag{6.29}$$

must hold for all u. If we select $u = u^{(3)}(\alpha)$,

$$D^{(j)}_{\mu'\mu}(\alpha) = e^{-i\mu\alpha}\delta_{\mu'\mu} \tag{6.30}$$

so that

$$M_{\mu\mu'}(e^{-i\mu'\alpha} - e^{-i\mu\alpha}) = 0 \ . \tag{6.31}$$

Since this condition must hold for all α, $M_{\mu\mu'} = 0$ unless $\mu = \mu'$. Thus Eq. (6.29) reduces to

$$D^{(j)}_{\mu\mu'}(u)(M_{\mu\mu} - M_{\mu'\mu'}) = 0 \ .$$

But one can always find an element u of $SU(2)$ such that $D^{(j)}_{\mu\mu'}(u) \neq 0$. Thus, $M_{\mu\mu} = M_{\mu'\mu'}$ and M is a multiple of the unit matrix. This establishes the irreducibility of $D^{(j)}(u)$.

We recall that u and $-u$ give rise to the same rotation. Now

$$D^{(j)}_{\mu'\mu}(-u) = (-1)^{2j} D^{(j)}_{\mu'\mu}(u) \ . \tag{6.32}$$

Thus, for $j = 1, 2, 3, \ldots$ the transformations $D^{(j)}(u)$ form a single- valued representation of $SO(3)$.

The character of $D^{(j)}$ is obtained from Eq. (6.30) after making the remark that all rotations by α are in the same conjugate class. Thus

$$\chi^{(j)}(\alpha) = \sum_{\mu=-j}^{j} e^{-i\mu\alpha} = \frac{\sin(j+\frac{1}{2})\alpha}{\sin\frac{\alpha}{2}}\,. \tag{6.33}$$

When $j = \frac{1}{2}$, a rotation by α about the z-axis is described by

$$u^{(3)}(\alpha) = \begin{pmatrix} e^{-i\alpha/2} & 0 \\ 0 & e^{i\alpha/2} \end{pmatrix} = \sigma_0 \cos\frac{\alpha}{2} - i\sigma_3 \sin\frac{\alpha}{2}\,. \tag{6.34}$$

We note that if $\alpha = 2\pi$, $u^{(3)}(2\pi) = -1$. Thus, for vectors belonging to the representation $D^{(\frac{1}{2})}$ a rotation by 2π is not the identity operation but rather changes the sign of the vectors. The same conclusion holds for all half-integral representations. We shall return to this subject in more detail later when we deal with the physical applications of group theory. Following Bethe we introduce the operation $\bar{E}$ as a rotation by 2π, E being a rotation by 4π. Since $\bar{E}$ simply changes the sign of vectors belonging to half-integral representations, it commutes with all orthogonal operations. Denoting also by $\bar{E}$ the group $\{E, \bar{E}\}$ of order 2, we form $\bar{E} \times SO(3)$ which is, then, isomorphic to $SU(2)$.

Problem. Derive expressions for $D^{(1)}(\alpha,\beta,\gamma)$ and $D^{(3/2)}(\alpha,\beta,\gamma)$.

Solution:

$$D^{(1)}(\alpha,\beta,\gamma) = \begin{pmatrix} \frac{1}{2}e^{-i(\alpha+\gamma)}(1+\cos\beta) & -\frac{1}{\sqrt{2}}e^{-i\alpha}\sin\beta & \frac{1}{2}e^{-i(\alpha-\gamma)}(1-\cos\beta) \\ \frac{1}{\sqrt{2}}e^{-i\gamma}\sin\beta & \cos\beta & -\frac{1}{\sqrt{2}}e^{i\gamma}\sin\beta \\ \frac{1}{2}e^{i(\alpha-\gamma)}(1-\cos\beta) & \frac{1}{\sqrt{2}}e^{i\alpha}\sin\beta & \frac{1}{2}e^{i(\alpha+\gamma)}(1+\cos\beta) \end{pmatrix}$$

$$D^{(3/2)}(\alpha,\beta,\gamma) = \begin{pmatrix} a^3 & \sqrt{3}a^2b & \sqrt{3}ab^2 & b^3 \\ -\sqrt{3}a^2b^* & a(|a|^2-2|b|^2) & b(2|a|^2-|b|^2) & \sqrt{3}a^*b^2 \\ \sqrt{3}ab^{*2} & -b^*(2|a|^2-|b|^2) & a^*(|a|^2-2|b|^2) & \sqrt{3}a^{*2}b \\ -b^{*3} & \sqrt{3}a^*b^{*2} & -\sqrt{3}a^{*2}b^* & a^{*3} \end{pmatrix}$$

6.7 Clebsch-Gordan coefficients

Let $\Gamma^{(i)}$ and $\Gamma^{(j)}$ be two irreducible, unitary representations of a finite group $H = \{e, a, b, \cdots\}$ generated by $\{\phi_\kappa^{(i)}\}$ and $\{\phi_{\kappa'}^{(j)}\}$, two sets of orthonormal vectors spanning the subspaces of $\Gamma^{(i)}$ and $\Gamma^{(j)}$, respectively. The indices κ and κ' range from 1 to ℓ_i and ℓ_j, respectively where ℓ_i and ℓ_j are the dimension of $\Gamma^{(i)}$ and $\Gamma^{(j)}$.

The set $\{\phi_\kappa^{(i)}\phi_{\kappa'}^{(j)}\}$ defined in the tensor product of the spaces of $\Gamma^{(i)}$ and $\Gamma^{(j)}$ generates the direct product representation

$$\Gamma^{(i\times j)} = \Gamma^{(i)} \times \Gamma^{(j)}$$

of the two representations. In general, the resulting representation of H is reducible:

$$\Gamma^{(i\times j)} = \Gamma^{(i)} \times \Gamma^{(j)} = \sum_{k=1}^{N_\Gamma} b_{ijk}\Gamma^{(k)} \ . \tag{7.1}$$

From Eq. (1.31) we obtain the coefficients b_{ijk} in the form

$$\begin{aligned} b_{ijk} &= \frac{1}{h}\sum_{a\in H}\chi^{(k)*}(a)\chi^{(i)}(a)\chi^{(j)}(a) \\ &= \frac{1}{h}\sum_{\mu=1}^{N_c} n_\mu\chi^{(k)*}(C_\mu)\chi^{(i)}(C_\mu)\chi^{(j)}(C_\mu) \ . \end{aligned} \tag{7.2}$$

Clearly $b_{ijk} = b_{jik}$.

Quite often, it is not enough to find this reduction but it is imperative to obtain those linear combinations of $\phi_\kappa^{(i)}\phi_{\kappa'}^{(j)}$ which belong to the several irreducible representations of the group. If all non-vanishing coefficients b_{ijk} are equal to unity, the symmetry adapted vectors are obtained free from ambiguity. When this is the case for all $\Gamma^{(i)} \times \Gamma^{(j)}$ we say that the group is simply reducible. Of course, the combinations of vectors for irreducible representations of dimension two or higher can be transformed by a change of bases.

Suppose that one or more of the b_{ijk} is greater than unit. Then, there are b_{ijk} sets of functions of the form

$$\psi_{\nu_k,\lambda}^{(k)} \qquad \nu_k = 1,2,\ldots b_{ijk} \ ; \ \lambda = 1,2,\ldots \ell_k$$

which are linear combinations of the $\phi_\kappa^{(i)}\phi_{\kappa'}^{(j)}$ and belong to the λ row of $\Gamma^{(k)}$. Any linear combination of the functions for fixed k and λ also belong to the λ row of $\Gamma^{(k)}$ so that there is a considerable amount of arbitrariness in their selection. We make the convention that they be selected in such a manner that they are orthogonal and normalized. Thus, there is a unitary transformation of the form

$$\psi_{\nu_k,\lambda}^{(k)} = \sum_{\kappa,\kappa'} \phi_\kappa^{(i)}\phi_{\kappa'}^{(j)} < ij;\kappa\kappa'|ij;k\nu_k\lambda > \ . \tag{7.3}$$

The coefficients $< ij;\kappa\kappa'|ij;k\nu_k\lambda >$ are called the Clebsch-Gordan coefficients. The sums over κ and κ' range from 1 to ℓ_i and from 1 to ℓ_j, respectively. The transformation (7.3) can be inverted to give

$$\phi_\kappa^{(i)}\phi_{\kappa'}^{(j)} = \sum_{k=1}^{N_\Gamma}\sum_{\nu_k=1}^{b_{ijk}}\sum_{\lambda=1}^{\ell_k} \psi_{\nu_k,\lambda}^{(k)} < ij;k\nu_k\lambda|ij;\kappa\kappa' > \ . \tag{7.4}$$

Let $D^{(j_1)}$ and $D^{(j_2)}$ be two irreducible representations of $SU(2)$ (or of $SO(3)$). The direct product of $D^{(j_1)}$ and $D^{(j_2)}$ is

$$D^{(j_1)} \times D^{(j_2)} = \sum_{J=|j_1-j_2|}^{j_1+j_2} D^{(J)} \tag{7.5}$$

We shall soon prove this result, but after accepting it, we conclude that $SO(3)$ is a simply reducible group. To prove this we remark that the character of $D^{(j_1)} \times D^{(j_2)}$ is

$$\begin{aligned} \chi^{(j_1 \times j_2)}(\varphi) &= \chi^{(j_1)}(\varphi)\chi^{(j_2)}(\varphi) \\ &= \sum_{m_1=-j_1}^{j_1} \sum_{m_2=-j_2}^{j_2} e^{-i(m_1+m_2)\varphi} \ . \end{aligned}$$

We show that this sum is equal to

$$\sum_{J=|j_1-j_2|}^{j_1+j_2} \sum_{M=-J}^{J} e^{iM\varphi} \ .$$

To prove this we proceed by direct calculation. Without loss of generality we can suppose that $j_1 \geq j_2$. Then

$$\begin{aligned} \sum_{J=j_1-j_2}^{j_1+j_2} \sum_{M=-J}^{J} e^{-iM\varphi} &= \frac{1}{\sin\frac{\varphi}{2}} \sum_{J=j_1-j_2}^{j_1+j_2} \sin\left(J+\frac{1}{2}\right)\varphi \\ &= \frac{1}{\sin\frac{\varphi}{2}} Im \sum_{J=j_1-j_2}^{j_1+j_2} e^{i(J+\frac{1}{2})\varphi} \\ &= \frac{1}{\sin\frac{\varphi}{2}} Im \left[e^{\frac{i\varphi}{2}} e^{i(j_1-j_2)\varphi} \frac{e^{i(2j_2+1)\varphi}-1}{e^{i\varphi}-1} \right] \\ &= \frac{\sin(j_1+\frac{1}{2})\varphi \sin(j_2+\frac{1}{2})\varphi}{\sin^2\frac{\varphi}{2}} \ . \end{aligned}$$

But this is precisely $\chi^{(j_1)}(\varphi)\chi^{(j_2)}(\varphi)$. This establishes the validity of Eq. (7.5). If $|j_1m_1>$ and $|j_2m_2>$ are orthonormal basis vectors for $D^{(j_1)}$ and $D^{(j_2)}$, respectively, where the first index gives the eigenvalue of $J^2_{1,2}$ while the second indicates that of $J_{z1,2}$, the basis vectors

$$|j_1m_1> \times |j_2m_2> = |j_1j_2;m_1m_2>$$

span the space of $D^{(j_1)} \times D^{(j_2)}$. We wish to find the linear combinations of these vectors corresponding to the eigenvalues of J^2 and J_z where

$$\boldsymbol{J} = \boldsymbol{J}_1 + \boldsymbol{J}_2 \ .$$

These linear combinations are of the form

$$|j_1j_2; JM> = \sum_{m_1=-j_1}^{j_1} \sum_{m_2=-j_2}^{j_2} |j_1j_2; m_1m_2>< j_1j_2; m_1m_2|j_1j_2; JM> \quad (7.6)$$

where the coefficients

$$< j_1j_2; m_1m_2|j_1j_2; JM >$$

are the Clebsch-Gordan coefficients for $SO(3)$. From (7.6) we obtain

$$|j_1j_2; m_1m_2 >= \sum_{J=|j_1-j_2|}^{j_1+j_2} \sum_{M=J}^{J} |j_1j_2; JM >< j_1j_2; JM|j_1j_2; m_1m_2 > . \quad (7.7)$$

There are several ways to determine the Clebsch-Gordan coefficients. Here we give a purely algebraic derivation. We consider two pairs of vectors (ξ_1, ξ_2) and (η_1, η_2) each of which transforms under the operations of $SU(2)$ as in Eqs. (6.24). The form

$$\xi_1\eta_2 - \xi_2\eta_1 \quad (7.8)$$

is invariant under all transformations of $SU(2)$. Let (x_1, x_2) be a vector whose components transform according to $u^{-1} = u^\dagger$, *i.e,*

$$\begin{aligned} x'_1 &= a^*x_1 - bx_2 , \\ \text{and} \qquad & \qquad (7.9) \\ x'_2 &= b^*x_1 + ax_2 , \end{aligned}$$

The forms $x_1\xi_1 + x_2\xi_2$ and $x_1\eta_1 + x_2\eta_2$ are invariant under all operations of $SU(2)$. Vectors corresponding to the $D^{(j_1)}$ and $D^{(j_2)}$ representations of $SU(2)$ are given by the expressions (6.25) which, in the present context we write as

$$\psi^{(j_1)}_{m_1} = \frac{\xi_1^{j_1+m_1}\xi_2^{j_1-m_1}}{[(j_1+m_1)!(j_1-m_1)!]^{\frac{1}{2}}} \quad (7.10)$$

and

$$\psi^{(j_2)}_{m_2} = \frac{\eta_1^{j_2+m_2}\eta_2^{j_2-m_2}}{[(j_2+m_2)!(j_2-m_2)!]^{\frac{1}{2}}} . \quad (7.11)$$

For a value of J between $|j_1 - j_2|$ and $j_1 + j_2$, *i.e.*, a member of the sequence

$$|j_1 - j_2| \ , \ |j_1 - j_2| + 1 \ , \ \ldots |j_1 + j_2| - 1 \ , \ j_1 + j_2$$

we construct the polynomial

$$F_J = (\xi_1\eta_2 - \xi_2\eta_1)^{j_1+j_2-J}(x_1\xi_1 + x_2\xi_2)^{j_1-j_2+J}(x_1\eta_1 + x_2\eta_2)^{j_2-j_1+J} \quad (7.12)$$

of degree $2J$ in x_1 and x_2. F_J is invariant under all transformations of $SU(2)$. Noting that all exponents are integral we expand F_J with the use of the binomial theorem to obtain

$$F_J = \sum_{\kappa\lambda\mu}(-1)^\kappa \begin{pmatrix} j_1+j_2-J \\ \kappa \end{pmatrix} \begin{pmatrix} j_1-j_2+J \\ \lambda \end{pmatrix} \begin{pmatrix} j_2-j_1+J \\ \mu \end{pmatrix}$$

$$x_1^{2J-\lambda-\mu} x_2^{\lambda+\mu} \xi_1^{2j_1-\kappa-\lambda} \xi_2^{\kappa+\lambda} \eta_1^{j_2-j_1+J+\kappa-\mu} \eta_2^{j_2+j_1-J-\kappa+\mu} . \tag{7.13}$$

The sums over κ, λ and μ extend over all integer values for which the factorials in the binomial coefficients are defined, *i.e*,

$$\begin{aligned} \kappa &= 0,1,2,\dots j_1+j_2-J , \\ \lambda &= 0,1,2,\dots j_1-j_2+J , \end{aligned} \tag{7.14}$$

and

$$\mu = 0,1,2,\dots j_2-j_1+J .$$

We now introduce new summation indices m_1 and m_2, ranging in the intervals $-j_1 \le m_1 \le j_1$ and $-j_2 \le m_2 \le j_2$ in steps of unity, defined by

$$m_1 = j_1 - \kappa - \lambda ,$$

and (7.15)

$$m_2 = J - j_1 + \kappa - \mu .$$

With these the sum in F_J becomes

$$F_J = \sum_\kappa \sum_{m_1,m_2}(-1)^\kappa \begin{pmatrix} j_1+j_2-J \\ \kappa \end{pmatrix} \begin{pmatrix} j_1-j_2+J \\ j_1-m_1-\kappa \end{pmatrix} \begin{pmatrix} j_2-j_1+J \\ j_2+m_2-\kappa \end{pmatrix}$$

$$\xi_1^{j_1+m_1} \xi_2^{j_1-m_1} \eta_1^{j_2+m_2} \eta_2^{j_2-m_2} x_1^{J+m_1+m_2} x_2^{J-m_1-m_2} . \tag{7.16}$$

Using the definitions (7.10) and (7.11) we rewrite F_J in the form

$$F_J = \sum_{M=-J}^{J} \Omega_M^{(J)} \chi_M^{(J)} \tag{7.17}$$

where

$$\Omega_M^{(J)} = \frac{x_1^{J+M} x_2^{J-M}}{[(J+M)!(J-M)!]^{\frac{1}{2}}} \tag{7.18}$$

and

$$\chi_M^{(J)} = \sum{}'_{m_1,m_2} C^{J,M}_{j_1 j_2; m_1 m_2} \psi_{m_1}^{(j_1)} \psi_{m_2}^{(j_2)} \tag{7.19}$$

The coefficients $C^{J,M}_{j_1 j_2; m_1 m_2}$ are given by

$$C^{J,M}_{j_1j_2;m_1m_2} = \sum_\kappa (-1)^\kappa \begin{pmatrix} j_1+j_2-J \\ \kappa \end{pmatrix} \begin{pmatrix} j_1-j_2+J \\ j_1-m_1-\kappa \end{pmatrix} \begin{pmatrix} j_2-j_1+J \\ j_2+m_2-\kappa \end{pmatrix}$$

$$[(J+M)!(J-M)!(j_1+m_1)!(j_1-m_1)!(j_2+m_2)!(j_2-m_2)!]^{\frac{1}{2}} \tag{7.20}$$

In Eq. (7.19) the sum over m_1 and m_2 is restricted to those values for which $m_1+m_2=M$, this being indicated by a prime on the summation symbol.

The vectors $\chi^{(J)}_M$ transform under $SU(2)$ according to $D^{(J)}$. We prove this statement by considering the invariant

$$(x_1\xi_1+x_2\xi_2)^{2J} = (2J)! \sum_{M=-J}^{J} \Omega^{(J)}_M \frac{\xi_1^{J+M}\xi_2^{J-M}}{[(J+M)!(J-M)!]^{\frac{1}{2}}} . \tag{7.21}$$

In order for the $\chi^{(J)}_M$ in Eq. (7.17) to give rise to the invariant F_J its $2J+1$ components must transform as

$$\frac{\xi_1^{J+M}\xi_2^{J-M}}{[(J+M)!(J-M)!]^{\frac{1}{2}}} .$$

as demonstrated by the invariant character of $(x_1\xi_1+x_2\xi_2)^{2J}$. We have established, therefore, that $\chi^{(J)}_M$ for $M=-J,-J+1,\dots J-1,J$ generate the representation $D^{(J)}$ of $SU(2)$. But the transformation (7.19) is not unitary. We can make it unitary by dividing by the square root of

$$\sum{}'_{m_1,m_2} \left| C^{J,M}_{j_1j_2;m_1m_2} \right|^2 \qquad ; \qquad (m_1+m_2=M)$$

This quantity is independent of m_1, m_2 and M so that we can calculate its value for the particular case in which $M=J$. In this case the sum over κ is restricted to $\kappa \le j_1-m_1$. But $m_2 = J-m_1$ so that the third binomial coefficient is defined only for $\kappa \ge j_1-m_1$. Thus, the sum over κ reduces to the single term with $\kappa = j_1-m_1$. Therefore

$$C^{J,J}_{j_1j_2;m_1m_2} = (-1)^{j_1-m_1}(j_1+j_2-J)![(2J)!(j_1-j_2+J)!(j_2-j_1+J)!]^{\frac{1}{2}}$$

$$\left[\begin{pmatrix} j_1+m_1 \\ j_2-m_2 \end{pmatrix} \begin{pmatrix} j_2+m_2 \\ j_1-m_1 \end{pmatrix} \right]^{\frac{1}{2}} . \tag{7.22}$$

where we used $m_1+m_2=J$.

We recall that

$$\begin{pmatrix} n \\ k \end{pmatrix} = \frac{n!}{k!(n-k)!} = (-1)^k \begin{pmatrix} k-n-1 \\ k \end{pmatrix} \tag{7.23}$$

and obtain

$$C^{J,J}_{j_1j_2;m_1m_2} = (-1)^{j_1-m_1}(j_1+j_2-J)![(2J)!(j_1-j_2+J)!(j_2-j_1+J)!]^{\frac{1}{2}} \left[(-1)^{j_1+j_2-J}\binom{j_2-j_1-J-1}{j_2-m_2}\binom{j_1-j_2-J-1}{j_1-m_1}\right]^{\frac{1}{2}} .$$

To calculate the sum of the squares of these quantities we use the identity

$$\binom{m+n}{k} = {\sum}'_{p,q}\binom{m}{p}\binom{n}{q} \tag{7.24}$$

where the sum is restricted to values of p and q such that $p+q=k$. This identity follows from

$$\begin{aligned}(1+x)^{m+n} = \sum_k \binom{m+n}{k} x^k &= \sum_{p,q}\binom{m}{p}\binom{n}{q} x^{p+q} \\ &= \sum_k {\sum}'_{p,q}\binom{m}{p}\binom{n}{q} x^k .\end{aligned} \tag{7.25}$$

We find that

$$\begin{aligned}&{\sum}'_{m_1,m_2}\left|C^{J,J}_{j_1j_2;m_1m_2}\right|^2 \\ &= \frac{(j_1+j_2-J)!(j_1-j_2+J)!(j_2-j_1+J)!(j_1+j_2+J+1)!}{2J+1} .\end{aligned} \tag{7.26}$$

Thus,

$$\psi^{(J)}_M = {\sum}'_{m_1,m_2}\psi^{(j_1)}_{m_1}\psi^{(j_2)}_{m_2} < j_1j_2;m_1m_2|j_1j_2;JM > \tag{7.27}$$

with

$$\begin{aligned}&< j_1j_2;m_1m_2|j_1j_2;JM >= \\ &\left[\frac{2J+1}{(j_1+j_2-J)!(j_1-j_2+J)!(j_2-j_1+J)!(j_1+j_2+J+1)!}\right]^{\frac{1}{2}} \\ &\sum_\kappa(-1)^\kappa\binom{j_1+j_2-J}{\kappa}\binom{j_1-j_2+J}{j_1-m_1-\kappa}\binom{j_2-j_1+J}{j_2+m_2-\kappa} \\ &[(J+M)!(J-M)!(j_1+m_1)!(j_1-m_1)!(j_2+m_2)!(j_2-m_2)!]^{\frac{1}{2}} .\end{aligned} \tag{7.28}$$

To investigate the symmetry properties of the Clebsch-Gordan coefficients we rewrite Eq. (7.28) in the form

$$< j_1 j_2; m_1 m_2 | j_1 j_2; j_3 m_3 >=$$

$$\left[\frac{(2j_3+1)(j_1+m_1)!(j_1-m_1)!(j_2+m_2)!(j_2-m_2)!(j_3+m_3)!(j_3-m_3)!}{(j_1+j_2-j_3)!(j_1-j_2+j_3)!(j_2-j_1+j_3)!(j_1+j_2+j_3+1)!}\right]^{\frac{1}{2}}$$

$$\sum_\kappa (-1)^\kappa \begin{pmatrix} j_1+j_2-j_3 \\ \kappa \end{pmatrix} \begin{pmatrix} j_1-j_2+j_3 \\ j_1-m_1-\kappa \end{pmatrix} \begin{pmatrix} j_2-j_1+j_3 \\ j_2+m_2-\kappa \end{pmatrix} \tag{7.29}$$

where $m_3 = m_1 + m_2$ and the sum over κ extends over all integer values for which the factorials in the binomial coefficients are well defined. If we exchange the indices 1 and 2, this is equivalent to interchanging ξ and η in the definition of F_J. This gives a factor of

$$(-1)^{j_1+j_2-j_3}$$

so that

$$< j_1 j_2; m_1 m_2 | j_1 j_2; j_3 m_3 >= (-1)^{j_1+j_2-j_3} < j_2 j_1; m_2 m_1 | j_2 j_1; j_3 m_3 > \ . \tag{7.30}$$

Similarly

$$< j_1 j_2; -m_1, -m_2 | j_1 j_2; j_3, -m_3 >= (-1)^{j_1+j_2-j_3} < j_1 j_2; m_1 m_2 | j_1 j_2; j_3 m_3 > \ . \tag{7.31}$$

A symmetric way to express many properties of the Clebsch-Gordan coefficients is with the use of the, so-called, $3j$ symbols defined by

$$< j_1 j_2; m_1 m_2 | j_1 j_2; j_3 m_3 >= (-1)^{j_1-j_2+m_3}(2j_3+1)^{\frac{1}{2}} \begin{pmatrix} j_1 & j_2 & j_3 \\ m_1 & m_2 & -m_3 \end{pmatrix} \tag{7.32}$$

We have

$$\begin{pmatrix} j_1 & j_2 & j_3 \\ m_1 & m_2 & m_3 \end{pmatrix} = (-1)^{-j_1+j_2+m_3}$$

$$\left[\frac{(j_1+m_1)!(j_1-m_1)!(j_2+m_2)!(j_2-m_2)!(j_3+m_3)!(j_3-m_3)!}{(j_1+j_2-j_3)!(j_1-j_2+j_3)!(-j_1+j_2+j_3)!(j_1+j_2+j_3+1)!}\right]^{\frac{1}{2}}$$

$$\sum_\kappa (-1)^\kappa \begin{pmatrix} j_1+j_2-j_3 \\ \kappa \end{pmatrix} \begin{pmatrix} j_1-j_2+j_3 \\ j_1-m_1-\kappa \end{pmatrix} \begin{pmatrix} -j_1+j_2+j_3 \\ j_2+m_2-\kappa \end{pmatrix} . \tag{7.33}$$

The $3j$ symbols satisfy the following conditions:

$$\begin{pmatrix} j_2 & j_1 & j_3 \\ m_2 & m_1 & m_3 \end{pmatrix} = (-1)^{j_1+j_2+j_3} \begin{pmatrix} j_1 & j_2 & j_3 \\ m_1 & m_2 & m_3 \end{pmatrix} , \tag{7.34}$$

$$\begin{pmatrix} j_1 & j_2 & j_3 \\ -m_1 & -m_2 & -m_3 \end{pmatrix} = (-1)^{j_1+j_2+j_3} \begin{pmatrix} j_1 & j_2 & j_3 \\ m_1 & m_2 & m_3 \end{pmatrix} , \tag{7.35}$$

and

$$\begin{pmatrix} j_1 & j_3 & j_2 \\ m_1 & m_3 & m_2 \end{pmatrix} = (-1)^{j_1+j_2+j_3} \begin{pmatrix} j_1 & j_2 & j_3 \\ m_1 & m_2 & m_3 \end{pmatrix} . \tag{7.36}$$

Combining these equations we conclude that the $3j$-symbols remain invariant under cyclic permutation of the indices 1, 2, 3.

6.8 Transformation of operators under rotations

Let A be a self-adjoint operator acting on the vectors of a Hilbert space. The vectors of the space are transformed according to the rule

$$\psi \to \psi' = D(R)\psi \tag{8.1}$$

where R is a rotation. We define the rotated expression of the linear operator A as an operator A' having the same relation with respect of the rotated vectors as A had with respect of the original vectors, *i.e.*,

$$< \phi'|A'|\psi' > \equiv < \phi|A|\psi > , \tag{8.2}$$

where ϕ and ψ are any two vectors in the space. Thus

$$A' = D(R)AD^{\dagger}(R) . \tag{8.3}$$

Thus, for a rotation by φ about an axis along $\hat{n}$ we have

$$A' = \exp(-i\varphi\hat{n}\cdot\boldsymbol{J})A\exp(i\varphi\hat{n}\cdot\boldsymbol{J}) \tag{8.4}$$

For an infinitesimal rotation by $\delta\varphi$ about $\hat{n}$

$$A' = A + \delta A = \exp(-i\delta\varphi\hat{n}\cdot\boldsymbol{J})A\exp(i\delta\phi\hat{n}\cdot\boldsymbol{J}) = \hat{A} - i\delta\phi[\hat{n}\cdot\boldsymbol{J}, A] + \ldots . \tag{8.5}$$

A is invariant under rotations about $\hat{n}$ if $\delta A = 0$, *i.e.*, if

$$[\hat{n}\cdot\boldsymbol{J}, A] = 0 . \tag{8.6}$$

A is said to be a scalar operator if it is invariant with respect to all rotations, *i.e.*, if

$$[\boldsymbol{J}, A] = 0 \ . \tag{8.7}$$

The transformation of a vector $\boldsymbol{V}$ is

$$\boldsymbol{V}' = R^{-1}\boldsymbol{V} \ . \tag{8.8}$$

Thus, under an infinitesimal rotation by $\delta\varphi$ about $\hat{n}$,

$$\boldsymbol{V}' = \boldsymbol{V} - i\delta\varphi[\hat{n}\cdot\boldsymbol{J}, \boldsymbol{V}] + \cdots = \boldsymbol{V} - \delta\varphi\hat{n}\times\boldsymbol{V} + \cdots \ . \tag{8.9}$$

Thus

$$[\hat{n}\cdot\boldsymbol{J}, \boldsymbol{V}] = -i\hat{n}\times\boldsymbol{V} \ . \tag{8.10}$$

In terms of Cartesian components, Eq. (8.10) is equivalent to

$$[J_i, V_j] = i\sum_k \epsilon_{ijk}V_k \tag{8.11}$$

Consider two sets of vectors $|njm>$ $(m = -j, -j+1, \cdots j-1, j)$ and $|n'j'm'>$ $(m' = -j', -j'+1, \cdots j'-1, j')$ generating the representations $D^{(j)}$ and $D^{(j')}$ of SO(3), respectively. The matrix elements of a scalar operator A, namely

$$< n'j'm'|A|njm >$$

vanish unless $m' = m$ and $j' = j$ and the quantities

$$< n'jm|A|njm >$$

are independent of the value of m. In fact, from

$$[J_z, A] = 0$$

we deduce that

$$< n'j'm'|J_zA - AJ_z|njm >= (m' - m) < n'j'm'|A|njm >= 0 \ .$$

From

$$[J^2, A] = 0$$

we obtain

$$< n'j'm'|J^2A - AJ^2|njm >= [j'(j'+1) - j(j+1)] < n'j'm'|A|njm >= 0 \ .$$

The independence of $< n'jm|A|njm >$ from the value of m follows from

$$[J_\pm, A] = 0 \ .$$

Consider, for example,

$$J_+A|njm >= AJ_+|njm >= \sqrt{(j-m)(j+m+1)}A|nj, m+1 >$$

so that

$$< n'j, m+1|J_+A|njm >= \sqrt{(j-m)(j+m+1)} < n'j, m+1|A|nj, m+1 > .$$

But

$$\begin{aligned} < n'j, m+1|J_+A|njm > &= < njm|AJ_-|n'j, m+1 >^* \\ &= \sqrt{(j-m)(j+m+1)} < n'jm|A|njm > \end{aligned}$$

so that

$$< n'j, m+1|A|nj, m+1 >=< n'jm|A|njm > \ .$$

Matrix elements of vector operators are obtained using similar procedures. Rather than develop them we prefer to give the general result embodied in the Wigner-Eckart theorem discussed in the next section.

6.9 Tensor Operators. The Wigner-Eckart Theorem

In Sec. 8 we have shown that V_1, V_2, V_3 are the components of a vector if

$$V_i' = D(R)V_iD^\dagger(R) = \sum_j R_{ji}V_j \tag{9.1}$$

where R is an element of SO(3). A second rank tensor, whose components T_{ij} transform as the components of the product of two vectors U_iV_j, obeys the law of transformation

$$D(R)T_{ij}D^\dagger(R) = \sum_{k\ell} R_{ki}R_{\ell j}T_{k\ell} \ . \tag{9.2}$$

The components of a second rank tensor belong, therefore, to

$$D^{(1)} \times D^{(1)} = D^{(0)} + D^{(1)} + D^{(2)} \ . \tag{9.3}$$

The trace of the tensor, namely

$$Tr\boldsymbol{T} = T_{11} + T_{22} + T_{33} \tag{9.4}$$

belongs to $D^{(0)}$ *i.e.*, it is a scalar. The three quantities $T_{ij} - T_{ji}$ for $i \neq j$ belong to $D^{(1)}$ and the five quantities

$$T_{ij} + T_{ji} - \frac{2}{3}\delta_{ij}Tr\boldsymbol{T} \tag{9.5}$$

belong to $D^{(2)}$. Note that the trace of the components in Eq. (9.5) vanishes so that only two of the diagonal terms ($i = j$) are independent.

The transformations of tensors of higher rank are described in a similar manner. But it is better to introduce tensors in terms of components that, themselves, generate irreducible representations of the rotation group in 3-space. For example, instead of the components (9.5) it is more convenient to employ

$$\begin{aligned} T_0^{(2)} &= \sqrt{\frac{2}{3}}(2T_{33} - T_{11} - T_{22}) \ , \\ T_{\pm 1}^{(2)} &= \mp(T_{31} + T_{13}) - i(T_{23} + T_{32}) \ , \end{aligned} \tag{9.6}$$

and

$$T_{\pm 2}^{(2)} = T_{11} - T_{22} \pm i(T_{12} + T_{21}) \ .$$

We note that these quantities transform as the spherical harmonics $r^2 Y_2^m(\theta \ , \phi)$, $m = -2, -1, 0, 1, 2$, namely

$$\sqrt{\frac{2}{3}}(2z^2 - x^2 - y^2) \ ,$$

$$\mp 2z(x \pm iy) \ ,$$

and

$$(x \pm iy)^2 \ .$$

In general, we define the irreducible components of a tensor of rank k as the set of operators

$$T_\kappa^{(k)} \quad , \quad \kappa = -k, -k+1, \dots k-1, k$$

transforming according to

$$D(R) T_\kappa^{(k)} D^\dagger(R) = \sum_{\kappa'=-k}^{k} D_{\kappa'\kappa}^{(k)}(R) T_{\kappa'}^{(k)} \ . \tag{9.7}$$

The operators $T_\kappa^{(k)}$ acting on the subspace generated by the set $|njm>$ belonging to $D^{(j)}$ give rise to the $(2k+1)(2j+1)$-dimensional subspace generated by

$$T_\kappa^{(k)}|njm> \quad , \quad \kappa = -k, -k+1, \dots k; m = -j, -j+1, \dots j \ . \tag{9.8}$$

Now

$$\begin{aligned} D(R) T_\kappa^{(k)}|njm> &= D(R) T_\kappa^{(k)} D^\dagger(R) D(R)|njm> \\ &= \sum_{\kappa'=-k}^{k} \sum_{m'=-j}^{j} D_{\kappa'\kappa}^{(k)}(R) D_{m'm}^{(j)}(R) T_{\kappa'}^{(k)}|njm'> \ . \end{aligned} \tag{9.9}$$

The vectors in the set (9.8) generate the direct product representation

$$D^{(k)} \times D^{(j)} = \sum_{j''=|j-k|}^{j+k} D^{(j'')} \ . \tag{9.10}$$

We project from the set $\{T_\kappa^{(k)}|njm>\}$, $\kappa = -k, -k+1, \ldots k$; $m = -j, -j+1, \ldots j$ a vector belonging to the m''-row of $D^{(j'')}$ using the Clebsch-Gordan coefficients:

$$|n'j''m''> W_{n'n}(k,j,j'') = \sum_{\kappa=-k}^{k} \sum_{m=-j}^{j} T_\kappa^{(k)}|njm><jk;m\kappa|jk;j''m''> \ . \tag{9.11}$$

We recall that all Clebsch-Gordan coefficients for which $m'' \neq m+\kappa$ vanish. Thus, the coefficient of proportionality $W_{n'n}(k,j,j'')$ does not depend on the "angular" variables m, κ and m''. The orthogonality of the Clebsch-Gordan coefficients allows us to solve Eq.s (9.11) for the quantities (9.8):

$$T_\kappa^{(k)}|njm>= \sum_{j''=|j-k|}^{j+k} \sum_{m''=-j''}^{j''} W_{n'n}(k,j,j'')|n'j''m''><jk;j''m''|jk;m\kappa> \tag{9.12}$$

The inner product with $|n'j'm''>$ gives

$$<n'j'm'|T_\kappa^{(k)}|njm>= W_{n'n}(k,j,j') <jk;j'm'|jk;m\kappa> \ . \tag{9.13}$$

This result, known as the Wigner-Eckart theorem, shows that the calculation of the $(2j'+1)(2k+1)(2j+1)$ matrix elements of $\boldsymbol{T}^{(k)}$ between vectors of $D^{(j)}$ and $D^{(j')}$ is reduced to the determination of the single quantity $W_{n'n}(k,j,j')$ independent of κ, m and m'. It is traditional to write this result in the form

$$\boxed{<n'j'm'|T_\kappa^{(k)}|njm>= (2j'+1)^{-1/2}<n'j'\|T^{(k)}\|nj><jk;j'm'|jk;m\kappa>} \tag{9.14}$$

The quantity $<nj' \parallel T^{(k)} \parallel nj>$ is called the reduced matrix element of $T^{(k)}$ with respect to the subspaces generated by the sets $|n'j'm'>$ and $|njm>$.

In terms of the 3-j symbols we have

$$<n'j'm'|T_\kappa^{(k)}|njm>= (-1)^{j'-m'}<n'j'\|T^{(k)}\|nj> \begin{pmatrix} j' & k & j \\ -m' & \kappa & m \end{pmatrix} . \tag{9.15}$$

The commutation rules of the irreducible components $T_\kappa^{(k)}$ of a tensor of rank k with the operators J_1, J_2, J_3 are obtained from Eq. (9.7). For an infinitesimal rotation by $\delta\varphi$ about an axis parallel to the unit vector $\hat{n}$, Eq. (9.7) becomes

$$\begin{aligned} e^{-i\delta\varphi\hat{n}\cdot\mathbf{J}} T_\kappa^{(k)} e^{i\delta\varphi\hat{n}\cdot\mathbf{J}} &= \sum_{\kappa'=-k}^{k} D_{\kappa'\kappa}^{(k)}(R) T_{\kappa'}^{(k)} \\ &= \sum_{\kappa'=-k}^{k} <k\kappa'|e^{-i\delta\varphi\mathbf{J}\cdot\hat{n}}|k\kappa> T_{\kappa'}^{(k)} \ . \end{aligned}$$

This yields

$$T_\kappa^{(k)} - i\delta\varphi[\hat{n}\cdot\boldsymbol{J}, T_\kappa^{(k)}] + \cdots = \sum_{\kappa'} <k\kappa'|1 - i\delta\varphi\hat{n}\cdot\boldsymbol{J} + \cdots|k\kappa> T_{\kappa'}^{(k)}$$

or

$$[\hat{n}\cdot \boldsymbol{J}, T^{(k)}_\kappa] = \sum_{\kappa'} < k\kappa'|\hat{n}\cdot\boldsymbol{J}|k\kappa > T^{(k)}_{\kappa'} \tag{9.16}$$

If the vectors $|k\kappa>$ are eigenvectors of J_3 we find

$$[J_3, T^{(k)}_\kappa] = \kappa T^{(k)}_\kappa \tag{9.17}$$

and

$$[J_\pm, T^{(k)}_\kappa] = [(k\mp\kappa)(k\pm\kappa+1)]^{\frac{1}{2}} T^{(k)}_{\kappa\pm 1} , \tag{9.18}$$

where $J_\pm = J_1 \pm iJ_2$.

The products of the irreducible components of two tensors, $U^{(k_1)}_{\kappa_1}$ and $V^{(k_2)}_{\kappa_2}$ generate the representation

$$D^{(k_1)} \times D^{(k_2)} = \sum_{k=|k_1-k_2|}^{k_1+k_2} D^{(k)} .$$

Using the Clebsch-Gordan coefficients we can project out the components of a tensor of rank k ($|k_1-k_2| \leq k \leq k_1+k_2$) as follows

$$T^{(k)}_\kappa = \sum_{\kappa_1=-k_1}^{k_1} \sum_{\kappa_2=-k_2}^{k_2} < k_1k_2;\kappa_1\kappa_2|k_1k_2;k\kappa > U^{(k_1)}_{\kappa_1} V^{(k_2)}_{\kappa_2} . \tag{9.19}$$

If $k_1 = k_2$ we can form the scalar

$$\boldsymbol{U}\cdot\boldsymbol{V} = (-1)^k(2k+1)^{\frac{1}{2}} \sum_{\kappa_1,\kappa_2} < kk;\kappa_1\kappa_2|kk;00 > U^{(k)}_{\kappa_1} V^{(k)}_{\kappa_2} \tag{9.20}$$

Since only terms with $\kappa_1+\kappa_2 = 0$ occur in the sum this scalar can be rewritten in the form

$$\boldsymbol{U}\cdot\boldsymbol{V} = \sum_\kappa (-1)^\kappa U^{(k)}_\kappa V^{(k)}_{-\kappa} . \tag{9.21}$$

To prove this we obtain the Clebsch-Gordan coefficient in Eq. (9.20), namely

$$< kk;\kappa,-\kappa|kk;00 >= \frac{(-1)^{k-\kappa}}{\sqrt{2k+1}} . \tag{9.22}$$

Substitution in Eq. (9.20) yields Eq. (9.21).

As an example we show that the definition (9.21) is consistent with the usual definition in the case of the scalar product of two vectors, $\boldsymbol{U}$ and $\boldsymbol{V}$. The irreducible components of $\boldsymbol{U}$ are

$$U_1 = -\frac{1}{\sqrt{2}}(U_x+iU_y) \quad , \quad U_o = U_z \, , \; U_{-1} = \frac{1}{\sqrt{2}}(U_x-iU_y) . \tag{9.23}$$

A similar set of equations gives the irreducible components of $\boldsymbol{V}$. Then

$$\begin{aligned} \boldsymbol{U}\cdot\boldsymbol{V} &= \frac{1}{2}(U_x+iU_y)(V_x-iV_y) + U_zV_z + \frac{1}{2}(U_x-iU_y)(V_x+iV_y) \\ &= U_xV_x + U_yV_y + U_zV_z . \end{aligned} \tag{9.24}$$

6.10 Further Developments

We have obtained the irreducible representations of $SO(3)$, namely $D^{(j)}$ where j takes the values $j = 0, \frac{1}{2}, 1, \frac{3}{2} \ldots$. For each integral value of j, $D^{(j)}(R)$ is an irreducible representation of $SO(3)$. For j equal to a half-integer, when R is a rotation by 2π about an arbitrary axis, $D^{(j)}(R)$ is not the identity but rather its negative. In such a case for any rotation R, there are two matrices $D^{(j)}(R)$ differing in sign. We say that such representations are double valued. For j integral the representations are said to be single valued or faithful. As already mentioned, we can introduce an operation $\bar{E}$ to represent a rotation by 2π about an arbitrary axis and regard the group as the direct product of $SO(3)$ and $\{E, \bar{E}\}$ with $\bar{E}^2 = E$.

The orthogonal group $O(3)$ is

$$O(3) = SO(3) \times C_i \ . \tag{10.1}$$

For each irreducible representation $D^{(j)}$ of $SO(3)$ there are two irreducible representations of $O(3)$ denoted by $D_+^{(j)}$ and $D_-^{j)}$. If $R \in SO(3)$, then

$$D_+^{(j)}(iR) = D^{(j)}(R) \tag{10.2}$$

and

$$D_-^{(j)}(iR) = -D^{(j)}(R) \ . \tag{10.3}$$

The operators

$$J_i = -i\epsilon_{ijk}x_j\partial_k \tag{10.4}$$

obey the commutation relations

$$[J_i, J_j] = i\epsilon_{ijk}J_k \tag{10.5}$$

verified by the generators of $SO(3)$, J_1, J_2 and J_3. Thus, a set of vectors $|\ell m>$ $(m = -\ell, -\ell+1, \ldots \ell)$ generating $D^{(\ell)}$ can be constructed satisfying the differential equations

$$J_3|\ell m> = m|\ell m> \ , \tag{10.6}$$

and

$$J_\pm|\ell m> = \sqrt{(\ell \mp m)(\ell \pm m + 1)}|\ell, m \pm 1> \ . \tag{10.7}$$

Since the resulting vectors, satisfying these differential equations whose solutions are functions of position, thus periodic with respect to rotations by 2π, the $D^{(\ell)}$ correspond to integral values of ℓ. This leads to the conclusion that the half-integral irreducible representations $D^{(j)}$ cannot be generated by functions of position but by the *spinons* ξ_1 and ξ_2 already introduced.

We note that the operators J_i defined in Eqs. (10.5) satisfy

$$P_i \boldsymbol{J} P_i^{-1} = \boldsymbol{J} \ . \tag{10.8}$$

Denoting ξ_1 and ξ_2, satisfying $J_z\xi_1 = \frac{1}{2}\xi_1$, $J_z\xi_2 = -\frac{1}{2}\xi_2$ by ξ_+ and ξ_-, respectively we write

$$J_z\xi_\pm = \pm\frac{1}{2}\xi_\pm \ . \tag{10.9}$$

Now

$$P_iJ_zP_i^{-1}P_i\xi_\pm = \pm\frac{1}{2}P_i\xi_\pm \tag{10.10}$$

so that we may assume that $P_i\xi_\pm = \xi_\pm$.

If C_n is a rotation by $2\pi/n$ about an arbitrary axis, C_n^n is a rotation by 2π so that for half-integral representations it is $\bar{E}$ rather than E. Since $P_i^{-1} = P_i$, $P_i^2 = E$ and we write

$$i^2 = E \tag{10.11}$$

For every operation R in $O(3)$ or any of its subgroups we define

$$\bar{R} = \bar{E}R \ . \tag{10.12}$$

Clearly $\bar{E}$ commutes with all elements of $O(3)$.

Let σ be a reflection in a mirror plane. We can always express σ as a rotation by π around the normal to the plane followed by inversion about the intersection of the axis of rotation and the plane, *i.e.*,

$$\sigma = iC_2 \ . \tag{10.13}$$

The square of σ is

$$\sigma^2 = i^2C_2^2 = \bar{E} \ . \tag{10.14}$$

We recall that, if C_2 and C_2' are rotations by π about intersecting orthogonal axes, then they commute, *i.e.*,

$$C_2C_2' = C_2'C_2 \ . \tag{10.15}$$

For spinors, this is no longer the case. In fact, for $j = \frac{1}{2}$, rotations by π about the x-axis and y-axis of a Cartesian coordinate system are represented by the operators

$$P_{C_2} = \exp\left(-\frac{i\pi}{2}\sigma_x\right) = -i\sigma_x \quad , \quad P_{C_2'} = \exp\left(-\frac{i\pi}{2}\sigma_y\right) = -i\sigma_y \tag{10.16}$$

and

$$P_{C_2}P_{C_2'} = -\sigma_x\sigma_y = -i\sigma_z \tag{10.17}$$

while

$$P_{C_2'}P_{C_2} = -\sigma_y\sigma_x = i\sigma_z \ . \tag{10.18}$$

Thus

$$C_2C_2' = \bar{C}_2'C_2 \ , \tag{10.19}$$

or

$$\bar{C}_2' = C_2C_2'C_2^{-1} \ . \tag{10.20}$$

This shows that, if C_2 and C_2' are elements of a group, then C_2' and $\bar{C}_2'$ belong to the same conjugate class. Multiplying both sides of Eq. (10.19) by the inversion i about the intersection of the axes of C_2 and C_2' we obtain

$$C_2\sigma' = \bar{\sigma}'C_2 \tag{10.21}$$

where $\sigma' = iC_2'$ is a mirror reflection in a plane perpendicular to the axis of C_2'. We conclude that σ' and $\bar{\sigma}'$ are in the same conjugate class if C_2, a rotation by π about an axis in the plane of reflection is an element of the group. Equation (10.21) also implies that

$$C_2\sigma' = \sigma'\bar{C}_2 \tag{10.22}$$

meaning that C_2 and $\bar{C}_2$ are conjugate if the group contains a mirror plane through the axis of C_2. Multiplying Eq. (10.21) by i we obtain

$$\sigma\sigma' = \sigma'\bar{\sigma}\ . \tag{10.23}$$

Thus σ and $\bar{\sigma}$ are in the same class if the group contains a mirror plane σ' normal to σ.

Let $\Gamma(R)$ be a representation of $O(3)$ or of one of its subgroups. Then

$$Tr\Gamma(\bar{R}) = \begin{cases} \chi(R) & \text{for } j \text{ integer} \\ -\chi(R) & \text{for } j \text{ half-integer}\ . \end{cases} \tag{10.24}$$

If R and $\bar{R}$ are in the same conjugate class, in a half-integer representation

$$\chi(\bar{R}) = \chi(R) = -\chi(R) = 0\ . \tag{10.25}$$

If R is a rotation by φ

$$\chi^{(j)}(\varphi) = \frac{\sin(j+\frac{1}{2})\varphi}{\sin\frac{\varphi}{2}} \tag{10.26}$$

and

$$\chi^{(j)}(\varphi + 2\pi) = (-1)^{2j}\chi^{(j)}(\varphi)\ . \tag{10.27}$$

For $j = \frac{1}{2}, \frac{3}{2}, \ldots,$ $\chi^{(j)}(\varphi) = 0$ only if $\varphi = \pi$. Therefore, when establishing conditions for R and $\bar{R}$ to be in the same class it is enough to limit consideration to two-fold rotations and to mirror planes, the latter being the only improper rotations derived from two-fold rotations.

These results are summarized in the following rules due to Opechowski:

(i) C_2 and $\bar{C}_2$ are in the same conjugate class of a double group if the group contains a two-fold axis normal to that of C_2 or a reflection plane σ containing the axis of C_2.

(ii) σ and $\bar{\sigma}$ are in the same class of a double group if the group contains a two-fold rotation about an axis in the plane σ or a reflection plane σ' normal to σ.

As an example we consider the group

$$\bar{T}_d = T_d \times \bar{E}\ . \tag{10.28}$$

According to Opechowski's rules C_2 and $\bar{C}_2$ as well as σ_d and $\bar{\sigma}_d$ are in the same conjugate classes. Thus, the class structure of $\bar{T}_d$ is

$$\bar{T}_d = \{E, \bar{E}, 8C_3, 8\bar{C}_3, (3C_2, 3\bar{C}_2), (6\sigma_d, 6\bar{\sigma}_d), 6S_4, 6\bar{S}_4\} \ . \tag{10.29}$$

There are eight conjugate classes and, hence, eight inequivalent irreducible representations. Five irreducible representations, corresponding to those of T_d are irreducible representation of $\bar{T}_d$ with $\bar{E} = E$. The additional representations Γ_6, Γ_7 and Γ_8 have dimensions ℓ_6, ℓ_7 and ℓ_8. By virtue of the celebrated theorem

$$\ell_6^2 + \ell_7^2 + \ell_8^2 = 24 \ .$$

Thus, $\ell_6, \ell_7 = 2$ and $\ell_8 = 4$.

We must be careful at the outset to properly specify the elements in the different classes. The class $8C_3$ contains

$$4C_3 \quad , \quad 4C_3^{-1}$$

because C_3^2 is a barred operation. In fact $C_3^3 = \bar{E}$ so that

$${C_3}^2 = \bar{E}C_3^{-1} = \overline{C_3^{-1}} \ .$$

The element S_4 of T_d is a rotation by $\pi/2$ followed by a reflection in a plane perpendicular to the axis of rotation, *i.e.*,

$$S_4 = \sigma_h C_4 \ .$$

We recall that while S_4 is an element of T_d, σ_h and C_4 are not. Now from

$$\sigma_h = iC_2$$

it follows that

$$S_4 = iC_4^3 \ .$$

We now adopt the following convention: if C_2 and C_φ are rotations about the same axis

$$C_2 C_\varphi = \begin{cases} \bar{C}_{\varphi+\pi} & \text{if} \quad 0 < \varphi \leq \pi \\ C_{\varphi+\pi} & \text{if} \quad -\pi < \varphi \leq 0 \ . \end{cases}$$

This convention is consistent with

$$C_2^2 = \bar{E} \ .$$

Then, it follows that

$$\sigma_h C_4 = iC_2 C_4$$

is a barred operation, namely

$$\bar{S}_4 = iC_4^3 = i\bar{E}C_4^{-1} = \overline{iC_4^{-1}} \ .$$

Thus

$$S_4 = iC_4^{-1} \; ,$$

and, hence

$$6S_4 = (3iC_4 \; , \; 3iC_4^{-1}) \tag{10.30}$$

and

$$6\bar{S}_4 = (\overline{3iC_4} \; , \; \overline{3iC_4^{-1}}) \; . \tag{10.31}$$

The action of P_{S_4} on the space spanned by ξ_+ and ξ_- is

$$P_{S_4} = P_i \exp(-i\pi\sigma_z/4) = \frac{1}{\sqrt{2}} P_i(1 - i\sigma_z) \; , \tag{10.32}$$

taking one of the operations, namely with C_4 along $\hat{z}$, so that

$$P_{S_4}\xi_\pm = \frac{1}{\sqrt{2}}(1 \pm i)\xi_\pm \; . \tag{10.33}$$

Thus the trace of S_4 in the representation Γ_6 generated by the spinons $\xi_\pm$ is

$$\chi(S_4) = \sqrt{2} \; . \tag{10.34}$$

For C_3 about $(\hat{x} + \hat{y} + \hat{z})/\sqrt{3} = \hat{n}$,

$$P_{C_3} = e^{-\frac{i\pi}{3}\hat{n}\cdot\boldsymbol{\sigma}} = \frac{1}{2} - \frac{i}{2}\sqrt{3}\hat{n}\cdot\boldsymbol{\sigma} \; . \tag{10.35}$$

The trace of C_3 in Γ_6 is thus

$$\chi(C_3) = 1 \; .$$

Thus, $\xi_\pm$ generate the representation

E	$\bar{E}$	$8C_3$	$8\bar{C}_3$	$(3C_2, 3\bar{C}_2)$	$(6\sigma_d, 6\bar{\sigma}_d)$	$6S_4$	$6\bar{S}_4$
2	-2	1	-1	0	0	$\sqrt{2}$	$-\sqrt{2}$

Theorem 1.5 shows that this representation is irreducible. We form $\Gamma_6 \times \Gamma_2$ whose character is

$$2 \;\; -2 \;\; 1 \;\; -1 \;\; 0 \;\; 0 \;\; -\sqrt{2} \;\; \sqrt{2}$$

which is also irreducible and is denoted by Γ_7. The character of $\Gamma_5 \times \Gamma_6$ is

$$6 \;\; -6 \;\; 0 \;\; 0 \;\; 0 \;\; 0 \;\; -\sqrt{2} \;\; \sqrt{2}$$

which contains Γ_7 once and a second irreducible representation whose character is

$$4 \;\; -4 \;\; -1 \;\; 1 \;\; 0 \;\; 0 \;\; 0 \;\; 0 \; .$$

It is denoted by Γ_8. Thus, the character table of $\bar{T}_d$ is

$\bar{T}_d$	E	$\bar{E}$	$8C_3$	$8\bar{C}_3$	$3C_2$ $3\bar{C}_2$	$6\sigma_d$ $6\bar{\sigma}_d$	$6S_4$	$6\bar{S}_4$	Basic Functions	
Γ_1	1	1	1	1	1	1	1	1	α	a
Γ_2	1	1	1	1	1	-1	-1	-1	β	a
Γ_3	2	2	-1	-1	2	0	0	0	γ_1, γ_2	a
Γ_4	3	3	0	0	-1	-1	1	1	$\delta_1, \delta_2, \delta_3$	a
Γ_5	3	3	0	0	-1	1	-1	-1	$\epsilon_1, \epsilon_2, \epsilon_3$	a
Γ_6	2	-2	1	-1	0	0	$\sqrt{2}$	$-\sqrt{2}$	ξ_+, ξ_-	c
Γ_7	2	-2	1	-1	0	0	$-\sqrt{2}$	$\sqrt{2}$	$\phi_{1/2}, \phi_{-1/2}$	c
Γ_8	4	-4	-1	1	0	0	0	0	$\psi_{3/2}, \psi_{1/2}, \psi_{-1/2}, \psi_{-3/2}$	c

$$\begin{aligned}
\psi_{3/2} &= -\frac{i}{\sqrt{2}}(\epsilon_1 + i\epsilon_2)\xi_+ \ , \ \psi_{1/2} = -\frac{i}{\sqrt{6}}(\epsilon_1 + i\epsilon_2)\xi_- + i\sqrt{\frac{2}{3}}\epsilon_3\xi_+ \\
\psi_{-1/2} &= \frac{i}{\sqrt{6}}(\epsilon_1 - i\epsilon_2)\xi_+ + i\sqrt{\frac{2}{3}}\epsilon_3\xi_- \ , \ \psi_{-3/2} = \frac{i}{\sqrt{2}}(\epsilon_1 - i\epsilon_2)\xi_- \\
\phi_{1/2} &= -\frac{i}{\sqrt{3}}(\epsilon_1 + i\epsilon_2)\xi_- - \frac{i}{\sqrt{3}}\epsilon_3\xi_+ \ , \ \phi_{-1/2} = -\frac{i}{\sqrt{3}}(\epsilon_1 - i\epsilon_2)\xi_+ \frac{i}{\sqrt{3}}\epsilon_3\xi_-
\end{aligned}$$

Chapter 7

Applications of the Theory of Group Representations in Quantum Mechanics

7.1 General Considerations

In chapter 2 we discussed the relation between symmetry principles and conservation laws in classical mechanics. We turn now to the similar questions in quantum theory. We recall that if $F(\boldsymbol{q}, \boldsymbol{p})$ is a dynamical variable in which $\boldsymbol{q}$ and $\boldsymbol{p}$ are operators obeying the Heisenberg commutation relations then, for an arbitrary state $\psi(\boldsymbol{q}, t)$ of the system

$$\frac{d}{dt} < \psi|\widehat{F}|\psi >= \frac{1}{i\hbar} < \psi|[\widehat{F}, \widehat{H}]|\psi > \tag{1.1}$$

so that the operation of taking the time derivative of the expectation value of a dynamical variable which is not explicitly time dependent is represented by

$$\widehat{\dot{F}} = \frac{1}{i\hbar}[\widehat{F}, \widehat{H}] \ . \tag{1.2}$$

Here $\widehat{H}$ is the Hamiltonian operator of the system. Thus, $\widehat{F}$ is a constant of the motion if it commutes with the Hamiltonian.

If $\widehat{H}$ is invariant with respect to arbitrary translations along an axis $\hat{e}$, then the projection of the momentum of the system is a constant of the motion. In fact, let P_a be the operator representing a translation by $\boldsymbol{a}$, parallel to $\hat{e}$. Since

$$P_a \widehat{H} P_a^{-1} = \widehat{H} \ , \tag{1.3}$$

for an arbitrary state of N particles

$$\begin{aligned} P_a\psi(\boldsymbol{r}_1,\boldsymbol{r}_2,\dots \boldsymbol{r}_N,t) &= \psi(\boldsymbol{r}_1-\boldsymbol{a},\boldsymbol{r}_2-\boldsymbol{a},\dots \boldsymbol{r}_N-\boldsymbol{a},t) \\ &= \exp\left(-\boldsymbol{a}\cdot\sum_{i=1}^{N}\nabla_i\right)\psi(\boldsymbol{r}_1,\boldsymbol{r}_2,\dots \boldsymbol{r}_N,t) \\ &= \exp\left(-\frac{i}{\hbar}\boldsymbol{a}\cdot\boldsymbol{P}\right)\psi(\boldsymbol{r}_1,\boldsymbol{r}_2\dots \boldsymbol{r}_N,t) \end{aligned} \tag{1.4}$$

where

$$\widehat{\boldsymbol{P}}=\sum_{i=1}^{N}\frac{\hbar}{i}\nabla_i=\sum_{i=1}^{N}\hat{\boldsymbol{p}}_i\ . \tag{1.5}$$

Therefore

$$P_a=\exp\left(\frac{i}{\hbar}\boldsymbol{a}\cdot\boldsymbol{P}\right)\ . \tag{1.6}$$

Combining Eqs. (1.3) and (1.6) we obtain

$$\left[\exp\left(-\frac{i}{\hbar}\boldsymbol{a}\cdot\boldsymbol{P}\right)\ ,\widehat{H}\right]=0 \tag{1.7}$$

Since $\boldsymbol{a}=a\hat{e}$ with a arbitrary,

$$[\boldsymbol{P}\cdot\hat{e},\widehat{H}]=0 \tag{1.8}$$

and $\boldsymbol{P}\cdot\hat{e}$ is a constant of the motion.

Let $\widehat{H}$ be invariant with respect to rotations about an axis along the unit vector $\hat{n}$. For any function $\psi(\boldsymbol{r}_1,\boldsymbol{r}_2,\dots \boldsymbol{r}_N,t)$, a rotation by $\delta\varphi$ about $\hat{n}$ transforms $\boldsymbol{r}_i$ into

$$\boldsymbol{r}_i'=\boldsymbol{r}_i+\delta\varphi\hat{n}\times\boldsymbol{r}_i \tag{1.9}$$

and ψ into

$$\begin{aligned} P_R\psi &= \psi(\boldsymbol{r}_1-\delta\varphi\hat{n}\times\boldsymbol{r}_1,\boldsymbol{r}_2-\delta\varphi\hat{n}\times r_2,\dots,\boldsymbol{r}_N-\delta\varphi\hat{n}\times\boldsymbol{r}_N,t) \\ &= \exp(-\delta\varphi\sum_i(\hat{n}\times\boldsymbol{r}_i)\cdot\nabla_i)\psi(\boldsymbol{r}_1,\dots \boldsymbol{r}_N,t) \\ &= \exp(-i\delta\varphi\hat{n}\cdot\boldsymbol{L})\psi(\boldsymbol{r}_1,\dots \boldsymbol{r}_N,t) \end{aligned} \tag{1.10}$$

where

$$\boldsymbol{L}=\frac{1}{i}\sum_{i=1}^{N}\boldsymbol{r}_i\times\nabla_i\ . \tag{1.11}$$

If H is invariant with respect to rotations about $\hat{n}$,

$$\left[e^{-i\delta\varphi\mathbf{L}\cdot\hat{n}},\widehat{H}\right]=0 \tag{1.12}$$

and, thus

$$[\boldsymbol{L}\cdot\hat{n},\widehat{H}]=0\ . \tag{1.13}$$

In general if

$$\left[e^{i\alpha\widehat{K}}, \widehat{H}\right] = 0 \tag{1.14}$$

where α is an arbitrary real number and $\widehat{K}$ a Hermitian operator, then

$$[\widehat{K}, \widehat{H}] = 0 \tag{1.15}$$

Thus, for every set of unitary operations depending on a continuous parameter which leave the Hamiltonian invariant we obtain a constant of the motion.

In applications of the theory of group representations in quantum mechanics the first step is to identify the group of transformations which leave the Hamiltonian $\widehat{H}$ invariant. We call it the group of $\widehat{H}$. Let $\{R\}$ be the group of geometrical transformation in the group of $\widehat{H}$. Associated with this group there is a group $\{P_R\}$ of unitary operators acting on the Hilbert space of the system. Now

$$P_R \widehat{H} P_R^{-1} = \widehat{H} \tag{1.16}$$

or

$$[P_R, \widehat{H}] = 0 \ . \tag{1.17}$$

Let ϕ be a normalized eigenvector of $\widehat{H}$ with eigenvalue E, *i.e.*,

$$\widehat{H}\phi = E\phi \ . \tag{1.18}$$

Then

$$\widehat{H} P_R \phi = E P_R \phi \tag{1.19}$$

so that $P_R\phi$ is also an eigenvector with eigenvalue E. The set $\{P_R\phi\}$ is a subspace of the Hilbert space of the system. Let $\phi_1, \phi_2, \ldots \phi_\ell$ be an orthonormal basis on this subspace. Then,

$$P_R \phi_k = \sum_{n=1}^{\ell} \phi_n \Gamma_{nk}(R) \tag{1.20}$$

where

$$\Gamma_{nk}(R) = <\phi_n|P_R|\phi_k> \ . \tag{1.21}$$

The matrices $\Gamma(R)$ defined by Eq. (1.21) form an ℓ-dimensional, unitary representation of $\{R\}$. The subspace spanned by $\phi_1, \phi_2 \ldots \phi_\ell$ will be called the space of the eigenvalue E. If there is no proper subspace of the space of E invariant under all operations of $\{P_R\}$, *i.e.*, if the representation Γ is irreducible, then we say that the ℓ-fold degeneracy of E is a necessary one. If this is not the case the degeneracy is called accidental. As an example we consider the p-states of a charged particle in a central potential $V(r)$. The Hamiltonian is

$$\widehat{H} = \frac{p^2}{2m} + V(r) \ . \tag{1.22}$$

Clearly the symmetry group of $\widehat{H}$ is $O(3)$. Thus the eigenvectors of $\widehat{H}$ are of the form $|E\ell m>$. For $\ell = 1$ ($m = 1, 0, -1$) the p states are of the form

$$p_{+1} = -\frac{i}{\sqrt{2}}(x+iy)f(r) \quad , \quad p_0 = izf(r) \quad , \quad p_{-1} = \frac{i}{\sqrt{2}}(x-iy)f(r)$$

where $f(r)$ is a function of the radius vector $r = |\boldsymbol{r}|$. The functions p_m ($m = 1, 0, -1$) generate the representation $D_-^{(1)}$ of $O(3)$ which is irreducible. In the case of the hydrogen atom

$$V(r) = -\frac{e^2}{r} \tag{1.23}$$

and the energy eigenvectors are $|n\ell m>$, $n = 1, 2, \ldots$; $\ell = 0, 1, \ldots n-1$, $m = -\ell, -\ell+1, \ldots \ell$. The states np ($|n1m>$) and ns $|n00>$) are degenerate. Consider again the p states above. There is no operation of $O(3)$ that can transform a p-state into an s-state, the latter being spherically symmetric. Thus, it is said that the np and ns degeneracy of the hydrogen atom is accidental. There is no room for accidents in physics. In this case it turns out that $O(3)$ does not include all symmetry elements of the Hamiltonian. We shall show later that the appropriate symmetry is $O(4)$ in momentum space. Thus we may say in general that associated with every energy eigenvalue of a quantum mechanical system there is an irreducible representation generated by the space of the eigenvectors corresponding to that eigenvalue. If one should find that the representation is reducible he should conclude that there has been a failure to properly identify the symmetry group of the Hamiltonian.

Let $\phi_k^{(i)}$ $k = 1, 2, \ldots \ell_i$ be a set of eigenvectors of $\widehat{H}$ with eigenvalue E_i ($\widehat{H}\phi_k^{(i)} = E_i\phi_k^{(i)}$) generating the irreducible representation $\Gamma^{(i)}$ of $\{P_R\}$, the group of $\widehat{H}$. If $\{P_R\}$ is a finite group and the matrices of the irreducible unitary representation are known, knowledge of one of the functions $\phi_k^{(i)}$ is enough to generate the others, called its partners. In fact

$$P_R\phi_k^{(i)} = \sum_{n=1}^{\ell i} \phi_n^{(i)}\Gamma_{nk}^{(i)}(R) \ . \tag{1.24}$$

Multiplying both sides of Eq. (1.24) by $\Gamma_{n'k'}^{(j)*}(R)$ and summing over R we obtain

$$\begin{aligned} \sum_R \Gamma_{n'k'}^{(j)*}(R)P_R\phi_k^{(i)} &= \sum_{n=1}^{\ell i}\phi_n^{(i)}\sum_R \Gamma_{n'k'}^{(j)*}(R)\Gamma_{nk}^{(i)}(R) \\ &= \frac{h}{\ell_i}\delta_{ij}\delta_{k'k}\phi_{n'}^{(i)} \end{aligned}$$

where we made use of the orthogonality theorem (6. 1.15). Hence,

$$\frac{\ell_j}{h}\sum_R \Gamma_{nk}^{(j)*}(R)P_R$$

acting on $\phi_k^{(i)}$ gives $\delta_{ij}\delta_{k'k}\phi_n^{(i)}$. We deduce that

$$\phi_n^{(i)} = \frac{\ell_i}{h} \sum_R \Gamma_{nk}^{(i)*} P_R \phi_k^{(i)} \; . \tag{1.25}$$

This operator is used when we wish to find the partners of one of the eigenvectors, say $\phi_k^{(i)}$ of $\widehat{H}$. The method requires the knowledge of the matrices $\Gamma^{(i)}(R)$ and is often not very practical. Equation (1.25) can be generalized to the case of infinite groups in which the sum over R is replaced by a suitable integration over continuous parameters. Consider an arbitrary vector F in the Hilbert space of a system. It is always possible to express F in the form

$$F = \sum_{i=1}^{N_\Gamma} \sum_{n=1}^{\ell_i} a_n^{(i)} f_n^{(i)} \tag{1.26}$$

where $f_n^{(i)}$ belongs to the nth row of the unitary irreducible representation $\Gamma^{(i)}$. Note that we do not assume that $f_n^{(i)}$ is an eigenvector of an operator such as $\widehat{H}$ but, rather, it is a linear combination of such eigenvectors.

The theory of group representations provides us with procedures to simplify the calculation of matrix elements of quantum mechanical operators. Let us consider a system whose symmetry group is G with irreducible representations $\Gamma^{(i)}, \Gamma^{(2)}, \ldots \Gamma^{(N_\Gamma)}$, and Ω_m a set of operators $(m = 1, 2, \ldots \ell_\Omega)$ belonging to the irreducible representation Γ_Ω of G and form matrix elements of Ω_m between vectors $\phi_n^{(i)}$ and $\phi_{n'}^{(j)}$ belonging to $\Gamma^{(i)}$ and $\Gamma^{(j)}$, respectively. The vectors

$$\Omega_m \phi_n^{(i)} \quad , \quad m = 1, 2, \ldots \ell_\Omega \quad ; \quad n = 1, 2, \ldots \ell_i$$

generate the $\ell_\Omega \ell_i$-dimensional representation $\Gamma_\Omega \times \Gamma^{(i)}$ of G. In general this representation is reducible, *i.e.*,

$$\Gamma_\Omega \times \Gamma^{(i)} = \sum_{j=1}^{N_\Gamma} n_j \Gamma^{(j)} \; . \tag{1.27}$$

We recall that G is simply reducible if all n_j are either zero or unity. We have shown in 6.7.4 that

$$\Omega_m \phi_n^{(i)} = \sum_{j=1}^{N_\Gamma} \sum_{\nu_j=1}^{n_j} \sum_{n'=1}^{\ell_j} \psi_{n',\nu_j}^{(j)} < \Gamma_\Omega i; jn', \nu_j | \Gamma_\Omega i; mn > \tag{1.28}$$

where the functions $\psi_{n',\nu_j}^{(j)}$ belong to $\Gamma^{(j)}$. Then

$$< \phi_{n'}^{(j)} | \Omega_n^{(i)} | \phi_n^{(i)} > = \sum_{\nu_j=1}^{n_j} < \phi_{n'}^{(j)} | \psi_{n',\nu_j}^{(j)} > < \Gamma_\Omega i; jn', \nu_j | \Gamma_\Omega i; mn > \; . \tag{1.29}$$

This result shows that to calculate the $\ell_j \ell_\Omega \ell_i$ matrix elements $< \phi_{n'}^{(j)} | \Omega_m | \phi_n^{(i)} >$ is is enough to calculate the n_j matrix elements $< \phi_{n'}^{(j)} | \psi_{n',\nu_j}^{(j)} >$. The remaining coefficients are the Clebsch-Gordan coefficients for the group G. This theorem is a generalization of the Wigner-Eckart theorem to groups which are not simply reducible.

7.2 Selection rules for quantum mechanical transitions in the presence of electromagnetic radiation

We consider the coupling between a quantum mechanical system composed of N charged particles of masses, charges and intrinsic magnetic moments m_i, q_i and $\boldsymbol{\mu}_i$, respectively and a plane electromagnetic wave of angular frequency ω described by the vector potential

$$\boldsymbol{A}(\boldsymbol{r}, t) = \Re e[\boldsymbol{A}_0 e^{i(\mathbf{q}\cdot\mathbf{r}-\omega t)}] \tag{2.1}$$

Here $\boldsymbol{q}$ is the wave vector of the wave and $\boldsymbol{A}_0$ a complex amplitude perpendicular to $\boldsymbol{q}$. By a suitable selection of the origin of the phase we can set

$$\boldsymbol{A}_0 = A_1 \hat{e}_1 + A_2 \hat{e}_2 e^{-i\delta} \ . \tag{2.2}$$

Here A_1 and A_2 are real amplitudes and $\hat{e}_1$ and $\hat{e}_2$ two orthogonal unit vectors at right angles to $\boldsymbol{q}$. Without loss of generality we can require A_1 and A_2 to be positive. The real number δ gives the phase difference between the two components of the wave field.

The kinetic energy of a particle of charge q and mass m is

$$\frac{1}{2m}\left(\boldsymbol{p} - \frac{q}{c}\boldsymbol{A}\right)^2 = \frac{p^2}{2m} - \frac{q}{2mc}(\boldsymbol{p}\cdot\boldsymbol{A} + \boldsymbol{A}\cdot\boldsymbol{p}) + \frac{q^2}{2mc^2}A^2$$

in the presence of the field. The coupling between the system of N particles and the electromagnetic wave is

$$H_{\text{I}} + H_{\text{II}}$$

where

$$H_{\text{I}} = -\sum_{i=1}^{N} \frac{q_i}{m_i c}\boldsymbol{p}_i \cdot \boldsymbol{A}(\boldsymbol{r}_i, t) - \sum_{i-1}^{N} \boldsymbol{\mu}_i \cdot (\nabla \times \boldsymbol{A}(\boldsymbol{r}_i, t)) \tag{2.3}$$

and

$$H_{\text{II}} = \sum_{i=1}^{N} \frac{q_i^2}{2m_i c^2}(\boldsymbol{A}(\boldsymbol{r}_i, t))^2 \ . \tag{2.4}$$

We have made use of the commutativity of $\boldsymbol{A}_i$ and $\boldsymbol{p}_i \cdot \hat{e}_{1,2}$ in writing Eq. (2.3). H_{I} is of first order in the amplitude of the electromagnetic wave and H_{II} is of second order.

The interaction $H_{\rm I}$ can be rewritten in the form

$$H_{\rm I} = -\frac{1}{2c}e^{-i\omega t}(A_1K_1 + A_2e^{-i\delta}K_2) - \frac{1}{2c}e^{i\omega t}(A_1K_1^\dagger + A_2e^{i\delta}K_2^\dagger) \tag{2.5}$$

where

$$K_{1,2} = \sum_{i=1}^{N}\left[\frac{q_i}{m_i}\hat{e}_{1,2}\cdot \boldsymbol{p}_i - ic(\hat{e}_{1,2}\times \boldsymbol{q})\cdot \boldsymbol{\mu}_i\right]e^{i\mathbf{q}\cdot\mathbf{r}_i}\ . \tag{2.6}$$

Let $\widehat{H}$ be the Hamiltonian of the quantum mechanical system and $\psi_\nu = |\nu>$ its eigenvectors in the absence of the electromagnetic wave. Then

$$\widehat{H}|\nu >= E_\nu|\nu >\ . \tag{2.7}$$

Let $\widehat{H}$ have full rotational symmetry as well as a center of inversion. Its symmetry is, therefore $O(3)$ and the energy eigenstates can be classified according to the irreducible representations $D_\pm^{(J)}$.

The first term in the interaction (2.5) gives rise to absorption of radiation while the system experiences a transition from an initial state $|\nu_o>$ to a final state $|\nu>$ such that

$$E_\nu - E_{\nu_0} = \hbar\omega\ . \tag{2.8}$$

The second term in Eq. (2.5) is responsible for emission of radiation.

We now prove a series of selection rules for such transitions in first order in the intensity of the radiation. The transition probabilities for absorption of radiation while the system goes from $|\nu_0>$ to $|\nu>$ is proportional to the square of the magnitude of the matrix element

$$< \nu|H_{\rm I}|\nu_0 >\ .$$

Thus we are led to the study of

$$< \nu|K|\nu_0 >\ .$$

We prove that transitions between states belonging to $D_\pm^{(0)}$ are not allowed. We say that $J = 0 \to J = 0$ transitions are strictly forbidden. In fact, if $|\nu_0>$ and $|\nu>$ belong to $D_\pm^0$, they are invariant under arbitrary rotations, *i.e.*,

$$e^{-i\varphi\mathbf{J}\cdot\hat{n}}\psi_{\nu,\nu_0} = \psi_{\nu,\nu_0} \tag{2.9}$$

for any φ and $\hat{n}$. Thus

$$< e^{-i\varphi\mathbf{J}\cdot\hat{n}}\psi_\nu|K|e^{-i\varphi\mathbf{J}\cdot\hat{n}}\psi_{\nu_0} >=< \psi_\nu|K|\psi_{\nu_o} >$$

where K is either K_1 or K_2. But this is equivalent to

$$< \psi_\nu|e^{i\varphi\mathbf{J}\cdot\hat{n}}Ke^{-i\varphi\mathbf{J}\cdot\hat{n}}|\psi_{\nu_0} >=< \psi_0|K|\psi_{\nu_0} >\ . \tag{2.10}$$

Taking $\varphi = \pi$ and $\hat{n} \parallel \boldsymbol{q}$

$$e^{i\pi\mathbf{J}\cdot\hat{q}}Ke^{-i\pi\mathbf{J}\cdot\hat{q}} = -K \tag{2.11}$$

since the transformation reverses $\hat{e} \cdot \boldsymbol{p}$ and $(\hat{e} \times \hat{q}) \cdot \boldsymbol{\mu}$. Thus

$$< \psi_\nu|K|\psi_{\nu_0} >= - < \psi_\nu|K|\psi_{\nu_0} >= 0 \ , \tag{2.12}$$

thereby establishing our result.

If a is the maximum diameter of the system of particles and the incident radiation has a wavelength much greater than a, $qa \ll 1$. Taking a position $\boldsymbol{r}_0$ within the system

$$e^{i\mathbf{q}\cdot\mathbf{r}_i} = e^{i\mathbf{q}\cdot\mathbf{r}_0} e^{i\mathbf{q}\cdot(\mathbf{r}_i - \mathbf{r}_0)}$$

so that K is multiplied by the phase factor $e^{i\mathbf{q}\cdot\mathbf{r}_0}$. We can thus take the origin of coordinates within the volume occupied by the quantum mechanical system, *i.e.*, we can set $\boldsymbol{r}_0 = 0$. The wave functions ψ_ν decrease exponentially with $|\boldsymbol{r}|$ as $|\boldsymbol{r}| \gg a$. Thus, in evaluating the matrix elements of K we can assume that $\boldsymbol{q} \cdot \boldsymbol{r}$ is small compared to unity. Therefore, we can expand the exponential in K and obtain

$$K(\boldsymbol{q}) = \sum_{\ell=0}^{\infty} K_\ell(\boldsymbol{q}) \tag{2.13}$$

where

$$K_0 = \sum_{i=1}^{N} \frac{q_i}{m_i} \hat{e} \cdot \boldsymbol{p}_i \tag{2.14}$$

and

$$K_\ell = \frac{i^\ell}{\ell!} \sum_{i=1}^{N} \left[\frac{q_i}{m_i} \hat{e} \cdot \boldsymbol{p}_i (\boldsymbol{q} \cdot \boldsymbol{r}_i)^\ell - \ell c (\hat{e} \times \boldsymbol{q}) \cdot \boldsymbol{\mu}_i (\boldsymbol{q} \cdot \boldsymbol{r}_i)^{\ell-1} \right] . \tag{2.15}$$

We define the electric dipole moment of the system by

$$\boldsymbol{d} = \sum_{i=1}^{N} q_i \boldsymbol{r}_i \ . \tag{2.16}$$

Now

$$[\boldsymbol{d}, \widehat{H}] = \sum_{i=1}^{N} q_i [\boldsymbol{r}_i, \widehat{H}] = i\hbar \sum_{i=1}^{N} \frac{q_i}{m_i} \boldsymbol{p}_i \ . \tag{2.17}$$

Thus

$$< \nu|K_0|\nu_0 >= \frac{1}{i\hbar} [\hat{e} \cdot \boldsymbol{d}, \widehat{H}]_{\nu\nu_0} = i\omega_{\nu\nu_0} < \nu|\hat{e} \cdot \boldsymbol{d}|\nu_0 > \ . \tag{2.18}$$

The matrix element of H_{I} responsible for absorption of radiation is

$$< \nu|H_{\mathrm{I}}|\nu_0 >= -\frac{i\omega_{\nu\nu_0}}{2c} \left(A_1 \hat{e}_1 + A_2 e^{-i\delta} \hat{e}_2 \right) \cdot < \nu|\boldsymbol{d}|\nu_0 > \tag{2.19}$$

in the electric dipole approximation which consists in neglecting K_ℓ for $\ell \geq 1$.

For plane polarized radiation with its electric vector along $\hat{e}_1$,

$$< \nu|H_{\mathrm{I}}|\nu_0 >= -\frac{i\omega_{\nu\nu_0}}{2c} A_1 \hat{e}_1 \cdot < \nu|\boldsymbol{d}|\nu_0 > \tag{2.20}$$

For cirtularly polarized radiation $A_2 = A_1 \quad , \quad \delta = \mp\frac{\pi}{2}$ where the upper sign corresponds to positive helicity (left circularly polarized radiation) and the lower one to negative helicity. Then

$$\begin{aligned} < \nu|H_I|\nu_0 > &= -\frac{i\omega_{\nu\nu_0}}{2c}A_1(\hat{e}_1 \pm i\hat{e}_2)\cdot < \nu|\boldsymbol{d}|\nu_0 > \\ &= -\frac{i\omega_{\nu\nu_0}A_1}{\sqrt{2}c}\hat{e}_\pm\cdot < \nu|\boldsymbol{d}|\nu_0 > \end{aligned} \tag{2.21}$$

where

$$\hat{e}_\pm = \frac{1}{\sqrt{2}}(\hat{e}_\pm \pm i\hat{e}_2) \ . \tag{2.22}$$

These vectors satisfy the following properties

$$\begin{aligned} \hat{e}^*_\pm &= \hat{e}_\mp \ , \\ \hat{e}_\pm \cdot \hat{e}_\pm &= 0 \ , \\ \hat{e}^*_\pm \cdot \hat{e}_\pm &= 1 \ . \end{aligned} \tag{2.23}$$

The irreducible components of the electric dipole moment are

$$d_{\pm 1} = \mp\frac{i}{\sqrt{2}}(d_x \pm i d_y) \quad , \quad d_0 = i d_z \tag{2.24}$$

The Wigner-Eckart theorem gives

$$< n'j'm'|d_\kappa|njm >= (-1)^{j'-m'} < n'j' \parallel \boldsymbol{d} \parallel nj > \begin{pmatrix} j' & 1 & j \\ -m' & \kappa & m \end{pmatrix} \ . \tag{2.25}$$

This shows that the matrix elements of d_κ vanish unless

$$|j' - j| = 0 \ , \ 1 \tag{2.26}$$

and

$$\kappa = m' - m \ . \tag{2.27}$$

Thus the selection rules for electric dipole transitions are

$$\Delta j = 0, \pm 1 \quad ; \quad \Delta m = 0, \pm 1 \tag{2.28}$$

but $\Delta j = 0$ for $j = j' = 0$ is strictly forbidden. If the system has a center of inversion, the energy eigenfunctions can be selected so that they have definite parity. Since

$$P_i \boldsymbol{d} P_i^{-1} = -\boldsymbol{d} \ , \tag{2.29}$$

$$< \psi_\nu|\boldsymbol{d}|\psi_{\nu_0} >= - < \psi_\nu|P_i\boldsymbol{d}P_i^{-1}|\psi_{\nu_0} >= - < P_i\psi_\nu|P_i\psi_{\nu_0} > \ . \tag{2.30}$$

Therefore, the matrix elements of $\boldsymbol{d}$ vanish between states of the same parity. Since for spinless particles the parity of $|n\ell m>$ is $(-1)^\ell$, the selection rules (2.28) and (2.30) reduce to

$$\Delta\ell = \pm 1 \quad , \quad \Delta m = 0, \pm 1 \; . \tag{2.31}$$

The next term in the expansion of K is

$$\begin{aligned} K_1(\boldsymbol{q}) &= i\sum_{i=1}^{N} \frac{q_i}{2m_i}(\hat{e}\cdot\boldsymbol{p}_i\boldsymbol{r}_i\cdot\boldsymbol{q} + \hat{e}\cdot\boldsymbol{r}_i\boldsymbol{p}_i\cdot\boldsymbol{q}) \\ &\quad - i\sum_{i=1}^{N}\frac{q_i}{2m_i}(\hat{e}\cdot\boldsymbol{r}_i\boldsymbol{p}_i\cdot\boldsymbol{q} - \hat{e}\cdot\boldsymbol{p}_i\boldsymbol{r}_i\cdot\boldsymbol{q}) - ic(\hat{e}\times\boldsymbol{q})\cdot\sum_{i=1}^{N}\boldsymbol{\mu}_i \\ &= i\hat{e}\cdot\sum_{i=1}^{N}\frac{q_i}{2m_i}(\boldsymbol{p}_i\boldsymbol{r}_i + \boldsymbol{r}_i\boldsymbol{p}_i)\cdot\boldsymbol{q} - i(\hat{e}\times\boldsymbol{q})\cdot\sum_{i=1}^{N}\left(\frac{q_i}{2m_i}\boldsymbol{r}_i\times\boldsymbol{p}_i + c\boldsymbol{\mu}_i\right) \; . \end{aligned} \tag{2.32}$$

Now

$$\sum_{i=1}^{N}\frac{q_i}{2m_i}(\boldsymbol{p}_i\boldsymbol{r}_i + \boldsymbol{r}_i\boldsymbol{p}_i) = -\frac{i}{2\hbar}\left[\sum_i q_i\boldsymbol{r}_i\boldsymbol{r}_i, \widehat{H}\right] \; . \tag{2.33}$$

We define the electric quadrupole tensor by

$$\boldsymbol{Q} = \sum_i q_i(3\boldsymbol{r}_i\boldsymbol{r}_i - \mathbf{1}r_i^2) \tag{2.34}$$

and the magnetic dipole moment operator by

$$\boldsymbol{\mu} = \sum_{i=1}^{N}\left(\frac{q_i}{2m_ic}\boldsymbol{r}_i\times\boldsymbol{p}_i + \boldsymbol{\mu}_i\right) \; . \tag{2.35}$$

Then

$$K_1(\boldsymbol{q}) = \frac{1}{6}\hat{e}\cdot\boldsymbol{Q}\cdot\boldsymbol{q} - ic(\hat{e}\times\boldsymbol{q})\cdot\boldsymbol{\mu} \; . \tag{2.36}$$

The irreducible components of $\boldsymbol{Q}$ are

$$\begin{aligned} Q_{\pm 2} &= -\frac{1}{2}(Q_{xx} - Q_{yy} \pm 2iQ_{xy}) \quad , \quad Q_{\pm 1} = \pm(Q_{zx} \pm iQ_{yz}) \; , \\ Q_0 &= -\frac{1}{\sqrt{6}}(2Q_{zz} - Q_{xx} - Q_{yy}) \; . \end{aligned} \tag{2.37}$$

The Wigner-Eckart theorem shows that

$$< n'j'm'|Q_\kappa|njm >= (-1)^{j'-m'} < n'j' \parallel \boldsymbol{Q} \parallel nj > \begin{pmatrix} j' & 2 & j \\ -m' & \kappa & m \end{pmatrix} . \tag{2.38}$$

We conclude that the matrix elements of $\boldsymbol{Q}$ vanish unless $|j' - j| = 0, 1, 2$ and $m' - m = \kappa = 0, \pm 1, \pm 2$. There is no change in parity between initial and final states and, as always, $j = 0$ to $j = 0$ transitions are forbidden.

For magnetic dipole transitions $\Delta j = 0$, ± 1 ; $\Delta m = 0, \pm 1$ and there is no change in parity; $j = 0 \rightarrow j = 0$ is forbidden.

7.3 Time reversal invariance in quantum mechanics

We have seen that a linear operator U which, acting on vectors of a Hilbert space, leaves inner products invariant satisfies

$$U^{-1} = U^{\dagger} \ . \tag{3.1}$$

Such operators are called unitary. Their fundamental property is that for any pair of vectors, ψ and ϕ in the space

$$< U\psi|U\phi > = < \psi|\phi > \ . \tag{3.2}$$

Not all operators used in quantum mechanics are either Hermitian or unitary. We consider now operators such that

$$| < \Omega\psi|\Omega\phi > | = | < \psi|\phi > | \ . \tag{3.3}$$

The reason for introducing such operators is that the measurable quantities in quantum theory are probabilities, not probability amplitudes. Thus, only the square of the magnitudes of inner products have a physical meaning. Wigner showed that operators obeying the requirement (3.3) are either unitary or anti-unitary. The operator Ω is anti-unitary if

$$< \Omega\psi|\Omega\phi > = < \psi|\phi >^* = < \phi|\psi > \ . \tag{3.4}$$

For any two vectors ψ and ϕ

$$\Omega(\psi + \phi) = \Omega\psi + \Omega\phi \ . \tag{3.5}$$

In fact

$$\begin{aligned} < \Omega\chi|\Omega(\psi+\phi) > &= < \psi + \phi|\chi > = < \psi|\chi > + \phi|\chi > \\ &= < \Omega\chi|\Omega\psi > + < \Omega\chi|\Omega\phi > \\ &= < \Omega\chi|\Omega\psi + \Omega\phi > \end{aligned}$$

for all χ. Thus Eq. (3.5) follows. However Ω is not linear because, in general

$$\Omega(c\phi) \neq c\Omega\phi \ .$$

In fact

$$\begin{aligned} < \Omega\psi|\Omega(c\phi) > &= < c\phi|\psi > = c^* < \phi|\psi > = c^* < \Omega\psi|\Omega\phi > \\ &= < \Omega\psi|c^*\Omega\phi > \end{aligned}$$

for all ψ. Thus

$$\Omega(c\phi) = c^*\Omega\phi \ . \tag{3.6}$$

Let $\{\varphi_i\}$ be an orthonormal basis in the Hilbert space under consideration, *i.e.*, every vector ϕ in the space is expressible as

$$\phi = \sum_i c_i \varphi_i$$

and

$$< \varphi_i | \varphi_j >= \delta_{ij} \ .$$

We define

$$\varphi_i' = \Omega \varphi_i \ .$$

Then

$$< \varphi_i' | \varphi_j' >=< \Omega\varphi_i | \Omega\varphi_j >=< \varphi_j | \varphi_i >= \delta_{ij} \ ,$$

so that $\{\varphi_i'\}$ is also an orthonormal set.

We suppose that Ω maps the Hilbert space of the system into itself. The null space of Ω consists of the zero vector alone. In fact if $\chi \neq 0$, $\Omega\chi = 0$ leads to a contradiction:

$$0 =< \Omega\chi | \Omega\chi >=< \chi | \chi > \neq 0 \ .$$

Thus Ω possesses an inverse. It follows that the set $\{\varphi_i'\}$ is a basis for the Hilbert space since for every ϕ,

$$\Omega^{-1}\phi = \sum c_i \varphi_i$$

so that

$$\phi = \sum_i c_i^* \varphi_i' \ .$$

In classical mechanics we defined the time reversed of a physical state as one in which the positions of all the particles in the system are kept invariant while all their momenta are reversed. If there are no velocity dependent forces, after the operation of time reversal is performed, the system retraces its original trajectory in configuration space with reversed momenta. Such systems are said to obey time reversal symmetry. If a magnetic field is present this is not the case unless the charged particles responsible for the field are part of the dynamical system under consideration.

We now consider time reversal in quantum mechanics. Let T be the quantum equivalent of the classical time reversal operation. For each vector ψ in the Hilbert space of the system we define a vector $\bar{\psi} = T\psi$ in the same space whose property is to leave positions invariant but reverse momenta. If $|\boldsymbol{r}'>$ is a state of a particle with its position well defined at $\boldsymbol{r}'$, *i.e.*, if

$$\hat{\boldsymbol{r}} | \boldsymbol{r}' >= \boldsymbol{r}' | \boldsymbol{r}' >$$

then

$$T | \boldsymbol{r}' >= | \boldsymbol{r}' > \ . \tag{3.7}$$

If $|\boldsymbol{p}'>$ is a state in which the momentum has the well defined value $\boldsymbol{p}'$, *i.e.*,

$$\hat{\boldsymbol{p}} | \boldsymbol{p}' >= | \boldsymbol{p}' >$$

then

$$T|\boldsymbol{p}'>=|-\boldsymbol{p}'> \ . \tag{3.8}$$

We require that

$$|<T\psi|T\phi>|=|<\psi|\phi>| \tag{3.9}$$

for any two vectors ψ and ϕ of the space. We shall show that T must be anti-unitary by demonstrating that the assumption that T is unitary is untenable. We know that T possesses an inverse T^{-1}. Now, from $\hat{\boldsymbol{p}}|\boldsymbol{p}'>=\boldsymbol{p}'|\boldsymbol{p}'>$ we deduce that

$$T\hat{\boldsymbol{p}}T^{-1}T|\boldsymbol{p}'>=\boldsymbol{p}'T|\boldsymbol{p}'>=\boldsymbol{p}'|-\boldsymbol{p}'>=-\hat{\boldsymbol{p}}T|\boldsymbol{p}'>$$

so that

$$T\hat{\boldsymbol{p}}T^{-1}=-\hat{\boldsymbol{p}} \ . \tag{3.10}$$

In a similar way we show that

$$T\hat{\boldsymbol{r}}T^{-1}=\hat{\boldsymbol{r}} \ . \tag{3.11}$$

The action of T on the angular momentum operator is

$$T\hat{\boldsymbol{r}}\times\hat{\boldsymbol{p}}T^{-1}=-\hat{\boldsymbol{r}}\times\hat{\boldsymbol{p}} \ . \tag{3.12}$$

Let $\widehat{H}$ be the Hamiltonian of the system. If the system obeys time reversal symmetry we have

$$T\widehat{H}T^{-1}=\widehat{H}$$

or

$$[\widehat{H},T]=0 \ , \tag{3.13}$$

where we assume that $\widehat{H}$ is not an explicit function of the time and that $\widehat{H}$ is quadratic in the momentum operators.

The evolution of the system starting from an arbitrary state ψ_0 is given by the solution

$$\psi=e^{-i\widehat{H}t/\hbar}\psi_0 \tag{3.14}$$

of the Schrödinger equation. If the system obeys time reversal symmetry, the evolution of $T\psi_0$ in the reversed time, *i.e.*,

$$e^{i\widehat{H}t/\hbar}T\psi_0$$

must be just $T\psi$, *i.e.*,

$$Te^{-i\widehat{H}t/\hbar}\psi_0=e^{i\widehat{H}t/\hbar}T\psi_0$$

for all ψ_0. Thus

$$Te^{-i\widehat{H}t/\hbar}=e^{i\widehat{H}t/\hbar}T \ , \tag{3.15}$$

or, equivalently,

$$T\left(1-i\frac{\widehat{H}t}{\hbar}+\cdots\right)=\left(1+i\frac{\widehat{H}t}{\hbar}+\cdots\right)T \ . \tag{3.16}$$

Equation (3.16) implies that

$$T\widehat{H} + \widehat{H}T = 0 \tag{3.17}$$

if T is unitary and

$$[T, \widehat{H}] = 0 \tag{3.18}$$

if it is anti-unitary. Equation (3.17) would imply

$$T\widehat{H}T^{-1} = -\widehat{H}$$

which is in direct contradiction with

$$T\frac{p^2}{2m}T^{-1} = \frac{p^2}{2m}$$

for the Hamiltonian of a free particle. Thus, T must be anti-unitary.

The nature of T makes it impossible to introduce it as an additional operation in the symmetry group of a quantum system and use the theorems developed for irreducible representations. We shall see, however, that as a separate symmetry, it can introduce additional degeneracies over and above those made necessary by those symmetry elements which are described by unitary operations.

For a spinless particle, T is simply equivalent to complex conjugation in the Schrödinger representation. Let H be the Hamiltonian for a particle in a field $U(\boldsymbol{r})$. Then, Eq. (3.18) is satisfied and

$$i\hbar\frac{\partial\psi}{\partial t} = H\psi \tag{3.19}$$

describes the evolution of the system. Taking the complex conjugate of both sides of Eq. (3.19)

$$-i\hbar\frac{\partial\psi^*}{\partial t} = H\psi^* \tag{3.20}$$

so that ψ^* describes the evolution of the system in time $t' = -t$. Denoting by K the operation of complex conjugation we have, therefore that

$$T = K\ . \tag{3.21}$$

If U is a central potential the eigenfunctions of H are of the form

$$\psi_{E\ell m} = R_{E\ell}(r)y_\ell^m(\vartheta, \varphi) \tag{3.22}$$

where

$$y_\ell^m(\vartheta, \varphi) = i^\ell Y_\ell^m(\vartheta, \varphi)\ .$$

We have

$$T\psi_{E\ell m} = (-1)^{\ell - m}\psi_{E\ell, -m}\ . \tag{3.23}$$

For particles of spin $\frac{1}{2}$, when the intrinsic angular momentum is

$$\frac{1}{2}\hbar\boldsymbol{\sigma}$$

where $\boldsymbol{\sigma}$ is a vector whose components are the Pauli matrices we must have

$$T\boldsymbol{\sigma}T^{-1} = -\boldsymbol{\sigma} \ . \tag{3.24}$$

We attempt to find T in the form

$$T = UK \tag{3.25}$$

where U is a unitary operator acting on the spin variable. Since $K^2 = E$,

$$T^{-1} = KU^{\dagger} \tag{3.26}$$

and Eq. (3.24) becomes

$$UK\boldsymbol{\sigma}KU^{\dagger} = -\boldsymbol{\sigma} \tag{3.27}$$

Now, in the usual representation of $\boldsymbol{\sigma}$, σ_1 and σ_3 are real while σ_2 is purely imaginary. Thus

$$K\sigma_1K = \sigma_1 \quad , \quad K\sigma_2K = -\sigma_2 \quad , \quad K\sigma_3K = \sigma_3 \tag{3.28}$$

so that we require U to be such that

$$U\sigma_1U^{\dagger} = -\sigma_1 \quad , \quad U\sigma_2U^{\dagger} = \sigma_2 \quad , \quad U\sigma_3U^{\dagger} = -\sigma_3 \ . \tag{3.29}$$

Conditions (3.29) are verified if U equals σ_2 multiplied by an arbitrary phase factor of modulus unit. Thus, we select

$$T = -i\sigma_2K \tag{3.30}$$

For the spin states ξ_+, ξ_- we have

$$T\xi_+ = \xi_- \quad , \quad T\xi_- = -\xi_+ \tag{3.31}$$

For spinless particles

$$T^2 = K^2 = E \ , \tag{3.32}$$

while for particles of spin $\frac{1}{2}$

$$T^2 = -\sigma_2^2 = -E \tag{3.33}$$

In general we can state that

$$T^2\psi = c\psi, \tag{3.34}$$

with $c = \pm 1$. Now

$$< T^2\psi|T^2\psi >= c^*c < \psi|\psi >$$

and

$$< T^2\psi|T^2\psi >=< T\psi|T\psi >=< \psi|\psi >$$

so that

$$|c|^2 = 1 \ .$$

Furthermore

$$T^3\psi = c^*T\psi$$

and

$$T^2(T\psi)=c(T\psi)$$

so that $c^* = c$. Thus c is real and can only be 1 or -1. We have seen that for a spinless particle $c = 1$ while for particles of spin $\frac{1}{2}, c = -1$.

Let ϕ be any state of a quantum system and define $\psi = T\phi$. We have

$$< T\psi|T\phi >=< \phi|\psi >=< \phi|T\phi >$$

But $T\psi = T^2\phi = c\phi$. This shows that

$$< \phi|T\phi >= c < \phi|T\phi > \ .$$

If $c = 1$, this is an identity but if $c = -1$, it shows that ϕ and $T\phi$ are orthogonal. If ϕ is an eigenvector of the Hamiltonian of a system for which $T^2 = -1$, ϕ and $T\phi$ are orthogonal and, hence, linearly independent. Furthermore, if $H\phi = E\phi$, then $TH\phi = ET\phi$. If the system obeys time reversal invariance $TH = HT$ so that $H(T\phi) = E(T\phi)$. In this case ϕ and $T\phi$ are degenerate states of H. We conclude that if $T^2 = -1$ and $[H,T] = 0$ all energy eigenstates have even-fold degeneracies. This result is known as Kramer's theorem.

As an example we consider a system of N electrons. The time reversal operator for such a system is

$$T = (-i)^N \sigma_{21}\sigma_{22}\ldots\sigma_{2N}K \ , \tag{3.35}$$

where σ_{2j} is the 2-component of $\boldsymbol{\sigma}$ for the jth electron. We obtain T^2 by simple multiplication as

$$T^2 = (-1)^N \ . \tag{3.36}$$

We conclude that systems containing an odd number of electrons can only have energy eigenvectors of even-fold degeneracy. The theorem cannot establish any consequence on systems containing an even number of electrons. This result is of particular interest in the quantum theory of magnetism.

To avoid future difficulties we remind the reader that the spin-orbit interaction

$$H_{SO} = \frac{\hbar}{4m^2c^2}\boldsymbol{\sigma}\cdot(\nabla U \times \boldsymbol{p}) \tag{3.37}$$

for an electron moving in a potential U is invariant under time reversal. This is, of course, due to the source of magnetic field being internal to the system, *i.e.*, it is the electron itself.

Consider now a system whose Hamiltonian $\widehat{H}$ is invariant under the operations of a group $\{P_R\}$ of orthogonal transformations. To each energy level there corresponds an irreducible representation of $\{P_R\}$. We shall assume that we have properly identified all symmetry elements of $\widehat{H}$ so that there is no question of there being "accidental" degeneracies.

If $\widehat{H}$ is time reversal invariant, additional degeneracies may occur. We turn now to the investigation of this question. Since T is not unitary it cannot be incorporated as an additional set of elements $\{TP_R\}$ in the group since $\{P_R\}$ consists entirely of unitary operators. However, it is possible to establish what additional degeneracies may occur because of time reversal invariance.

Let E be a ℓ-fold degenerate level of $\widehat{H}$ whose subspace of eigenvectors with eigenvalue E is spanned by the ℓ orthonormal vectors $\phi_n(n = 1, 2, \ldots \ell)$. If Γ is the irreducible representation associated with this level

$$P_R\phi_n = \sum_{n'} \Gamma_{n'n}(R)\phi_{n'} \ . \tag{3.38}$$

Since the operations P_R reflect the action of the orthogonal operations R on the states ϕ_n and this operation is real, P_R and T commute. Thus, acting on both sides of Eq. (3.38) with the time reversal operator T we obtain

$$P_R\bar{\phi}_n = \sum_{n'} \Gamma^*_{n'n}(R)\bar{\phi}_{n'} \tag{3.39}$$

where

$$\bar{\phi}_n = T\phi_n \ . \tag{3.40}$$

The states $\bar{\phi}_n$ generate the complex conjugate representation $\{\Gamma^*(R)\}$ which, as we saw in chapter 6, sec. 5, is like Γ, an irreducible representation of the group. The states $\bar{\phi}_n$, like the ϕ_n are orthonormal but need not span the same subspace of the Hilbert space of $\widehat{H}$.

We recall that complex conjugate representations can be of three types labeled a, b and c in chapter 6, sec. 5.

If $\Gamma(R)$ is of type a, then, by means of a unitary transformation it is possible to find an orthonormal basis $\{\phi_n\}$ for the subspace of Γ such that all $\Gamma_{n'n}(R)$ in Eq. (3.38) are real. Then, the bases $\{\phi_n\}$ and $\{\bar{\phi}_n\}$ generate the same irreducible representation in **identical** real and unitary form.

By virtue of theorem 3.1 in chapter 6, $< \bar{\phi}_{n'}|\phi_n >= 0$ if $n' \neq n$ and $< \bar{\phi}_n|\phi_n >$ is independent of n. The question: do the bases $\{\phi_n\}$ and $\{\bar{\phi}_n\}$ span the same or distinct subspaces of the Hilbert space of the system? arises in a natural way. If $T^2 = 1$, nothing can be said about $< \bar{\phi}_n|\phi_n >$. Since we have assumed that there are no "accidental" degeneracies, time reversal, in this case, does not introduce an additional degeneracy. This result is in contrast to that applying to systems for which $T^2 = -1$ because, then, ϕ_n and $\bar{\phi}_n$ are orthogonal:

$$< \bar{\phi}_n|\phi_n >= - < \bar{\phi}_n|T^2\phi_n >= - < \bar{\phi}_n|\phi_n >= 0 \ . \tag{3.41}$$

The sets $\{\phi_n\}$ and $\{\bar{\phi}_n\}$ are, therefore, linearly independent but correspond to the same eigenvalue of $\widehat{H}$: the degeneracy is doubled.

No additional proof of this result is required. But many authors proceed by defining the vectors

$$\psi_n = \frac{1}{\sqrt{2}}(\phi_n + \bar{\phi}_n) \tag{3.42}$$

and

$$\chi_n = \frac{i}{\sqrt{2}}(\phi_n - \bar{\phi}_n) \ . \tag{3.43}$$

We mention their argument for the sake of completeness. If $T^2 = 1$, ψ_n and χ_n are their own time reversed conjugates: $\overline{\psi}_n = T\psi_n = \psi_n$ and $\overline{\chi}_n = T\chi_n = \chi_n$. The operator T does not mix the set $\{\psi_n\}$ and $\{\chi_n\}$ and, since we suppose there are no "accidental" degeneracies, they must span the same subspace. Hence no new degeneracies are introduced by time reversal invariance. On the other hand, if $T^2 = -1$, $\overline{\psi}_n = T\psi_n = i\chi_n$ and $\overline{\chi}_n = T\chi_n = -i\psi_n$. Since ψ_n and $\overline{\psi}_n$ are orthogonal, the states $\{\psi_n\}$ and $\{\chi_n\}$ are linearly independent and hence, the degeneracy is doubled.

If Γ is of type b, then, since Γ and Γ^* are inequivalent, the vectors $\phi_1, \phi_2 \dots \phi_n$, $\bar{\phi}_1 . \bar{\phi}_2, \dots \bar{\phi}_n$ are all mutually orthogonal and, hence, linearly independent. Thus, independently of whether T^2 is 1 or -1, the degeneracy is doubled by time reversal invariance.

We now turn to the case in which Γ is of type c, when Γ and Γ^* are equivalent but cannot be made real by a unitary transformation. The equivalence of Γ and Γ^* implies the existence of a unitary matrix U satisfying

$$\Gamma(R) = U^\dagger \Gamma^*(R) U \tag{3.44}$$

for all R. The vectors

$$\overline{\omega}_n = \sum_{n'} U_{n'n} \bar{\phi}_{n'} \tag{3.45}$$

transform according to

$$P_R \overline{\omega}_n = \sum_{n'} (U^\dagger \Gamma^*(R) U)_{n'n} \overline{\omega}_{n'} = \sum_{n'} \Gamma_{n'n}(R) \overline{\omega}_{n'} \ . \tag{3.46}$$

Thus, by virtue of theorem 3.1 in chapter 6

$$< \phi_{n'} | \overline{\omega}_n >= 0$$

if $n' \neq n$ and $< \phi_n | \overline{\omega}_n >$ is independent of n.

Again we ask whether the bases $\{\phi_n\}$ and $\{\bar{\phi}_n\}$ (or equivalently $\{\omega_n\}$) span the same or distinct subspaces. We shall see that the situation here is reversed with respect to that for irreducible representation of type a, *i.e.*, the degeneracy is doubled if $T^2 = 1$ while no additional degeneracy results if $T^2 = -1$. An indication that this is so follows from the fact that it is not possible to find a basis $\{\psi_n\}$ for which $\overline{\psi}_n = \psi_n$ as in the case (a). In fact, if this were possible

$$< \psi_{n'} | P_R | \psi_n >=< T^2 \psi_{n'} | P_R | T^2 \psi_n >=< \overline{\psi}_{n'} | P_R | \overline{\psi}_n >^* =< \psi_{n'} | P_R | \psi_n >^*$$

and the representation would be equivalent to a real one contrary to the hypothesis.

The proof of our assertion follows from the fact that, according to the results of Sec. 5, chapter 6, the transformation U in Eq. (3.44) satisfies the additional constraint

$$\widetilde{U} = -U \ . \tag{3.47}$$

The bases $\{\phi_n\}$ and $\{\bar{\phi}_n\}$ span the same subspace if, and only if, the vectors ϕ_n and $\overline{\omega}_n$ are equal except perhaps for a phase factor of absolute value unity, *i.e.*, if

$$\phi_n = e^{i\delta} \sum_{n'} U_{n'n} \bar{\phi}_{n'} \ . \tag{3.48}$$

Acting with T on both sides of Eq. (3.48) we obtain

$$\bar{\phi}_n = \pm \sum_{n'} \bar{\phi}_{n'} (UU^*)_{n'n} \tag{3.49}$$

where the upper and lower signs correspond to $T^2 = 1$ and to $T^2 = -1$, respectively. But

$$UU^* = -\widetilde{U}U^* = -(U^\dagger U)^* = -E \tag{3.50}$$

so that Eq. (3.49) reduces to

$$\bar{\phi}_n = \mp \bar{\phi}_n \ . \tag{3.51}$$

Thus, the identification (3.48) is impossible if $T^2 = 1$ and possible if $T^2 = -1$. In the absence of accidental degeneracies, the degeneracy is doubled if $T^2 = 1$ but not if $T^2 = -1$.

The following table summarizes these results. In the first column we give the type of irreducible representation, in the second the result of the Frobenius-Schur test described in Eq. (5.18) in chapter 6 while in the third and fourth columns we indicate whether the degeneracy is doubled (2) or that no extra degeneracy (1) occurs because of time reversal invariance.

Γ	$\Sigma\chi(R^2)$	$T^2 = 1$	$T^2 = -1$
a	h	1	2
b	0	2	2
c	$-h$	2	1

7.4 The "accidental" degeneracy of the hydrogen atom

We have mentioned the degeneracy of the ns and np states of the non-relativistic hydrogen atom. The symmetry group of the Hamiltonian of a particle of mass m in a central potential $V(r)$ is

$$H = \frac{p^2}{2m} + V(r) \ . \tag{4.1}$$

Its symmetry group is clearly $O(3)$ and the eigenfunctions are of the form

$$\psi_{n\ell m} = R_{n\ell}(r) Y_\ell^m(\vartheta, \varphi) \ . \tag{4.2}$$

The ns states are spherically symmetric while the three np states transform as x, y, z, the components of the position $\boldsymbol{r}$ of the particle. There is no operation of

$O(3)$ that can transform a np state into a ns state. Thus we conclude that $E_{n\ell}$, the eigenvalue of $\psi_{n\ell m}$ is such that $E_{n0} \neq E_{n1}$. We know that this is not the case for the hydrogen atom whose Hamiltonian is

$$H = \frac{p^2}{2m} - \frac{e^2}{r} \, . \tag{4.3}$$

For this case the energy eigenvalues are

$$E_n = -\frac{me^4}{2\hbar^2 n^2} \tag{4.4}$$

independently of ℓ for $\ell = 0, 1 \ldots n-1$. The purpose of this section is to uncover the additional symmetry responsible for this degeneracy.

In classical mechanics the angular momentum $\boldsymbol{L}$ of a particle moving in a central potential $V(r)$ according to the classical Hamiltonian (4.1) is a constant of the motion. In fact, the equations of motion are

$$\dot{\boldsymbol{r}} = \frac{\partial H}{\partial \boldsymbol{p}} = \frac{\boldsymbol{p}}{m} \quad , \quad \dot{\boldsymbol{p}} = -\frac{\partial H}{\partial \boldsymbol{r}} = -\frac{1}{r}\frac{dV}{dr}\boldsymbol{r} \tag{4.5}$$

so that

$$\dot{\boldsymbol{L}} = \boldsymbol{r} \times \dot{\boldsymbol{p}} + \dot{\boldsymbol{r}} \times \boldsymbol{p} = 0 \, . \tag{4.6}$$

Thus the orbit of the particle is confined to a plane normal to $\boldsymbol{L}$. For motion in a Coulomb potential the orbit is closed; for bound states ($E < 0$) it is an ellipse whose semi-major axis remains invariant in time. This means that there is an additional vector quantity, normal to $\boldsymbol{L}$, which is a constant of the motion. From the equations of motion we obtain

$$\frac{d}{dt}\left(\frac{\boldsymbol{p} \times \boldsymbol{L}}{m}\right) = -\frac{1}{mr}\frac{dV}{dr}\boldsymbol{r} \times (\boldsymbol{r} \times \boldsymbol{p}) \tag{4.7}$$

and

$$\frac{d}{dt}(\boldsymbol{r}V) = \frac{\boldsymbol{p}}{m}\frac{d}{dr}(rV) + \frac{1}{mr}\frac{dV}{dr}\boldsymbol{r} \times (\boldsymbol{r} \times \boldsymbol{p}) \, . \tag{4.8}$$

Thus, the sum of the quantities on the left-hand side of Eqs (4.7) and (4.8) is a constant of the motion for Coulomb potentials. Taking $V = -e^2/r$, the vector

$$\boldsymbol{\epsilon} = -\frac{\boldsymbol{r}}{r} + \frac{\boldsymbol{p} \times \boldsymbol{L}}{me^2} \tag{4.9}$$

is a constant of the motion. The vector $\boldsymbol{\epsilon}$ is called the eccentricity vector or the Runge-Lenz vector. We note that $\boldsymbol{\epsilon}$ is in the plane of the orbit since

$$\boldsymbol{\epsilon} \cdot \boldsymbol{L} = 0 \, . \tag{4.10}$$

Multiplying Eq. (4.9) scalarly by $\boldsymbol{r}$ we obtain

$$\boldsymbol{\epsilon} \cdot \boldsymbol{r} = -r + \frac{L^2}{me^2} \tag{4.11}$$

and denoting by θ the angle between $\boldsymbol{r}$ and $\boldsymbol{\epsilon}$ we obtain the equation of the orbit in polar coordinates

$$r(1+\epsilon\cos\theta) = \frac{L^2}{me^2} \ . \tag{4.12}$$

The square of the magnitude of the eccentricity vector is

$$\epsilon^2 = 1 + \frac{2L^2}{me^4}\left(\frac{p^2}{2m} - \frac{e^2}{r}\right) = 1 + \frac{2L^2}{me^4}H \ . \tag{4.13}$$

The generalization of the eccentricity vector in quantum mechanics must take into account that $\boldsymbol{p}$ and $\boldsymbol{L}$ do not commute and that we require $\boldsymbol{\epsilon}$ to be a Hermitian operator. The following satisfies the requirement of being Hermitian and at the same time being equal to the classical variable (4.9) in the limit $\hbar \to 0$:

$$\boldsymbol{\epsilon} = -\frac{\boldsymbol{r}}{r} + \frac{\hbar}{2me^2}(\boldsymbol{p}\times\boldsymbol{L} - \boldsymbol{L}\times\boldsymbol{p}) \tag{4.14}$$

where now $\boldsymbol{L}$ is the angular momentum operator in units of $\hbar$. In quantum theory $\boldsymbol{\epsilon}$ is a constant of the motion, *i.e.*,

$$[\boldsymbol{\epsilon}, H] = 0 \ . \tag{4.15}$$

The following commutation relations can be proved (after a considerable amount of algebra):

$$[L_i, \epsilon_j] = i\epsilon_{ijk}\epsilon_k \ , \tag{4.16}$$

and

$$[\epsilon_i, \epsilon_j] = i\epsilon_{ijk}\left(-\frac{2H\hbar^2}{me^4}\right)L_k \ . \tag{4.17}$$

For bound states the eigenvalues of H are negative so that the operator

$$\left(-\frac{2H\hbar^2}{me^4}\right)^{\frac{1}{2}}$$

is well defined in the subspace of bound states of H. We, consequently, introduce the operators

$$K_i = \epsilon_i\left(-\frac{2H\hbar^2}{me^4}\right)^{-\frac{1}{2}} \ . \tag{4.18}$$

It is legitimate to do this because ϵ_i and H commute so that the operators K_i are Hermitian. Then

$$[L_i, K_j] = i\epsilon_{ijk}K_k \tag{4.19}$$

and

$$[K_i, K_j] = i\epsilon_{ijk}L_k \ . \tag{4.20}$$

We define the vector operators (under rotations only)

$$\boldsymbol{J}_1 = \frac{1}{2}(\boldsymbol{L} + \boldsymbol{K}) \tag{4.21}$$

and

$$J_2 = \frac{1}{2}(L - K) \tag{4.22}$$

satisfying the commutation relations

$$[J_{1i}, J_{1j}] = i\epsilon_{ij} J_{1k} \ , \tag{4.23}$$

$$[J_{2i}, J_{2j}] = i\epsilon_{ijk} J_{2k} \ , \tag{4.24}$$

and

$$[J_{1i}, J_{2j}] = 0 \ . \tag{4.25}$$

Thus each of the operators J_1 and J_2 generate a $SU(2)$ algebra and since they commute with each other the algebra of the direct product space obtained from each of the spaces generated is $SU(2) \times SU(2)$. Alternatively this algebra is identical to $SO(4)$ generated by the six operators

$$x^\mu p_\nu - x^\nu p_\mu \quad , \quad \mu, \nu = 0, 1, 2, 3 \ .$$

For the hydrogen atom there are restrictions to the accessible vectors of the space generated by J_1 and J_2. In fact we have

$$\epsilon \cdot L = L \cdot \epsilon = 0 \tag{4.26}$$

or, equivalently

$$K \cdot L = L \cdot K = 0 \ . \tag{4.27}$$

This result imposes the restriction

$$J_1^2 = J_2^2 \ . \tag{4.28}$$

Now, the eigenvalues of J_1^2 and J_2^2 are $j_1(j_1+1)$ and $j_2(j_2+1)$ where j_1 and j_2 take the values $0, \frac{1}{2}, 1, \frac{3}{2}, \ldots$. Since the restriction (4.28) is imposed we must have $j_1 = j_2 = j$ for all states. The equation analogous to (4.13) in the quantum theory is

$$\epsilon^2 = 1 + \frac{2H\hbar^2}{me^4}(L^2 + 1) \ . \tag{4.29}$$

This gives

$$\frac{me^4}{2\hbar^2}\frac{1}{H} = -K^2 - L^2 - 1 = -2(J_1^2 + J_2^2) - 1 \tag{4.30}$$

after multiplication by $-(me^4/2\hbar^2 H)$. Since both J_1^2 and J_2^2 take the values $j(j+1)$ the eigenvalues of the left hand side of Eq. (4.30) are

$$-(2j+1)^2$$

and the eigenvalues of H are

$$-\frac{me^4}{2\hbar^2}\frac{1}{(2j+1)^2} \quad , \quad j = 0, \frac{1}{2}, 1, \frac{3}{2}, \ldots$$

Defining

$$n = 2j + 1 ,$$

the possible values of n are $1, 2, 3, \ldots$ and we obtain the energies of the bound states of the hydrogen atom, namely

$$-\frac{me^4}{2\hbar^2}\frac{1}{n^2} \quad , \quad n = 1, 2, 3, ..$$

An alternative proof due to Fock is based on the transformation of the Schrödinger equation

$$\left(\frac{p^2}{2m} - \frac{e^2}{r}\right)\psi = E\psi$$

from its representation in coordinate space to a representation in momentum space. If $\psi(\boldsymbol{r})$ is the wave function in coordinate space, the wave function in momentum space is

$$\phi(\boldsymbol{p}) = (2\pi)^{-3/2}\int e^{-i\boldsymbol{r}\cdot\boldsymbol{p}}\psi(\boldsymbol{r})d\boldsymbol{r} \tag{4.31}$$

where we use atomic units ($\hbar = 1, m = 1, e = 1$). The Schrödinger equation

$$\left(-\frac{1}{2}\nabla^2 - \frac{1}{r}\right)\psi(\boldsymbol{r}) = E\psi(\boldsymbol{r}) \tag{4.32}$$

is transformed using the inverse Fourier transform

$$\psi(\boldsymbol{r}) = (2\pi)^{-3/2}\int e^{i\boldsymbol{r}\cdot\boldsymbol{p}}\phi(\boldsymbol{p})d\boldsymbol{p} \tag{4.33}$$

and that of the Coulomb field $\frac{1}{r}$ obtained from the Poisson equation

$$\nabla^2\frac{1}{r} = -4\pi\delta(\boldsymbol{r}) = -\frac{1}{2\pi^2}\int d\boldsymbol{p}e^{i\mathbf{r}\cdot\mathbf{p}} . \tag{4.34}$$

We have

$$\frac{1}{r} = \frac{1}{2\pi^2}\int \frac{e^{i\mathbf{r}\cdot\mathbf{p}}}{p^2}d\boldsymbol{p} \tag{4.35}$$

Substitution in Eq. (4.32) yields

$$\int d\boldsymbol{p}\phi(\boldsymbol{p})\left(\frac{p^2}{2} - E\right)e^{i\mathbf{r}\cdot\mathbf{p}} - \frac{1}{2\pi^2}\int d\boldsymbol{q}d\boldsymbol{p}'\phi(\boldsymbol{p}')\frac{1}{q^2}e^{i\mathbf{r}\cdot(\mathbf{p}'+\mathbf{q})} = 0$$

Replacing $\boldsymbol{p}' + \boldsymbol{q}$ by $\boldsymbol{p}$ in the second integral we obtain

$$\int e^{i\mathbf{r}\cdot\mathbf{p}}d\boldsymbol{p}\left[\left(\frac{p^2}{2} - E\right)\phi(\boldsymbol{p}) - \frac{1}{2\pi^2}\int \frac{\phi(\boldsymbol{p}')}{|\mathbf{p}' - \mathbf{p}|^2}d\boldsymbol{p}'\right] = 0$$

From the orthogonality of the functions $\exp(i\boldsymbol{r}\cdot\boldsymbol{p})$ we obtain

$$\left(\frac{p^2}{2}-E\right)\phi(\boldsymbol{p})-\frac{1}{2\pi^2}\int\frac{\phi(\boldsymbol{p}')}{|\boldsymbol{p}'-\boldsymbol{p}|^2}d\boldsymbol{p}'=0\ . \tag{4.36}$$

This is the Schrödinger equation of the hydrogen atom in the momentum representation. It differs from Eq. (4.32) in that it is an integral equation.

We recall that $E<0$ for bound states. Thus we may define p_0 by

$$E=-\frac{1}{2}p_0^2\ . \tag{4.37}$$

Let p_1, p_2 and p_3 be the Cartesian components of $\boldsymbol{p}$ and define

$$\xi_0=\frac{p_0^2-p^2}{p_0^2+p^2}\quad,\quad\xi_1=\frac{2p_0p_1}{p_0^2+p^2}\quad,\quad\xi_2=\frac{2p_0p_2}{p_0^2+p^2}\quad,\quad\xi_3=\frac{2p_0p_3}{p_0^2+p^2}\ . \tag{4.38}$$

Equation (4.36) can be rewritten in the form

$$\int\frac{\phi(\boldsymbol{p}')d\boldsymbol{p}'}{|\boldsymbol{p}'-\boldsymbol{p}|^2}=\pi^2(p^2+p_0^2)\phi(\boldsymbol{p})\ . \tag{4.39}$$

We note that

$$\sum_{\mu=0}^{3}\xi_\mu^2=1\ , \tag{4.40}$$

i.e., instead of considering ϕ as a function of the momentum we can regard it as a function of a point on the four-dimensional sphere of unit radius defined by Eq. (4.40).

The problem consists of finding the functions $\phi(\boldsymbol{p})$ satisfying the integral equation (4.39) subject to the condition of integrability of $|\phi(\boldsymbol{p})|^2$. We shall see that such solutions exists for a discrete set of values of p_0. To accomplish this we rewrite Eq. (4.39) in spherical coordinates in the four-dimensional space:

$$\begin{aligned}\xi_0&=\rho\cos\alpha_1\ ,\\ \xi_1&=\rho\sin\alpha_1\cos\alpha_2\ ,\\ \xi_2&=\rho\sin\alpha_1\sin\alpha_2\cos\alpha_3\ ,\\ \xi_3&=\rho\sin\alpha_1\sin\alpha_2\sin\alpha_3\ ,\end{aligned} \tag{4.41}$$

with $0\le\rho<\infty$, $0\le\alpha_1\le\pi$, $0\le\alpha_2\le\pi$, $0\le\alpha_3<2\pi$. The element of solid angle is

$$d\Omega=\sin^2\alpha_1\sin\alpha_2 d\alpha_1 d\alpha_2 d\alpha_3 \tag{4.42}$$

obtained from the Jacobian determinant

$$\frac{\partial(\xi_0,\xi_1,\xi_2,\xi_3)}{\partial(\rho,\alpha_1,\alpha_2,\alpha_3)}=\rho^3\sin^2\alpha_1\sin\alpha_2\ . \tag{4.43}$$

Equations (4.38) give

$$|\boldsymbol{\xi}' - \boldsymbol{\xi}|^2 = \sum_{\mu=0}^{3}(\xi'_\mu - \xi_\mu)^2 = \frac{4p_0^2|\boldsymbol{p}' - \boldsymbol{p}|^2}{(p_0^2+p^2)(p_0^2+p'^2)} \ . \tag{4.44}$$

From Eqs. (4.38)

$$\begin{aligned} p_i &= \frac{p_0^2+p^2}{2p_0}\xi_i \\ &= \frac{p_0\xi_i}{1+\xi_0} = \frac{p_0\xi_i}{1+\cos\alpha_1} \quad , \quad i = 1,2,3 \ . \end{aligned} \tag{4.45}$$

The volume element in momentum space is

$$d\boldsymbol{p} = dp_1dp_2dp_3 = \frac{\partial(p_1,p_2,p_3)}{\partial(\alpha_1,\alpha_2,\alpha_3)}d\alpha_1 d\alpha_2 d\alpha_3 = \frac{p_0^3}{(1+\cos\alpha_1)^3}d\Omega \ . \tag{4.46}$$

From Eqs. (4.45) we deduce that

$$1+\cos\alpha_1 = \frac{2p_0^2}{p^2+p_0^2} \tag{4.47}$$

so that

$$d\boldsymbol{p} = \frac{(p_0^2+p^2)^3}{8p_0^3}d\Omega \ . \tag{4.48}$$

Defining

$$\psi(\boldsymbol{\xi}) = (p_0^2+p^2)\phi(\boldsymbol{p}) \tag{4.49}$$

we transform Eq. (4.39) into

$$\psi(\boldsymbol{\xi}) = \frac{1}{2\pi^2 p_0}\int \frac{\psi(\boldsymbol{\xi}')d\Omega'}{|\boldsymbol{\xi}' - \boldsymbol{\xi}|^2} \ . \tag{4.50}$$

This equation is equivalent to the Schrödinger equation and it is clearly invariant with respect to rotations in the four-dimensional space $(\xi_0, \xi_1, \xi_2, \xi_3)$.

We recall that in Eq. (4.50) the point $\boldsymbol{\xi}$ is on the surface of the unit 4-sphere. To solve this equation we extend the domain of definition of $\psi(\boldsymbol{\xi})$. While on $|\boldsymbol{\xi}|^2 = 1$ the function must satisfy Eq. (4.50) we can select $\psi(\underset{\sim}{x})$ in such a way that, away from the unit sphere, it satisfied the generalized Laplace equation.

$$\sum_{\mu=0}^{3}\partial_\mu^2\psi = 0 \ . \tag{4.51}$$

The Laplacian operator in four dimensions is

$$\begin{aligned} \sum_{\mu=0}^{3}\partial_\mu^2 &= \frac{1}{\rho^3}\frac{\partial}{\partial\rho}\left(\rho^3\frac{\partial}{\partial\rho}\right) + \frac{1}{\rho^2\sin^2\alpha_1}\frac{\partial}{\partial\alpha_1}\left(\sin^2\alpha_1\frac{\partial}{\partial\alpha_1}\right) \\ &+ \frac{1}{\rho^2\sin^2\alpha_1\sin\alpha_2}\frac{\partial}{\partial\alpha_2}\left(\sin\alpha_2\frac{\partial}{\partial\alpha_2}\right) + \frac{1}{\rho^2\sin^2\alpha_1\sin^2\alpha_2}\frac{\partial^2}{\partial\alpha_3^2} \ . \end{aligned} \tag{4.52}$$

A spherically symmetric solution of Eq. (4.52) satisfied

$$\frac{d}{d\rho}\left(\rho^3 \frac{d\psi}{d\rho}\right) = 0 \tag{4.53}$$

so that

$$\psi = \frac{A}{\rho^2} + B \ , \tag{4.54}$$

where A and B are constants. The function

$$G(\underset{\sim}{x}, \underset{\sim}{x}') = \frac{1}{4\pi^2 |\underset{\sim}{x} - \underset{\sim}{x}'|^2} \tag{4.55}$$

satisfies

$$\sum_{\mu=1}^{3} \partial_\mu^2 G(\underset{\sim}{x}, \underset{\sim}{x}') = -\delta(\underset{\sim}{x} - \underset{\sim}{x}') = -\prod_{\mu=0}^{3} \delta(x_\mu - x'_\mu) \ . \tag{4.56}$$

Clearly, from the result (4.54) above, when $\underset{\sim}{x}' \neq \underset{\sim}{x}$,

$$\sum_{\mu=0}^{3} \partial_\mu^2 G(\underset{\sim}{x}, \underset{\sim}{x}') = 0 \ . \tag{4.57}$$

To prove that when $\underset{\sim}{x}$ approaches $\underset{\sim}{x}'$, $\Sigma\partial_\mu^2 G$ is the δ-function we trace a sphere of radius ϵ centered at $\underset{\sim}{x}'$. Then

$$\int\limits_{|\underset{\sim}{x} - \underset{\sim}{x}'| \leq \epsilon} \sum_{\mu=0}^{3} \partial_\mu^2 G d\underset{\sim}{x} = \sum_\mu \int\limits_{(\Sigma)} n_\mu \partial_\mu G(\underset{\sim}{x}, \underset{\sim}{x}') d\Sigma = \int \frac{d}{d\rho}\left(\frac{1}{4\pi\rho^2}\right) \rho^3 d\Omega = -1 \ .$$

Here we used the divergence theorem in 4-space. Consider now a point $\underset{\sim}{x} = (\rho, \alpha_1, \alpha_2, \alpha_3)$ inside the unit sphere ($\rho < 1$) and its inverse $\underset{\sim}{y} = (\frac{1}{\rho}, \alpha_1, \alpha_2, \alpha_3)$. The function

$$G(\underset{\sim}{x}, \underset{\sim}{x}') = \frac{1}{4\pi^2 |\underset{\sim}{x} - \underset{\sim}{x}'|^2} + \frac{1}{4\pi^2 \rho^2} \frac{1}{|\underset{\sim}{y} - \underset{\sim}{x}'|^2} \tag{4.58}$$

satisfies Eq. (4.56) inside the unit sphere. Now

$$|\underset{\sim}{x} - \underset{\sim}{x}'|^2 = \rho^2 + \rho'^2 - 2\sum_{\mu=0}^{3} x_\mu x'_\mu = \rho^2 + \rho'^2 - 2\rho\rho' \cos\omega$$

where $\cos\omega$ can be evaluated directly from Eqs. (4.41). Furthermore

$$\rho^2 |\underset{\sim}{y} - \underset{\sim}{x}'| = 1 + \rho^2\rho'^2 - 2\rho\rho' \cos\omega$$

When $\underset{\sim}{x}'$ is on the surface of the unit sphere, $\rho' = 1$ and

$$G(\underset{\sim}{x}, \underset{\sim}{x}') = \frac{1}{2\pi^2}(\rho^2 + 1 - 2\rho\cos\omega)^{-1} \ . \tag{4.59}$$

From the second Green identity in 4-space

$$\int\limits_{|\underset{\sim}{x}'|\leq 1} d\underset{\sim}{x}'[\psi(\underset{\sim}{x}')\sum_{\mu=0}^{3}\partial_\mu'^2 G(\underset{\sim}{x}, \underset{\sim}{x}') - G(\underset{\sim}{x}, \underset{\sim}{x}')\sum_{\mu=0}^{3}\partial_\mu'^2\psi(\underset{\sim}{x}')]$$

$$= \int\limits_{\Sigma(|\underset{\sim}{x}'|=1)} [\psi(\underset{\sim}{x}')\sum_{\mu=0}^{3}\partial_\mu' G - G(\underset{\sim}{x}, \underset{\sim}{x}')\sum_{\mu=0}^{3}\partial_\mu'\psi(\underset{\sim}{x}')]n_\mu' d\Sigma' \ .$$

Inside the sphere $\sum_{\mu=0}^{3}\partial_\mu'^2 G(\underset{\sim}{x}, \underset{\sim}{x}') = 0$ so that using Eq. (4.56) we find

$$\psi(\underset{\sim}{x}) = -\int d\Omega'\left[\psi(\underset{\sim}{x}')\frac{\partial G(\underset{\sim}{x}, \underset{\sim}{x}')}{\partial\rho'} - G(\underset{\sim}{x}, \underset{\sim}{x}')\frac{\partial\psi(\underset{\sim}{x}')}{\partial\rho'}\right]_{\rho'=1} \ . \tag{4.60}$$

By direct computation of

$$\frac{\partial G}{\partial\rho'}$$

from Eq. (4.58) we find, setting $\rho' = 1$, that

$$\left(\frac{\partial G}{\partial\rho'}\right)_{\rho'=1} = (-G)_{\rho'=1} \ .$$

Thus

$$\psi(\underset{\sim}{x}) = \frac{1}{2\pi^2}\int\limits_{\rho'=|\underset{\sim}{x}'|=1}\frac{d\Omega'}{|\underset{\sim}{x} - \underset{\sim}{x}'|^2}\left[\psi(\underset{\sim}{x}') + \frac{\partial\psi(\underset{\sim}{x}')}{\partial\rho'}\right] \ . \tag{4.61}$$

The expansion

$$\psi(\underset{\sim}{x}) = \sum_{n=1}^{\infty} c_n\rho^{n-1}\psi_n(\alpha_1, \alpha_2, \alpha_3) = \sum_{n=1}^{\infty} c_n\psi_n(\underset{\sim}{x}) \tag{4.62}$$

is valid for all $|\underset{\sim}{x}| < |\underset{\sim}{x}'| = 1$ where we have defined $\psi_n(\underset{\sim}{x}) = \rho^{n-1}\psi_n(\alpha_1, \alpha_2, \alpha_3)$.

Clearly

$$\left[\psi_n(\underset{\sim}{x}) + \frac{\partial\psi_n(\underset{\sim}{x})}{\partial\rho}\right]_{\rho=1} = n\psi_n(\boldsymbol{\xi}) \tag{4.63}$$

so that

$$\psi_n(\boldsymbol{\xi}) = \frac{n}{2\pi^2} \int \frac{\psi_n(\boldsymbol{\xi}')d\Omega'}{|\boldsymbol{\xi}' - \boldsymbol{\xi}|^2} \tag{4.64}$$

for every integer $n \geq 1$. Hence $p_0 = \frac{1}{n}$ and we obtain the energy eigenvalues

$$E_n = -\frac{1}{2}p_0^2 = -\frac{1}{2n^2} \; . \tag{4.65}$$

The differential equation satisfied by $\psi_n(\boldsymbol{\xi})$ is obtained using Eq. (4.52). We have

$$\begin{aligned}(n^2-1)\psi_n(\boldsymbol{\xi}) \quad &+ \quad \frac{1}{\sin^2\alpha_1}\left[\frac{\partial}{\partial\alpha_1}\left(\sin^2\alpha_1\frac{\partial\psi_n}{\partial\alpha_1}\right) + \frac{1}{\sin\alpha_2}\frac{\partial}{\partial\alpha_2}\left(\sin\alpha_2\frac{\partial\psi_n}{\partial\alpha_2}\right)\right. \\ &+ \quad \left.\frac{1}{\sin^2\alpha_2}\frac{\partial^2\psi_n}{\partial\alpha_3^2}\right] = 0 \; ,\end{aligned} \tag{4.66}$$

since $\psi_n(\underset{\sim}{x}) \propto \rho^{n-1}$.

Non-singular solutions in α_2 and α_3 are proportional to the spherical harmonics

$$Y_\ell^m(\alpha_2, \alpha_3)$$

so that expanding $\psi_n(\boldsymbol{\xi})$ in the form

$$\psi_n(\boldsymbol{\xi}) = \psi_n(\alpha_1, \alpha_2, \alpha_3) = \sum_{\ell=0}^{\infty}\sum_{m=-\ell}^{\ell} \chi_{n\ell}(\alpha_1)Y_\ell^m(\alpha_2, \alpha_3) \tag{4.67}$$

we obtain the differential equation

$$\left[n^2 - 1 - \frac{\ell(\ell+1)}{\sin^2\alpha_1} + \frac{1}{\sin^2\alpha_1}\frac{d}{d\alpha_1}\left(\sin^2\alpha_1\frac{d}{d\alpha_1}\right)\right]\chi_{n\ell}(\alpha_1) = 0 \tag{4.68}$$

for the function $\chi_{n\ell}(\alpha_1)$. We make the change of variable

$$z = \frac{1}{2}(1 - \cos\alpha_1) \tag{4.69}$$

transforming the differential equation into

$$z(1-z)\frac{d^2\chi_{n\ell}}{dz^2} + \frac{3}{2}(1-2z)\frac{d\chi_{n\ell}}{dz} + \left(n^2 - 1 - \frac{\ell(\ell+1)}{4z(1-z)}\right)\chi_{n\ell} = 0 \; . \tag{4.70}$$

This differential equation is of a type which often occurs in mathematical physics. See, *e.g.*, S. Rodriguez, *"The Special Functions of Mathematical Physics"*, pp. 118–121. Changing the dependent variable to $w_{n\ell}$ related to $\chi_{n\ell}$ by

$$\chi_{n\ell} = z^\alpha(1-z)^\beta w_{n\ell} \tag{4.71}$$

it can be reduced to the equation for the hyper-geometric functions. In fact, the substitution (4.71) leads to

$$z(1-z)w''_{n\ell} + \Big[2(\alpha-(\alpha+\beta)z) + \frac{3}{2}(1-2z)\Big]w'_{n\ell}$$
$$+\Big[\frac{\alpha}{2z}(2\alpha+1) + \frac{\beta(2\beta+1)}{2(1-z)} - (\alpha+\beta)(\alpha+\beta+2)$$
$$+n^2 - 1 - \frac{\ell(\ell+1)}{4z} - \frac{\ell(\ell+1)}{4(1-z)}\Big]w_{n\ell} = 0\ .$$

Selecting $\alpha = \beta = \ell/2$ the equation reduces to

$$z(1-z)w''_{n\ell} + \left(\ell+\frac{3}{2}\right)(1-2z)w'_{n\ell} + (n^2-(\ell+1)^2)w_{n\ell} = 0\ . \qquad (4.72)$$

This is of the form

$$z(1-z)w'' + \{c-(a+b+1)z\}w' - abw = 0\ , \qquad (4.73)$$

with

$$\begin{aligned} a &= n+\ell+1\ , \\ b &= -n+\ell+1\ , \\ c &= \ell+\frac{3}{2}\ . \end{aligned} \qquad (4.74)$$

The solutions of this equation are

$$F(a,b,c;z) = 1+\frac{ab}{c}z+\cdots+\frac{a(a+1)\cdots(a+k-1)b(b+1)\cdots(b+k-1)}{k!c(c+1)\cdots(c+k-1)}z^k+\cdots \qquad (4.75)$$

and a second solution singular at $z=0$. Furthermore the function resulting from the solution is singular at $z=1$ unless it is a polynomial. This tells us that the physically acceptable solutions are those for which b is a negative integer or zero, *i.e.*,

$$b \le 0\ .$$

Thus

$$\ell \le n-1\ ,$$

showing that the degeneracy of the state with energy $E_n = -(1/2n^2)$ is n^2.

7.5 Selection rules for two-photon processes

In Sec. 7.2. we have studied selection rules for one-photon processes. In this section we discuss the selection rules for two-photon processes, namely, scattering of radiation by atoms, molecules or crystals and two-photon absorption or emission. We first review the theory of electron-photon interactions. We then discuss the theory of scattering of radiation and leave that of two-photon emission or absorption as an exercise for the reader.

We consider a system composed of N charged particles of masses m_i and electric charges q_i $(i = 1, 2, \cdots, N)$. The Hamiltonian operator describing such a system is of the form

$$\mathrm{H}_S = \sum_{i=1} \frac{\mathbf{p}_i^2}{2m_i} + U \tag{5.1}$$

where U includes all interactions between the particles. We assume the system is contained in a finite volume so that we can disregard retardation effects in the interaction U.

We now suppose that this system is in the presence of an electromagnetic field described by a vector potential operator $\mathbf{A}(\mathbf{r})$. The Hamiltonian of the system, the radiation field and their mutual interaction is

$$\mathrm{H} = \sum_{i=1}^{N} \frac{1}{2m_i}\Big(\mathbf{p}_i - \frac{q_i}{c}\mathbf{A}(\mathbf{r}_i)\Big)^2 + U - \sum_{i=1}^{N} \boldsymbol{\mu}_i \cdot \Big(\nabla \times \mathbf{A}(\mathbf{r}_i)\Big) + \mathrm{H}_R, \tag{5.2}$$

where H_R is the Hamiltonian of the radiation and $\boldsymbol{\mu}_i$ the magnetic moment of the ith particle. We expand the kinetic energy term to obtain

$$\mathrm{H} = \mathrm{H}_0 + \mathrm{H}_I + \mathrm{H}_{II}. \tag{5.3}$$

with

$$\mathrm{H}_0 = \mathrm{H}_S + \mathrm{H}_R. \tag{5.4}$$

$$\mathrm{H}_I = -\sum_{i=1}^{N} \frac{q_i}{2m_i c}\Big(\mathbf{A}(\mathbf{r}_i) \cdot \mathbf{p}_i + \mathbf{p}_i \cdot \mathbf{A}(\mathbf{r}_i)\Big) - \sum_{i=1}^{N} \boldsymbol{\mu}_i \cdot \Big(\nabla \times \mathbf{A}(\mathbf{r}_i)\Big), \tag{5.5}$$

and

$$\mathrm{H}_{II} = -\sum_{i=1}^{N} \frac{q_i^2}{2m_i c^2}\Big(\mathbf{A}(\mathbf{r}_i)\Big)^2. \tag{5.6}$$

We shall refer to H_I and H_{II} as the paramagnetic and diamagnetic interactions, respectively. We expand $\mathbf{A}(\mathbf{r})$ in terms of plane waves in the form

$$\mathbf{A}(\mathbf{r}) = \sum_{q\mu} \left(\frac{2\pi\hbar}{V\omega_q}\right)^{1/2} c\, \hat{\mathbf{e}}_{q\mu}\Big(a_{q\mu} \exp(i\,\mathbf{q}\cdot\mathbf{r}) + a_{q\mu}^{\dagger} \exp(-i\,\mathbf{q}\cdot\mathbf{r})\Big), \tag{5.7}$$

where we use periodic boundary conditions with a basic period in the form of a cube of volume V; ω_q is the frequency of an electromagnetic wave of wave vector $\mathbf{q}$, $a_{q\mu}$ and $a_{q\mu}^{\dagger}$ are destruction and creation operators for photons of wave vector $\mathbf{q}$ and polarization $\hat{\mathbf{e}}_{q\mu}$ ($\mu = 1, 2$ with $\mathbf{q} \cdot \hat{\mathbf{e}}_{q\mu} = 0$). To simplify the writing we define the current density operator

$$\mathbf{J}(\mathbf{r}) = \sum_{i} \left[\frac{q_i}{2m_i}\Big(\mathbf{p}_i\, \delta(\mathbf{r} - \mathbf{r}_i) + \delta(\mathbf{r} - \mathbf{r}_i)\, \mathbf{p}_i\Big)\right] + c\, \nabla \times \mathbf{M}(\mathbf{r}) \tag{5.8}$$

where the magnetization $\mathbf{M}(\mathbf{r})$ is given by

$$\mathbf{M}(\mathbf{r}) = \sum_i \boldsymbol{\mu}_i \, \delta(\mathbf{r} - \mathbf{r}_i). \tag{5.9}$$

The Fourier transform of $\mathbf{J}(\mathbf{r})$ is

$$\begin{aligned}\mathbf{J}(\mathbf{q}) &= \int_{(V)} d\mathbf{r} \; \mathbf{J}(\mathbf{r}) \exp(-i \, \mathbf{q} \cdot \mathbf{r}) \\ &= \sum_i \Big[\frac{q_i}{2m_i} \Big(\mathbf{p}_i \exp(-i \, \mathbf{q} \cdot \mathbf{r}_i) + \exp(-i \, \mathbf{q} \cdot \mathbf{r}_i) \, \mathbf{p}_i \Big) \\ &\quad + i \, c \, \mathbf{q} \times \boldsymbol{\mu}_i \exp(-i \, \mathbf{q} \cdot \mathbf{r}_i) \Big]. \end{aligned} \tag{5.10}$$

We note that since $\hat{\mathbf{e}}_{q\mu}$ and $\mathbf{q}$ are at right angles, $\hat{\mathbf{e}}_{q\mu} \cdot \mathbf{p}$ and $\exp(-i \, \mathbf{q} \cdot \mathbf{r}_i)$ are commuting operators. Thus,

$$\hat{\mathbf{e}}_{q\mu} \cdot \mathbf{J}(\mathbf{q}) = \sum_i \left[\frac{q_i}{m_i} \, \hat{\mathbf{e}}_{q\mu} \cdot \mathbf{p}_i \exp(-i \, \mathbf{q} \cdot \mathbf{r}_i) + i \, c \, (\hat{\mathbf{e}}_{q\mu} \times \mathbf{q}) \cdot \boldsymbol{\mu}_i \exp(-i \, \mathbf{q} \cdot \mathbf{r}_i) \right]. \tag{5.11}$$

We express H_I in the form

$$\mathrm{H}_I = -\sum_{q\mu} \left(\frac{2\pi\hbar}{V\omega_q} \right)^{1/2} \hat{\mathbf{e}}_{q\mu} \cdot \Big(a_{q\mu} \, \mathbf{J}^\dagger(\mathbf{q}) + a_{q\mu}^\dagger \, \mathbf{J}(\mathbf{q}) \Big). \tag{5.12}$$

A. The Raman scattering cross section

We designate the eigenstates of the scattering system (H_S) with symbols such as $\nu, \nu_0, \cdots$ and those of the radiation field (H_R) by $r, r_0, \cdots$. A Raman process is defined as one in which the initial state of the radiation field, r_0, differs from the final state, r, in that a photon characterized by ($\mathbf{q}$, $\hat{\mathbf{e}}$) has been replaced by a second with wave vector and polarization given by ($\mathbf{q}', \hat{\mathbf{e}}'$) while the scattering system has experienced a transition from ν_0 to ν. Let E_ν and E_{ν_0} be the energies of the scattering system in states ν and ν_0, respectively.

The probability that in time t this transition has taken place is

$$\begin{aligned}\mathrm{P}_{r_0\nu_0 \to r\nu}(t) &= \Bigg| < r\nu \, | - \frac{i}{\hbar} \int_0^t \mathrm{H}_{II}(t_1) \, dt_1 \\ &\quad - \frac{1}{\hbar^2} \int_0^t dt_2 \int_0^{t_2} dt_1 \; \mathrm{H}_I(t_2) \, \mathrm{H}_I(t_1) \mid r_0\nu_0 > \Bigg|^2, \end{aligned} \tag{5.13}$$

where H_{II} is taken in first order and H_I in second order. The operators $\mathrm{H}_I(t)$ and $\mathrm{H}_{II}(t)$ are those in Eqs. (5.17) and (5.6) expressed in the interaction representation, *i.e.*,

$$\mathrm{H}_I(t) = \exp\left(\frac{i}{\hbar} \mathrm{H}_0 t \right) \, \mathrm{H}_I \, \exp\left(-\frac{i}{\hbar} \mathrm{H}_0 t \right). \tag{5.14}$$

Defining

$$\omega_{\nu\nu_0} = (E_\nu - E_{\nu_0})/\hbar \tag{5.15}$$

we have

$$< r\nu \mid \mathrm{H}_{II}(t_1) \mid r_0\nu_0 >= \exp\Big(i\,(\omega_{\nu\nu_0} + \omega' - \omega)t_1\Big) < r\nu \mid \mathrm{H}_{II} \mid r_0\nu_0 >, \tag{5.16}$$

and

$$< r\nu \,|\, \mathrm{H}_I(t_2)\,\mathrm{H}_I(t_1) \,|\, r_0\nu_0 > = \sum_{r'\nu'} < r\nu \,|\, \mathrm{H}_I(t_2) \,|\, r'\nu' >< r'\nu' \,|\, \mathrm{H}_I(t_1) \,|\, r_0\nu_0 > . \tag{5.17}$$

Since H_I contains raising and lowering operators for photon mode occupation numbers only linearly, the intermediate states r' of the radiation field can only be $\mid n-1, n' >$ or $\mid n, n'+1 >$, where n is the photon population of the incident radiation and n' the initial population of the scattered radiation. The resulting possibilities are represented schematically by the diagrams in Fig. 7.1.

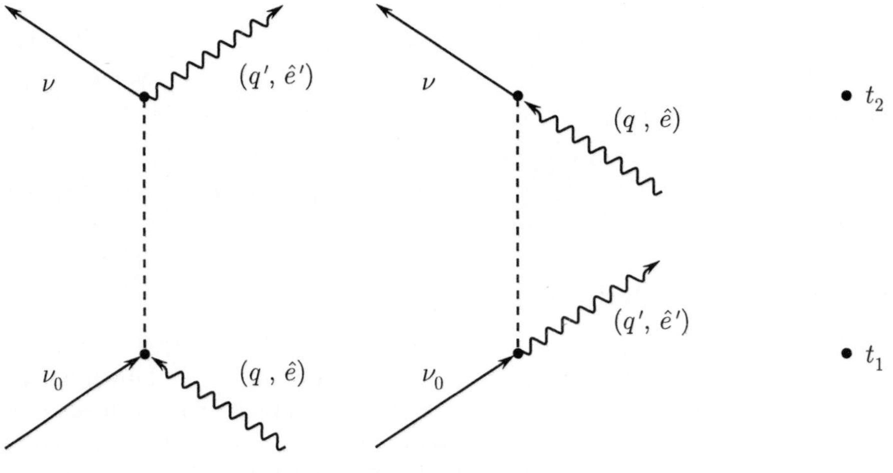

Figure 7.1

We obtain

$$< r\nu \mid \mathrm{H}_I(t_2)\,\mathrm{H}_I(t_1) \mid r_0\nu_0 >$$

$$= \frac{2\pi\hbar}{V}\left[\frac{n(n'+1)}{\omega\omega'}\right]^{1/2}$$

$$\Bigg(\hat{e}'\cdot\mathbf{J}_{\nu\nu'}(\mathbf{q}')\;\hat{e}\cdot\mathbf{J}^\dagger_{\nu'\nu_0}(\mathbf{q})\,\exp\Big(i(\omega_{\nu\nu'}+\omega')t_2\Big)\,\exp\Big(i(\omega_{\nu'\nu_0}-\omega)t_1\Big)$$

$$+\,\hat{e}\cdot\mathbf{J}^\dagger_{\nu\nu'}(\mathbf{q})\;\hat{e}\cdot\mathbf{J}_{\nu'\nu_0}(\mathbf{q}')\,\exp\Big(i(\omega_{\nu\nu'}-\omega)t_2\Big)\,\exp\Big(i(\omega_{\nu'\nu_0}+\omega')t_1\Big)\Bigg).$$

We need the expression

$$
\begin{aligned}
&-\frac{1}{\hbar^2}\int_0^t dt_2 \int_0^{t_2} dt_1 < r\nu \mid \mathrm{H}_I(t_2)\,\mathrm{H}_I(t_1) \mid r_0\nu_0 > \\
&= \frac{2\pi\hbar}{V}\left[\frac{n(n'+1)}{\omega\omega'}\right]^{1/2} \Big(\hat{\mathbf{e}}'\cdot \mathbf{J}_{\nu\nu'}(\mathbf{q}')\;\hat{\mathbf{e}}\cdot \mathbf{J}^{\dagger}_{\nu'\nu_0}(\mathbf{q}) \\
&\left(\frac{\exp\Big(i(\omega_{\nu\nu_0}+\omega'-\omega)t\Big)-1}{(\omega_{\nu'\nu_0}-\omega)(\omega_{\nu\nu_0}+\omega'-\omega)} + \frac{\exp\Big(i(\omega_{\nu\nu'}+\omega')t\Big)-1}{(\omega_{\nu'\nu_0}-\omega)(\omega_{\nu\nu'}+\omega')}\right) \\
&+\hat{\mathbf{e}}\cdot \mathbf{J}^{\dagger}_{\nu\nu'}(\mathbf{q})\;\hat{\mathbf{e}}'\cdot \mathbf{J}_{\nu'\nu_0}(\mathbf{q}')\left(\frac{\exp\Big(i(\omega_{\nu\nu_0}+\omega'-\omega)t\Big)-1}{(\omega_{\nu'\nu_0}+\omega')(\omega_{\nu\nu_0}+\omega'-\omega)}\right. \\
&\left. - \frac{\exp\Big(i(\omega_{\nu\nu'}-\omega)t\Big)-1}{(\omega_{\nu'\nu_0}+\omega')(\omega_{\nu\nu'}-\omega)}\right)\Big).
\end{aligned}
\tag{5.18}
$$

The diamagnetic contribution to the transition amplitude is

$$
\begin{aligned}
&-\frac{i}{\hbar}\int_0^t < r\nu \mid \mathrm{H}_{II}(t') \mid r_0\nu_0 > dt' \\
&= -\frac{2\pi}{V}\left[\frac{n(n'+1)}{\omega\omega'}\right]^{1/2} \hat{\mathbf{e}}\cdot\hat{\mathbf{e}}' < \nu \mid \sum_i \frac{q_i^2}{m_i}\exp\Big(i\,(\mathbf{q}-\mathbf{q}')\cdot \mathbf{r}_i\Big) \mid \nu_0 > \\
&\times \frac{\exp\Big(i(\omega_{\nu\nu_0}+\omega'-\omega)t\Big)-1}{(\omega_{\nu\nu_0}+\omega'-\omega)}.
\end{aligned}
\tag{5.19}
$$

In Eq. (5.18) we neglect the terms containing $\exp\Big(i(\omega_{\nu\nu'}+\omega')t\Big)$ and $\exp\Big(i(\omega_{\nu\nu'}-\omega)t\Big)$. These contributions become large when $\omega' = \omega_{\nu'\nu}$ and $\omega = \omega_{\nu\nu'}$, respectively. For a scattering process in which $\omega_{\nu\nu_0}+\omega'-\omega = 0$ the first gives rise to a secular transition when $\omega = \omega_{\nu'\nu_0}$ while the second requires $\omega' = \omega_{\nu_0\nu'}$. We assume that ν_0 is the ground state of the scattering system so that ω' cannot be equal to $\omega_{\nu_0\nu'}$. There remains the possibility that $\omega = \omega_{\nu'\nu_0}$, i.e., that absorption of radiation of angular frequency ω can occur. If such a transition is not possible, the terms we neglect give rise to oscillatory contributions to the probability amplitude and can legitimately be neglected. Thus, for the Raman process considered,

$$
\mathrm{P}_{r_0\nu_0\to r\nu}(t) = \frac{4\pi^2}{\hbar^2V^2}\left[\frac{n(n'+1)}{\omega\omega'}\right]\left|\frac{\exp\Big(i(\omega_{\nu\nu_0}+\omega'-\omega)t\Big)-1}{\omega_{\nu\nu_0}+\omega'-\omega}\right|^2 |\mathrm{M}_{\nu\nu_0}|^2,
\tag{5.20}
$$

where

$$\mathrm{M}_{\nu\nu_0} = \sum_{\nu'}\left[\frac{\hat{e}'\cdot\mathbf{J}_{\nu\nu'}(\mathbf{q}')\ \hat{e}\cdot\mathbf{J}^{\dagger}_{\nu'\nu_0}(\mathbf{q})}{\omega_{\nu'\nu_0}-\omega}+\frac{\hat{e}\cdot\mathbf{J}^{\dagger}_{\nu\nu'}(\mathbf{q})\ \hat{e}'\cdot\mathbf{J}_{\nu'\nu_0}(\mathbf{q}')}{\omega_{\nu'\nu_0}+\omega'}\right]$$
$$-\,\hbar\ \hat{e}\cdot\hat{e}' < \nu \mid \sum_i \frac{q_i^2}{m_i}\exp\Big(i\,(\mathbf{q}-\mathbf{q}')\cdot\mathbf{r}_i\Big) \mid \nu_0 > . \qquad (5.21)$$

The time dependent term in the transition probability (5.20) is equal to

$$\frac{\sin^2\frac{1}{2}(\omega_{\nu\nu_0}+\omega'-\omega)t}{\left[\frac{1}{2}(\omega_{\nu\nu_0}+\omega'-\omega)\right]^2} \approx 2\pi t\ \delta(\omega_{\nu\nu_0}+\omega'-\omega) \qquad (5.22)$$

for sufficiently long times. The transition probability per unit time is, therefore

$$\mathrm{w}_{r_0\nu_0\to r\nu}(t) = \frac{1}{t}\ \mathrm{P}_{r_0\nu_0\to r\nu}(t) = \frac{8\pi^3 n(n'+1)}{\hbar^2 V^2\omega\omega'}\ |\mathrm{M}_{\nu\nu_0}|^2\ \delta(\omega_{\nu\nu_0}+\omega'-\omega). \quad (5.23)$$

The differential scattering cross section for the scattered photon to propagate within a solid angle $d\Omega'$ around the direction $\mathbf{q}'$ is

$$d\sigma = \sum_{\mathbf{q}\in d\Omega'} \frac{8\pi^3 n(n'+1)}{\hbar^2 V^2\omega\omega'}\ |\mathrm{M}_{\nu\nu_0}|^2\ \delta(\omega_{\nu\nu_0}+\omega'-\omega)\ \frac{V}{nv}.$$

where v is the velocity of light in the medium. Now the sum over $\mathbf{q}'$ can be approximated by the integral

$$\frac{V d\Omega'}{(2\pi)^3}\int q'^2 dq' = \frac{V d\Omega'}{(2\pi)^3}\int\frac{\omega'^2 d\omega'}{v^3}$$

so that

$$\frac{d\sigma}{d\Omega'} = \left(\frac{\omega'}{\omega\hbar^2 v^4}\right)|\mathrm{M}_{\nu\nu_0}|^2(n'+1). \qquad (5.24)$$

B. The electric dipole approximation

The electric dipole approximation is accurate when the wavelengths of the incident and scattered photons is long compared to the dimensions of the scattering system. We note that a translation by $\mathbf{n}$ of all the particles in the scattering system amounts to multiplying the scattering amplitude $\mathrm{M}_{\nu\nu_0}$ by the phase factor $\exp(i\ (\mathbf{q}-\mathbf{q}')\cdot\mathbf{n})$ so that the cross section is not altered by this transformation. We can then select the origin of the $\mathbf{r}_i$ within the scattering system of maximum diameter, say, D. For $|\mathbf{r}_i| > D$ the wave functions $\mid \nu >$ and $\mid \nu_0 >$ decrease exponentially so that the factors $\exp(i\ \mathbf{q}\cdot\mathbf{r}_i)$ and $\exp(-i\ \mathbf{q}'\cdot\mathbf{r}_i)$ can be approximated by unity when the wavelength of the photons involved is much longer than D. Under these conditions $\mathbf{J}(\mathbf{q})$ can be approximated by

$$\mathbf{J} = \sum_i \frac{q_i}{m_i}\ \mathbf{p}_i = \mathbf{J}^{\dagger} \qquad (5.25)$$

and, since

$$[\mathbf{r}_i,\ \mathrm{H}_S] = i\hbar\,\frac{\mathbf{p}_i}{m_i}$$

we obtain

$$i\hbar\,\mathbf{J} = \sum_i q_i\,[\mathbf{r}_i,\ \mathrm{H}_S] = [\mathbf{d},\ \mathrm{H}_S] \tag{5.26}$$

where

$$\mathbf{d} = \sum_i q_i \mathbf{r}_i \tag{5.27}$$

is the electric dipole moment of the scattering system. But

$$\mathbf{J}_{\nu\nu'} = i\,\omega_{\nu\nu'}\,\mathbf{d}_{\nu\nu'} \tag{5.28}$$

so that the scattering amplitude is approximated by

$$\mathrm{M}_{\nu\nu_0} = \sum_{\nu'}\left(\frac{\hat{e}'\cdot\mathbf{d}_{\nu\nu'}\ \hat{e}\cdot\mathbf{d}_{\nu'\nu_0}}{\omega-\omega_{\nu'\nu_0}} - \frac{\hat{e}\cdot\mathbf{d}_{\nu\nu'}\ \hat{e}'\cdot\mathbf{d}_{\nu'\nu_0}}{\omega'+\omega_{\nu'\nu_0}}\right)\omega_{\nu\nu'}\omega_{\nu'\nu_0} - \hbar\,\hat{e}\cdot\hat{e}'\,\delta_{\nu\nu_0}\sum_i\frac{q_i^2}{m_i}. \tag{5.29}$$

To simplify this expression we consider the sum

$$\sum_{\nu'}(\omega_{\nu\nu'}\omega_{\nu'\nu_0}+\omega\omega')\left(\frac{\hat{e}'\cdot\mathbf{d}_{\nu\nu'}\ \hat{e}\cdot\mathbf{d}_{\nu'\nu_0}}{\omega-\omega_{\nu'\nu_0}} - \frac{\hat{e}\cdot\mathbf{d}_{\nu\nu'}\ \hat{e}'\cdot\mathbf{d}_{\nu'\nu_0}}{\omega'+\omega_{\nu'\nu_0}}\right) \tag{5.30}$$

and note that

$$\omega_{\nu\nu'}\omega_{\nu'\nu_0}+\omega\omega' = (\omega-\omega_{\nu'\nu_0})(\omega'+\omega_{\nu'\nu_0}). \tag{5.31}$$

Thus, the expression (5.30) is identical to

$$\begin{aligned}
&\sum_{\nu'}\left[(\omega'+\omega_{\nu'\nu_0})(\hat{e}'\cdot\mathbf{d}_{\nu\nu'}\ \hat{e}\cdot\mathbf{d}_{\nu'\nu_0}) - (\omega-\omega_{\nu'\nu_0})(\hat{e}\cdot\mathbf{d}_{\nu\nu'}\ \hat{e}'\cdot\mathbf{d}_{\nu'\nu_0})\right]\\
&\quad = \omega\sum_{\nu'}(\hat{e}'\cdot\mathbf{d}_{\nu\nu'}\ \hat{e}\cdot\mathbf{d}_{\nu'\nu_0} - \hat{e}\cdot\mathbf{d}_{\nu\nu'}\ \hat{e}'\cdot\mathbf{d}_{\nu'\nu_0})\\
&\quad\quad + \sum_{\nu'}(\omega_{\nu'\nu_0}\ \hat{e}\cdot\mathbf{d}_{\nu\nu'}\ \hat{e}'\cdot\mathbf{d}_{\nu'\nu_0} - \omega_{\nu\nu'}\ \hat{e}'\cdot\mathbf{d}_{\nu\nu'}\ \hat{e}'\cdot\mathbf{d}_{\nu'\nu_0})
\end{aligned}$$

where we replaced $\omega'+\omega_{\nu'\nu_0}$ with $\omega-\omega_{\nu\nu'}$. The first sum vanishes exactly since it equals

$$\omega\,\hat{e}'\cdot(\mathbf{d}\ \mathbf{d})_{\nu\nu_0}\cdot\hat{e} - \omega\,\hat{e}\cdot(\mathbf{d}\ \mathbf{d})_{\nu\nu_0}\cdot\hat{e}' \equiv 0$$

To evaluate the second sum we use Eq. (5.28) to obtain

$$\begin{aligned}
&\sum_{\nu'}(\,\omega_{\nu'\nu_0}\ \hat{e}\cdot\mathbf{d}_{\nu\nu'}\ \hat{e}'\cdot\mathbf{d}_{\nu'\nu_0} - \omega_{\nu\nu'}\ \hat{e}'\cdot\mathbf{d}_{\nu\nu'}\ \hat{e}\cdot\mathbf{d}_{\nu'\nu_0}\,)\\
&\quad = \frac{1}{i}\sum_{\nu'}(\,\hat{e}\cdot\mathbf{d}_{\nu\nu'}\ \hat{e}'\cdot\mathbf{J}_{\nu'\nu_0} - \hat{e}'\cdot\mathbf{J}_{\nu\nu'}\ \hat{e}\cdot\mathbf{d}_{\nu'\nu_0}\,)\\
&\quad = \frac{1}{i}(\,\hat{e}\cdot\mathbf{d}\ \hat{e}'\cdot\mathbf{J} - \hat{e}'\cdot\mathbf{J}\ \hat{e}\cdot\mathbf{d}\,)_{\nu\nu_0}\\
&\quad = \frac{1}{i}\left[\hat{e}\cdot\mathbf{d},\ \hat{e}'\cdot\mathbf{J}\right]_{\nu\nu_0}.
\end{aligned}$$

But the commutator of **d** and **J** is given by its components, namely

$$\begin{aligned}[\mathrm{d}_l,\ \mathrm{J}_k] &= \sum_{i,j}\left[q_i\, x_{il},\ \frac{q_j}{m_j}\, p_{jk}\right] \\ &= \sum_{i,j}\frac{q_i q_j}{m_j}\,[x_{il},\ p_{jk}], \quad i,j=1,2,...,N.\ :\quad l,k=x,y,z.\end{aligned}$$

However

$$[x_{il},\ p_{jk}] = i\hbar\ \delta_{ij}\delta_{lk}$$

so that

$$[\mathrm{d}_l,\ \mathrm{J}_k] = i\hbar\sum_i \frac{q_i^2}{m_i}\delta_{lk}.$$

and

$$\frac{1}{i}[\hat{e}\cdot \boldsymbol{d}, \hat{e}'\cdot \boldsymbol{J}] = \hbar\hat{e}\cdot\hat{e}'\sum_i \frac{q^2}{m_i}\ .$$

Combining these results we deduce that

$$\sum_{\nu'}(\omega_{\nu\nu'}\omega_{\nu'\nu_0}+\omega\omega')\left(\frac{\hat{\mathbf{e}}'\cdot\mathbf{d}_{\nu\nu'}\ \hat{\mathbf{e}}\cdot\mathbf{d}_{\nu'\nu_0}}{\omega-\omega_{\nu'\nu_0}}-\frac{\hat{\mathbf{e}}\cdot\mathbf{d}_{\nu\nu'}\ \hat{\mathbf{e}}'\cdot\mathbf{d}_{\nu'\nu_0}}{\omega'+\omega_{\nu'\nu_0}}\right)=\hbar\ \hat{e}\cdot\hat{e}'\ \delta_{\nu\nu_0}\sum_i\frac{q_i^2}{m_i}. \tag{5.32}$$

Thus, Eqs. (5.29) and (5.32) give

$$\mathrm{M}_{\nu\nu_0} = -\omega\omega'\left(\frac{\hat{\mathbf{e}}'\cdot\mathbf{d}_{\nu\nu'}\ \hat{\mathbf{e}}\cdot\mathbf{d}_{\nu'\nu_0}}{\omega-\omega_{\nu'\nu_0}}-\frac{\hat{\mathbf{e}}\cdot\mathbf{d}_{\nu\nu'}\ \hat{\mathbf{e}}'\cdot\mathbf{d}_{\nu'\nu_0}}{\omega'+\omega_{\nu'\nu_0}}\right). \tag{5.33}$$

The cross section is, thus, expressed in the form

$$\frac{d\sigma}{d\Omega'} = \varepsilon_\infty^2\left(\frac{\omega'^3\omega}{\hbar^2c^4}\right)\left|\sum_{\nu'}\left(\frac{\hat{\mathbf{e}}\cdot\mathbf{d}_{\nu\nu'}\ \hat{\mathbf{e}}'\cdot\mathbf{d}_{\nu'\nu_0}}{\omega'+\omega_{\nu'\nu_0}}-\frac{\hat{\mathbf{e}}'\cdot\mathbf{d}_{\nu\nu'}\ \hat{\mathbf{e}}\cdot\mathbf{d}_{\nu'\nu_0}}{\omega-\omega_{\nu'\nu_0}}\right)\right|^2(n'+1), \tag{5.34}$$

where ε_∞ is the optical dielectric constant of the medium.

C. The high frequency limit

We investigate the form of the cross section in Eq. (5.34) when ω is sufficiently large so that most of the scattering amplitude originates from matrix elements of the electric dipole moment between states of energies much less than $\hbar\omega$. To be true the sum over intermediate states extends to higher energy states. The high frequency limit, to be derived below, is valid when the matrix elements $\mathbf{d}_{\nu'\nu_0}$ and $\mathbf{d}_{\nu\nu'}$ decrease rapidly with increasing $E_{\nu'}$. We also suppose that $\omega_{\nu\nu_0} \ll \omega$.

Under these conditions we separate the sum over ν' into two parts, the first, which we suppose to be predominant extends over states such that $E_{\nu'} - E_\nu \ll \hbar\omega$ while the second includes all other intermediate states. These may become important when $\omega \sim \omega_{\nu'\nu_0}$, i.e., in the resonant Raman effect. We now obtain an expansion of the cross section (5.34) in powers of ω^{-1}. We replace $\omega' + \omega_{\nu'\nu_0}$ with $\omega - \omega_{\nu\nu'}$ and expand to obtain

$$\begin{aligned}\frac{d\sigma}{d\Omega'} &= \frac{\varepsilon_\infty^2 \omega'^3}{\omega\hbar^2 c^4}(n'+1)\Bigg|\sum_{\nu'}\Bigg(\left(1+\frac{\omega_{\nu\nu'}}{\omega}+\frac{{\omega_{\nu\nu'}}^2}{\omega^2}+\cdots\right)\hat{\mathbf{e}}\cdot\mathbf{d}_{\nu\nu'}\;\hat{\mathbf{e}}'\cdot\mathbf{d}_{\nu'\nu_0} \\ &\quad -\left(1+\frac{\omega_{\nu'\nu_0}}{\omega}+\frac{{\omega_{\nu'\nu_0}}^2}{\omega^2}+\cdots\right)\hat{\mathbf{e}}'\cdot\mathbf{d}_{\nu\nu'}\;\hat{\mathbf{e}}\cdot\mathbf{d}_{\nu'\nu_0}\Bigg)\Bigg|^2 .\end{aligned}$$

Now

$$\sum_{\nu'}\left(\hat{\mathbf{e}}\cdot\mathbf{d}_{\nu\nu'}\;\hat{\mathbf{e}}'\cdot\mathbf{d}_{\nu'\nu_0}-\hat{\mathbf{e}}'\cdot\mathbf{d}_{\nu\nu'}\;\hat{\mathbf{e}}\cdot\mathbf{d}_{\nu'\nu_0}\right)=0$$

if we extend the sum over ν' over all states of the scattering system. The next term in the expansion of the amplitude has already been calculated in Sec. B. It is

$$-\frac{\hbar\;\hat{\mathbf{e}}\cdot\hat{\mathbf{e}}'}{\omega}\delta_{\nu\nu 0}\sum_i\frac{q^2}{m_i} .$$

Thus

$$\begin{aligned}\frac{d\sigma}{d\Omega'} &= \frac{\varepsilon_\infty^2(n'+1)}{\hbar^2 c^4}\left(\frac{\omega'}{\omega}\right)^3\Bigg|-\hbar\;\hat{\mathbf{e}}'\cdot\hat{\mathbf{e}}\sum_i\frac{q_i^2}{m_i}\,\delta_{\nu\nu_0} \\ &\quad -\frac{1}{\omega}\sum_{\nu'}\left({\omega_{\nu\nu'}}^2\,\hat{\mathbf{e}}\cdot\mathbf{d}_{\nu\nu'}\;\hat{\mathbf{e}}'\cdot\mathbf{d}_{\nu'\nu_0}-{\omega_{\nu'\nu_0}}^2\,\hat{\mathbf{e}}\cdot\mathbf{d}_{\nu\nu'}\;\hat{\mathbf{e}}'\cdot\mathbf{d}_{\nu'\nu_0}\right) \\ &\quad +\cdots\Bigg|^2 .\end{aligned} \tag{5.35}$$

Thus, neglecting all but the first term in the expansion we obtain

$$\frac{d\sigma}{d\Omega'}=\frac{\varepsilon_\infty^2\,(n'+1)}{c^4}\,(\hat{\mathbf{e}}\cdot\hat{\mathbf{e}}')^2\left|\sum_i\frac{q_i^2}{m_i}\right|^2 . \tag{5.36}$$

This is the result one obtains taking only the diamagnetic term in the $\mathbf{A}\cdot\mathbf{p}$ approach described in detail in Sec. A. (see Eq. (5.21) in the case $\exp(i\;\mathbf{q}\cdot\mathbf{r})\approx 1$.) For scattering by a free electron we obtain (set $n'=0$)

$$\frac{d\sigma}{d\Omega'}=\left(\frac{e^2}{mc^2}\right)^2(\hat{\mathbf{e}}\cdot\hat{\mathbf{e}}')^2 . \tag{5.37}$$

For an incident unpolarized beam of angular frequency ω and scattering at angle θ

$$\frac{d\sigma}{d\Omega'}=\frac{1}{2}\left(\frac{e^2}{mc^2}\right)^2(1+\cos^2\theta) \tag{5.38}$$

so that

$$\sigma = \frac{8\pi}{3}\left(\frac{e^2}{mc^2}\right)^2, \tag{5.39}$$

the well known Thomson scattering cross section of X-rays by free electrons. Equation (5.38) is obtained by adding the contributions for two orthogonal polarization $\hat{\mathbf{e}}'$ for given $\hat{\mathbf{e}}$. For each choice of $\hat{\mathbf{e}}$ only one polarization of the scattered radiation occurs, hence the factor of 1/2 in Eq. (5.38).

D. Further discussion of the relation between the $\mathbf{A}\cdot\mathbf{p}$ and $\mathbf{d}\cdot\mathbf{E}$ interactions in the electric dipole approximation

In the long wavelength limit we can replace the vector potential of the radiation field, given in Eq. (5.7), with

$$\mathbf{A} = c\sum_{q\mu}\left(\frac{2\pi\hbar}{V\omega_q}\right)^{1/2}\hat{\mathbf{e}}_{q\mu}\,(a_{q\mu}+a^\dagger_{q\mu}). \tag{5.40}$$

In this form $\mathbf{A}$ is independent of position within the diameter of the scattering system, i.e., we can regard the fields as uniform in space. The interaction between the scattering system and the radiation is specified by the Hamiltonian (5.2) now in the form

$$\mathrm{H} = \sum_{i=1}^{N}\frac{1}{2m_i}\left(\mathbf{p}_i - \frac{q_i}{c}\mathbf{A}(\mathbf{r}_i)\right)^2 + \mathrm{U} + \mathrm{H}_R \tag{5.41}$$

with

$$\mathrm{H}_R = \sum_{q\mu}\hbar\omega_q\left(a^\dagger_{q\mu}a_{q\mu}+\frac{1}{2}\right). \tag{5.42}$$

We perform a gauge transformation using the unitary operator

$$\mathrm{S} = \exp\Bigl(-\frac{i}{\hbar c}\sum_{i=1}^{N}q_i\,\mathbf{r}_i\cdot\mathbf{A}\Bigr) = \exp\Bigl(-\frac{i}{\hbar c}\mathbf{d}\cdot\mathbf{A}\Bigr). \tag{5.43}$$

Then, if E_ν and $\mid \nu >$ are the eigenvalues and eigenvectors of the scattering system, i.e., if

$$\mathrm{H}_S \mid \nu > = E_\nu \mid \nu >, \tag{5.44}$$

then

$$\mathrm{S\,H\,S^{-1}S} \mid \nu\ > = E_\nu\,\mathrm{S} \mid \nu\ > . \tag{5.45}$$

We denote $\mathrm{S\,H\,S^{-1}}$ by $\tilde{\mathrm{H}}_S$ and $\mathrm{S}\mid \nu >$ by $\mid \tilde{\nu} >$ so that the Schrödinger equation is equivalent to

$$\tilde{\mathrm{H}}_S \mid \tilde{\nu} > = E_\nu \mid \tilde{\nu} > . \tag{5.46}$$

Now, since S depends only on $\sum_i q_i \mathbf{r}_i$ and $a_{q\mu}$ and $a^\dagger_{q\mu}$ it commutes with all position operators. The momentum operators are transformed as follows:

$$\begin{aligned} \mathrm{S}\,\mathbf{p}_i\,\mathrm{S}^{-1} &= \exp\left(-\frac{i}{\hbar c}\mathbf{d}\cdot\mathbf{A}\right)\,\mathbf{p}_i\,\exp\left(\frac{i}{\hbar c}\mathbf{d}\cdot\mathbf{A}\right) \\ &= \mathbf{p}_i + \exp\left(-\frac{i}{\hbar c}\mathbf{d}\cdot\mathbf{A}\right)\left[\mathbf{p}_i,\ \exp\left(\frac{i}{\hbar c}\mathbf{d}\cdot\mathbf{A}\right)\right]. \end{aligned}$$

But

$$\begin{aligned} \left[\boldsymbol{p}_i, \exp\left(\frac{i}{\hbar c}\boldsymbol{d}\cdot\boldsymbol{A}\right)\right] &= \left[\boldsymbol{p}_i, \exp\left(\frac{iq_i}{\hbar c}\boldsymbol{r}_i\cdot\boldsymbol{A}\right)\right]\exp\left(\frac{i}{\hbar c}\sum_{j\neq i} q_j\boldsymbol{r}_j\cdot\boldsymbol{A}\right) \\ &= -i\hbar\exp\left(\frac{iq_i}{\hbar c}\boldsymbol{r}_i\cdot\boldsymbol{A}\right)\left(\frac{iq_i}{\hbar c}\boldsymbol{A}\right)\exp\left(\frac{i}{\hbar c}\sum_{j\neq i} q_j\boldsymbol{r}_j\cdot\boldsymbol{A}\right) \\ &= \frac{q_i}{c}\boldsymbol{A} \end{aligned}$$

Since $\mathbf{A}$ commutes with S

$$\mathrm{S}\left(\mathbf{p}_i - \frac{q_i}{c}\mathbf{A}\right)\mathrm{S}^{-1} = \mathbf{p}_i, \tag{5.47}$$

and, consequently,

$$\frac{1}{2m_i}\,\mathrm{S}\sum_{i=1}^{N}\left(\mathbf{p}_i - \frac{q_i}{c}\mathbf{A}\right)^2\mathrm{S}^{-1} = \sum_{i=1}^{N}\frac{\mathbf{p}_i^2}{2m_i}. \tag{5.48}$$

In general

$$\left[a_{q\mu},\ a^\dagger_{q'\mu'}\right] = \delta_{qq'}\delta_{\mu\mu'}. \tag{5.49}$$

For any photon mode, say one corresponding to the operators a and $a^\dagger$ we have

$$\left[a,\ a^\dagger\right] = 1.$$

Then

$$\left[a^2,\ a^\dagger\right] = 2a,$$

$$\left[a^n,\ a^\dagger\right] = na^{n-1},$$

In a similar way

$$\left[a,\ a^{\dagger^n}\right] = na^{\dagger^{n-1}},$$

If $F(a, a^\dagger)$ is an arbitrary function of a and $a^\dagger$

$$\left[a,\ F(a,a^\dagger)\right] = \frac{\partial F}{\partial a^\dagger} \tag{5.50}$$

and

$$\left[a^\dagger, F(a,a^\dagger)\right] = -\frac{\partial F}{\partial a}\,. \tag{5.51}$$

Using these results we obtain

$$\left[a_{q\mu},\ \mathrm{S}^{-1}\right] = \mathrm{S}^{-1}\ i\ \mathbf{d}\cdot\hat{\mathbf{e}}_{q\mu}\left(\frac{2\pi}{V\hbar\omega_q}\right)^{1/2} \tag{5.52}$$

and

$$\left[a^\dagger_{q\mu},\ \mathrm{S}^{-1}\right] = -\mathrm{S}^{-1}\ i\ \mathbf{d}\cdot\hat{\mathbf{e}}_{q\mu}\left(\frac{2\pi}{V\hbar\omega_q}\right)^{1/2}\,. \tag{5.53}$$

We are now in a position to calculate $\tilde{\mathrm{H}}_R$

$$\widetilde{H}_R = \mathrm{S}\ \mathrm{H}_R\ \mathrm{S}^{-1} = \sum_{q\mu}\hbar\omega_{q\mu}\left(\mathrm{S}\ a^\dagger_{q\mu}a_{q\mu}\mathrm{S}^{-1} + \frac{1}{2}\right)\,. \tag{5.54}$$

But

$$\begin{aligned}\mathrm{S}\ a^\dagger_{q\mu}\ a_{q\mu}\ \mathrm{S}^{-1} &= a^\dagger_{q\mu}a_{q\mu} + \mathrm{S}\left[a^\dagger_{q\mu}a_{q\mu},\ \mathrm{S}^{-1}\right]\\ &= a^\dagger_{q\mu}a_{q\mu} + \mathrm{S}\ a^\dagger_{q\mu}\left[a_{q\mu},\ \mathrm{S}^{-1}\right] + \mathrm{S}[a^\dagger_{q\mu},\mathrm{S}^{-1}]a_{q\mu}\\ &= a^\dagger_{q\mu}a_{q\mu} + \left(\frac{2\pi}{V\hbar\omega_q}\right)^{1/2} i\ \mathbf{d}\cdot\hat{\mathbf{e}}_{q\mu}(\mathrm{S}\ a^\dagger_{q\mu}\mathrm{S}^{-1} - a_{q\mu})\end{aligned} \tag{5.55}$$

Furthermore

$$\begin{aligned}\mathrm{S}\ a^\dagger_{q\mu}\ \mathrm{S}^{-1} &= a^\dagger_{q\mu} + \mathrm{S}\left[a^\dagger_{q\mu},\ \mathrm{S}^{-1}\right]\\ &= a^\dagger_{q\mu} - i\ \mathbf{d}\cdot\hat{\mathbf{e}}_{q\mu}\left(\frac{2\pi}{V\hbar\omega_q}\right)^{1/2}\,.\end{aligned} \tag{5.56}$$

Combining Eqs. (5.55) and (5.56) we obtain

$$\mathrm{S}\ a^\dagger_{q\mu}\ a_{q\mu}\ \mathrm{S}^{-1} = a^\dagger_{q\mu}a_{q\mu} + i\left(\frac{2\pi}{V\hbar\omega_q}\right)^{1/2}\mathbf{d}\cdot\hat{\mathbf{e}}_{q\mu}(a^\dagger_{q\mu} - a_{q\mu}) + \frac{2\pi}{V\hbar\omega_q}(\mathbf{d}\cdot\hat{\mathbf{e}}_{q\mu})^2. \tag{5.57}$$

The electric field is obtained from

$$\mathbf{E} = -\frac{1}{c}\frac{\partial\mathbf{A}}{\partial t},$$

where it is understood that the time derivative of any operator F is obtained from the quantum-mechanical equation of motion

$$\frac{dF}{dt} = \frac{1}{i\hbar}\left[F,\, \mathrm{H}_R\right].$$

Thus

$$\begin{aligned}\mathbf{E} &= \frac{i}{\hbar c}\left[c\sum_{q\mu}\left(\frac{2\pi\hbar}{V\omega_q}\right)^{1/2}\hat{e}_{q\mu}(a^{\dagger}_{q\mu}+a_{q\mu}),\ \sum_{q'\mu'}\hbar\omega_{q'\mu'}\left(a^{\dagger}_{q'\mu'}a_{q'\mu'}+\frac{1}{2}\right)\right]\\ &= i\sum_{q\mu}\left(\frac{2\pi\hbar\omega_q}{V}\right)^{1/2}\hat{e}_{q\mu}(a_{q\mu}-a^{\dagger}_{q\mu}).\end{aligned} \tag{5.58}$$

Combining Eqs. (5.54), (5.57), and (5.58) we obtain

$$\tilde{\mathrm{H}}_R = \mathrm{H}_R - \mathbf{d}\cdot\mathbf{E} + \frac{2\pi}{V}\sum_{q\mu}(\mathbf{d}\cdot\hat{e}_{q\mu})^2. \tag{5.59}$$

Finally,

$$\tilde{\mathrm{H}} = \sum_{i=1}^{N}\frac{\mathbf{p}_i^2}{2m_i} + \mathrm{U} + \mathrm{H}_R - \mathbf{d}\cdot\mathbf{E} + \frac{2\pi}{V}\sum_{q\mu}(\mathbf{d}\cdot\hat{e}_{q\mu})^2. \tag{5.60}$$

The quantity

$$\frac{2\pi}{V}\sum_{q\mu}(\mathbf{d}\cdot\hat{e}_{q\mu})^2$$

is a self-energy of the electric dipole moment. The important point is that the only term in $\tilde{\mathrm{H}}$ containing the field operators of the radiation is $\mathbf{d}\cdot\mathbf{E}$ which is linear in the creation and destruction operators for photons. Thus, in order to evaluate the amplitude of a Raman process we take the $-\mathbf{d}\cdot\mathbf{E}$ interaction in second order perturbation theory. Therefore, there is no term similar to the diamagnetic contribution in the $\mathbf{A}\cdot\mathbf{p}$ approach. Of course, as already shown in Sec. B, the two procedures yield identical results in the electric dipole approximation valid when the wavelength of the radiation is much larger than the maximum diameter of the scattering system.

E. Selection rules for Raman scattering in the electric dipole approximation

Equation (5.34) can be rewritten as

$$\frac{d\sigma}{d\Omega'} = \left(\frac{e^2\epsilon_\infty}{mc^2}\right)^2 |<\nu|\hat{e}'\cdot\boldsymbol{\alpha}\cdot\hat{e}|\nu_0>|^2(n'+1) \tag{5.61}$$

where $<\nu|\boldsymbol{\alpha}|\nu_0>$ is a second rank tensor whose components are

$$
\begin{aligned}
<\nu|\alpha_{ij}|\nu_0> &= \frac{m(\omega'^3\omega)^{\frac{1}{2}}}{e^2}\sum_{\nu'}\left(\frac{<\nu|d_i|\nu'><\nu'|d_j|\nu_0>}{E_{\nu'}-E_{\nu_0}-\hbar\omega}\right. \\
&+ \left.\frac{<\nu|d_j|\nu'><\nu'|d_i|\nu_0>}{E_{\nu'}-E_\nu+\hbar\omega}\right) .
\end{aligned} \tag{5.62}
$$

We introduce the vector operator $\boldsymbol{b}$ satisfying the condition

$$[\boldsymbol{b}, H_S] + \hbar\omega\boldsymbol{b} = \boldsymbol{d} . \tag{5.63}$$

Taking matrix elements of both sides of Eq. (5.63) between eigenstates of H_s we obtain

$$<\nu|\boldsymbol{b}|\nu'>= \frac{<\nu|\boldsymbol{d}|\nu'>}{E_{\nu'}-E_\nu+\hbar\omega} . \tag{5.64}$$

Substituting the result of Eq. (5.64) in Eq. (5.61) and using the completeness of the states $|\nu'>$ we obtain

$$<\nu|\alpha_{ij}|\nu_0>= \frac{m(\omega'^3\omega)^{\frac{1}{2}}}{e^2} <\nu|b_jd_i - d_ib_j|\nu_0> .$$

The operators d_i and b_j do not commute. Since $\boldsymbol{d}$ and $\boldsymbol{b}$ are polar vector operators α_{ij} is a second rank tensor. Furthermore, if the system characterized by the Hamiltonian H_S possesses a center of inversion, the states $|\nu>$ have a definite parity. Thus, the matrix elements of α_{ij} vanish unless the states $|\nu_0>$ and $|\nu>$ have the same parity. This selection rule shows that for systems with inversion symmetry Raman and absorption spectroscopies are complementary tools for the investigation of the properties of these systems. This property is called the "rule of mutual exclusion".

7.6 Symmetry classification of molecular and crystal vibrations

A conservative mechanical system characterized by coordinates $q_1, q_2, \ldots q_n$ obeys the equations of motion

$$\frac{d}{dt}\frac{\partial L}{\partial \dot{q}_i} - \frac{\partial L}{\partial q_i} = 0 \tag{6.1}$$

already discussed in Sec. 2.1. Let $\{q_i^0\} = \{\boldsymbol{q}^0\}$ be a position in coordinate space for which the potential $V(q_1, q_2, \ldots q_n)$ is a minimum. For small deviations from $\boldsymbol{q}^0$ we can expand the potential in a power series in $q_i - q_i^0$. We can, for simplicity, redefine q_i^0 to be zero and take $V(q_1^0, q_2^0 \ldots q_n^0)$ to be also zero. Then

$$
\begin{aligned}
V(q_1, q_2 \ldots q_n) &= \sum_{i,j}\frac{1}{2}q_iq_j\left(\frac{\partial^2 V}{\partial q_i \partial q_i}\right)_0 + \cdots \\
&= \sum_{i,j}\frac{1}{2}C_{ij}q_iq_j + \cdots
\end{aligned} \tag{6.2}
$$

where

$$\left(\frac{\partial^2 V}{\partial q_i \partial q_j}\right)_0 = C_{ij} \tag{6.3}$$

are the second derivatives of the potential evaluated at the point q^0 which we call the position of stable equilibrium. Linear terms in q_i in the expansion vanish identically since V is an extremum at equilibrium. The kinetic energy $T(\boldsymbol{q}, \dot{\boldsymbol{q}})$ can also be expanded in a power series in the coordinate q_i and the generalized velocities $\dot{q}_i$. Now, taking the positions $\boldsymbol{r}_i$ of the particles to be functions of the generalized coordinates but not explicitly of the time, the velocities $\dot{\boldsymbol{r}}_i$ are linear functions of the $\dot{q}_i$. Thus, keeping terms to second powers in small quantities we have

$$L = \frac{1}{2}\sum_{ij}(A_{ij}\dot{q}_i\dot{q}_j - C_{ij}q_iq_j)$$

where A_{ij} and C_{ij} are constants. The quadratic forms

$$\sum_{i,j} A_{ij}\dot{q}_i\dot{q}_j$$

and

$$\sum_{i,j} C_{ij}q_iq_j$$

are positive definite and the matrices (A_{ij}) and (B_{ij}) are symmetric. A theorem in linear algebra states that A_{ij} and B_{ij} can be diagonalized simultaneously by an orthogonal transformation S:

$$q_i = \sum_j S_{ij}Q_j$$

so that

$$L = \frac{1}{2}\sum_i(a_i\dot{Q}_i^2 - c_iq_i^2)$$

The equations of motion are

$$\ddot{Q}_i = -\omega_i^2 Q_i$$

where $\omega_i = (c_i/a_i)^{\frac{1}{2}}$. The coordinates Q_i are called the normal modes of the system. The general solution of the equations of motion is

$$q_i = \sum_j Q_j^0 S_{ij}\cos(\omega_j - t + \varphi_j)\ .$$

If $Q_i^0 = 0$ for all i except j, then

$$q_i(t) = Q_j^0 S_{ij}\cos(\omega_j t + \varphi_j)\ .$$

Suppose that the q_i are Cartesian components of the displacements of the particles composing the system. If a symmetry element R of the system in its

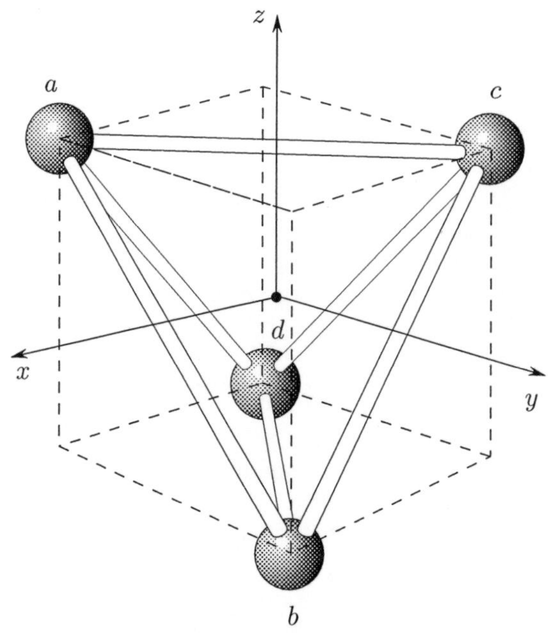

Figure 7.2

equilibrium state transforms it into itself the coordinates q_i are transformed into q_i' but both the kinetic and potential energies as functions of the q_i (and of the q_i') are invariant. The transformation generates a matrix whose dimension is three times the number of particles present. The set of all matrices thus obtained for all operations R of the group form a representation of the group, which is, in general, reducible.

As an example we consider the modes of vibration of the methane molecule CH_4. (See figure 7.2.) The symmetry group of transformation around the carbon atom is T_d. Since this molecule has five atoms the set of displacements of the atoms generate a fifteen-dimensional representation of the T_d. Each displacement by itself, being a vector, generates the Γ_5 representation of T_d. But a contribution to the trace of a particular operation R in the fifteen-dimensional representation only obtains for those atoms that remain fixed under R. The number of atoms of the methane molecule which remain fixed are 5 for E, 2 for C_3, 1 for C_2, 3 for σ_d and 1 for S_4. Thus the character of the representation Γ_t generated by the fifteen displacements is

E	$8C_3$	$3C_2$	$6\sigma_d$	$6S_4$
15	0	-1	3	-1

The reduction of Γ_t is

$$\Gamma_t = \Gamma_1 + \Gamma_3 + \Gamma_4 + 3\Gamma_5 \; .$$

However, a uniform displacement of all five atoms is not a vibration and generates the vector representation Γ_5. A rigid rotation of the molecule as whole corresponds to displacements generating the Γ_4 representation of T_d. Thus, the vibrational modes of the molecule are classified according to

$$\Gamma_v = \Gamma_1 + \Gamma_3 + 2\Gamma_5 \; .$$

This shows that the vibrational spectrum of methane has four different frequencies. Of these one is non-degenerate (Γ_1), one is doubly degenerate (Γ_3) and the two Γ_5 vibrations are, each, three-fold degenerate. Of these the two Γ_5 modes are infrared active and the Γ_1, Γ_3 and the two Γ_5 modes are Raman active. In fact, the Γ_5 modes transform as the components of a vector but also as the yz, zx and xy components of a second rank tensor. In the same manner the basis vectors Γ_1 and Γ_2 of the two-fold degenerate Γ_3 mode transform as additional components of a second rank tensor. We note that in this case the rule of mutual exclusion of infrared and Raman activity does not apply because of the absence of inversion symmetry.

The theory of group representations allows us to obtain additional selection rules for Raman transitions. Let $\alpha_{ij} \quad (i, j = x, y, z)$ be the components of the Raman tensor referred to as the cubic axes of the tetrahedral arrangement of the atoms in the methane molecule and let us consider Raman transitions from the Γ_1 ground state to excited states of symmetries Γ_1, Γ_3 or Γ_5. For a $\Gamma_1 \to \Gamma_1$ transition only the component $\alpha_{xx} + \alpha_{yy} + \alpha_{zz}$ of the Raman tensor is different from zero. Thus, $\alpha_{yz} = \alpha_{zx} = \alpha_{xy} = 0$, and $\alpha_{xx} = \alpha_{yy} = \alpha_{zz}$. Thus the Raman tensor for $\Gamma_1 \to \Gamma_1$ transitions is of the

$$\boldsymbol{\alpha} = \begin{pmatrix} a & 0 & 0 \\ 0 & a & 0 \\ 0 & 0 & a \end{pmatrix} .$$

This means that such transitions are observed only if the polarization of the scattered radiation is in the same direction as that of the incident radiation. For a $\Gamma_1 \to \Gamma_3$ Raman transition the Raman tensors are again diagonal. The Raman tensor for $\Gamma_1 \to \gamma_1$ transitions have components $\alpha_{xx} = \alpha_{yy}$ and $\alpha_{zz} = -2\alpha_{xx}$ because γ_1 behaves as $2z^2 - x^2 - y^2$ while γ_2 varies as $\sqrt{3}(x^2 - y^2)$. Thus, in this case

$$\boldsymbol{\alpha} = \begin{pmatrix} -b & 0 & 0 \\ 0 & -b & 0 \\ 0 & 0 & 2b \end{pmatrix}$$

In the same way, for a $\Gamma_1 \to \gamma_2$ transition

$$\boldsymbol{\alpha} = \begin{pmatrix} \sqrt{3} & 0 & 0 \\ 0 & -\sqrt{3}b & 0 \\ 0 & 0 & 0 \end{pmatrix}$$

by virtue of theorem 3.1 which, in the present context, implies that $< \gamma_1|2\alpha_{zz} - \alpha_{xx} - \alpha_{yy}|\Gamma_1 >=< \gamma_2|\sqrt{3}(\alpha_{xx} - \alpha_{yy})|\Gamma_1 >$. The Raman tensors for $\Gamma_1 \to \epsilon_i$ $(i = 1, 2, 3)$ transitions are

$$\boldsymbol{\alpha} = \begin{pmatrix} 0 & 0 & 0 \\ 0 & 0 & c \\ 0 & c & 0 \end{pmatrix} , \quad \boldsymbol{\alpha} = \begin{pmatrix} 0 & 0 & c \\ 0 & 0 & 0 \\ c & 0 & 0 \end{pmatrix} , \quad \boldsymbol{\alpha} = \begin{pmatrix} 0 & c & 0 \\ c & 0 & 0 \\ 0 & 0 & 0 \end{pmatrix} ,$$

respectively.

7.7 Application of the theory of group representations to the study of electron states in crystalline solids

Crystals possess translational symmetry. If we consider an infinite crystal we define the primitive translations $\boldsymbol{a}_1, \boldsymbol{a}_2$ and $\boldsymbol{a}_3$ as three non-coplanar vectors such that the necessary and sufficient condition for the crystal to remain invariant under a translation $\boldsymbol{n}$ is that $\boldsymbol{n}$ be expressible in the form

$$\boldsymbol{n} = n_1\boldsymbol{a}_1 + n_2\boldsymbol{a}_2 + n_3\boldsymbol{a}_3 \tag{7.1}$$

where n_1, n_2 and n_3 are integers. The operations $\{T_n\}$ defined by

$$T_n\psi(\boldsymbol{r}) = \psi(\boldsymbol{r} - \boldsymbol{n}) \tag{7.2}$$

form an Abelian group since

$$T_nT_{n'} = T_{n+n'} = T_{n'}T_n \ . \tag{7.3}$$

In addition to translations crystals possess symmetry properties such as rotations and improper rotations about a point. Let $\{R\}$ be the set of such operations. We have already seen that there are restrictions on the possible symmetry operations leaving a crystal invariant. We remark that the vectors $\boldsymbol{a}_1, \boldsymbol{a}_2$ and $\boldsymbol{a}_3$ can be selected in a number of different ways but the volume of the parallelepiped formed by them is always the same. This volume is called the volume of the primitive cell v_0 and is given by $v_0 = \boldsymbol{a}_1 \cdot (\boldsymbol{a}_2 \times \boldsymbol{a}_3)$. Here we label the primitive translations in such a manner that $v_0 > 0$. The group of orthogonal operations $\{R\}$ about a point is called the point group of the crystal if the origen is selected so that $\{R\}$ is maximal. We shall always assume that we can select this origen in this manner. The group $\{R\}$ is called the point group of the crystal.

The combination of R and T_n gives the transformation $\{R, \boldsymbol{n}\}$ such that $\boldsymbol{r} \to \boldsymbol{r}' = R\boldsymbol{r} + \boldsymbol{n}$, *i.e.*,

$$\{R, \boldsymbol{n}\}\psi(\boldsymbol{r}) = \psi(R^{-1}(\boldsymbol{r} - \boldsymbol{n})) \ . \tag{7.4}$$

Let $\{S, \boldsymbol{n}'\}$ be a second operation consisting of a translation by $\boldsymbol{n}'$ and a transformation S about a fixed point. Then

$$\{R, \boldsymbol{n}\}\{S, \boldsymbol{n}\}\boldsymbol{r} = \{R, \boldsymbol{n}\}(S\boldsymbol{r} + \boldsymbol{n}') = RS\boldsymbol{r} + R\boldsymbol{n}' + \boldsymbol{n} \ . \tag{7.5}$$

We note that

$$\{S, \boldsymbol{n}'\}\{R, \boldsymbol{n}\}\boldsymbol{r} = SR\boldsymbol{r} + S\boldsymbol{n} + \boldsymbol{n}' \ , \tag{7.6}$$

so that

$$\{R, \boldsymbol{n}\}\{S, \boldsymbol{n}'\} = \{RS, R\boldsymbol{n}' + \boldsymbol{n}\} \ . \tag{7.7}$$

Comparison of Eqs (7.5) and (7.6) shows that

$$\{R, \boldsymbol{n}\}\{S, \boldsymbol{n}'\} \neq \{S, \boldsymbol{n}'\}\{R, \boldsymbol{n}\}$$

in general.

The inverse of $\{R, \boldsymbol{n}\}$ is

$$\{R, \boldsymbol{n}\}^{-1} = \{R^{-1}, -R^{-1}\boldsymbol{n}\} \ . \tag{7.8}$$

Clearly the set of operations $\{R, \boldsymbol{n}\}$ forms a group. The set of operations leaving a crystal invariant is called the space group of the crystal. One may imagine that all space groups are of the form $\{\{R, \boldsymbol{n}\}\}$ but this is not the case. There are 73 space groups for which this is the case; these are called symorphic space groups. An example is the group of the zinc-blende structure. There are 157 non-symorphic space groups. Diamond has a non-symorphic space group.

In chapter 5 we introduced the reciprocal vectors $\boldsymbol{\tau}_1, \boldsymbol{\tau}_2$ and $\boldsymbol{\tau}_3$. It is convenient to describe the triply periodic function $f(\boldsymbol{r})$ with periods $\boldsymbol{a}_1, \boldsymbol{a}_2$ and $\boldsymbol{a}_3$ in a more compact fashion. An arbitrary point $\boldsymbol{r}$ is expressible as

$$\boldsymbol{r} = x_1\boldsymbol{a}_1 + x_2\boldsymbol{a}_2 + x_3\boldsymbol{a}_3 \tag{7.9}$$

where

$$x_i = \boldsymbol{\tau}_i \cdot \boldsymbol{r} \quad , \qquad i = 1, 2, 3 \ . \tag{7.10}$$

The function $f(\boldsymbol{r})$ can now be viewed as a function of x_1, x_2 and x_3, periodic in each of these variables with period unity. Thus, under suitable conditions of continuity and differentiability

$$f(\boldsymbol{r}) = \sum_{n_1=-\infty}^{\infty} \sum_{n_2=-\infty}^{\infty} \sum_{n_3=-\infty}^{\infty} f_{n_1 n_2 n_3} e^{2\pi i(n_1 x_1 + n_2 x_2 + n_3 x_3)} \tag{7.11}$$

but

$$n_1 x_1 + n_2 x_2 + n_3 x_3 = (n_1\boldsymbol{\tau}_1 + n_2\boldsymbol{\tau}_2 + n_3\boldsymbol{\tau}_3) \cdot \boldsymbol{r} \tag{7.12}$$

We define the set of vectors

$$\boldsymbol{G} = 2\pi(n_1\boldsymbol{\tau}_1 + n_2\boldsymbol{\tau}_2 + n_3\boldsymbol{\tau}_3) = n_1\boldsymbol{b}_1 + n_2\boldsymbol{b}_2 + n_3\boldsymbol{b}_3 \tag{7.13}$$

with $n_i = 0, \pm 1, \pm 1, \ldots$. We call these vectors the reciprocal vectors of the structure. In some instances it is preferable to use $\boldsymbol{b}_i = 2\pi \boldsymbol{\tau}_i$ while in others we prefer to use the $\boldsymbol{\tau}_i'$s. Equation (7.11) can be rewritten as

$$f(\boldsymbol{r}) = \sum_G f_G e^{i\mathbf{G}\mathbf{r}} \ . \tag{7.14}$$

The functions $\exp(i\mathbf{G}\cdot\mathbf{r})$ are orthogonal, *i.e.*,

$$\int_{(v_0)} e^{i(\mathbf{G}-\mathbf{G}')\cdot\mathbf{r}} d\boldsymbol{r} = v_0 \delta_{G,G'} \ . \tag{7.15}$$

Thus,

$$f_G = \frac{1}{v_0} \int_{v_0} f(\boldsymbol{r}) e^{-i\mathbf{G}\cdot\mathbf{r}} d\boldsymbol{r} \ . \tag{7.16}$$

It is often desirable to use finite crystals rather than infinite ones in theoretical descriptions. If the crystal is large and we are only concerned with bulk properties this leads to no errors. From the mathematical point of view we take a finite crystal containing the primitive cells within a parallelepiped of edges $N_0\boldsymbol{a}_1, N_0\boldsymbol{a}_2, N_0\boldsymbol{a}_3$ where N_0 is a large integer. We now imagine replicas of this crystal so arranged that they, together with the initial crystal, fill the whole space. We suppose that all properties of the extended system, so constructed, are the same as viewed from $\boldsymbol{r}$ as from $N_0\boldsymbol{n} + \boldsymbol{r}$ where $\boldsymbol{n}$ is an arbitrary translational of the lattice. We assume, therefore, that $T_n^{N_0} \equiv E$, the unit operation. Thus

$$T_{a_i}^{N_0} = E \quad , \quad i = 1, 2, 3 \ . \tag{7.17}$$

With this assumption, $\{T_n\}$ is a finite Abelian group of order $N = N_0^3$.

These considerations allow us to obtain the irreducible representations of the translation group. Since it is an Abelian group of order N, its elements are each in a separate conjugate class. Thus, there are N inequivalent irreducible representations all of which are one-dimensional. If $e^{-i\kappa_i}$ is the character associated with T_{a_i}, then, since $T_{a_i}^{N_0} = E$, we have

$$e^{-iN_0\kappa_i} = 1$$

so that

$$\kappa_i = \frac{2\pi\ell_i}{N_0}$$

where ℓ_i is an integer. The character of the element $T_n = \prod_{i=1}^3 (T_{a_i})^{n_i}$ is

$$e^{-i(n_1\kappa_1 + n_2\kappa_2 + n_3\kappa_3)} \ .$$

We now define the vector

$$\boldsymbol{k} = \kappa_1\boldsymbol{\tau}_1 + \kappa_2\boldsymbol{\tau}_2 + \kappa_3\boldsymbol{\tau}_3 = \frac{1}{2\pi}(\kappa_1\boldsymbol{b}_1 + \kappa_2\boldsymbol{b}_2 + \kappa_3\boldsymbol{b}_3)$$

and write

$$n_1\kappa_1 + n_2\kappa_2 + n_3\kappa_3 = \boldsymbol{k}\cdot\boldsymbol{n}\ .$$

Thus, to each irreducible representation we associate a vector $\boldsymbol{k}$ or, equivalently, the set of three numbers $\kappa_1, \kappa_2, \kappa_3$. In other words, the vectors $\boldsymbol{k}$ define the irreducible representations of the group. The character of T_n in the representation Γ_k associated with $\boldsymbol{k}$ is

$$\chi_k(\boldsymbol{n}) = e^{-i\mathbf{k}\cdot\mathbf{n}}\ . \tag{7.18}$$

We note that the vectors $\boldsymbol{k}$ and $\boldsymbol{k}'$ differing by a vector $\boldsymbol{G}$ of the reciprocal lattice give the same representation, *i.e.*,

$$\chi_{k'}(\boldsymbol{n}) = e^{-ik'\cdot\mathbf{n}} = e^{-i(\mathbf{k}+\mathbf{G})\cdot\mathbf{n}} = \chi_k(\boldsymbol{n})\ .$$

Therefore, to avoid redundancy we restrict the values of κ_i to the ranges $0 \leq \kappa_i < 2\pi$ or to the following values of $\boldsymbol{k}$:

$$\boldsymbol{k} = \frac{1}{N_0}(\ell_1\boldsymbol{b}_1 + \ell_2\boldsymbol{b}_2 + \ell_3\boldsymbol{b}_3) \tag{7.19}$$

with

$$\ell_i = 0, 1, 2, \ldots N_0 - 1\ .$$

Thus, there are $N = N_0^3$ distinct values of $\boldsymbol{k}$. Instead of restricting the values of $\boldsymbol{k}$ to the ranges

$$0 \leq \boldsymbol{k}\cdot\boldsymbol{a}_i < 2\pi \tag{7.20}$$

it is often preferred to use the ranges determined by

$$-\pi \leq \boldsymbol{k}\cdot\boldsymbol{a}_i < \pi\ ,\ i = 1, 2, 3\ . \tag{7.21}$$

The set of values of $\boldsymbol{k}$ satisfying these conditions is called the fundamental Brillouin zone.

The character table of the group of translations can be written in the form

$$\begin{array}{cc} & T_n \\ \Gamma_k & e^{-i\mathbf{k}\cdot\mathbf{n}} \end{array} \tag{7.22}$$

That these representations are irreducible follows from

$$\sum_n |e^{-i\mathbf{k}\cdot\mathbf{n}}|^2 = N\ . \tag{7.23}$$

Their non-equivalence is a consequence of the orthogonality relation

$$\sum_n e^{i\mathbf{k}'\cdot\mathbf{n}}e^{-i\mathbf{k}\cdot\mathbf{n}} = 0 \tag{7.24}$$

for $\boldsymbol{k}$ and $\boldsymbol{k}'$ distinct ($\boldsymbol{k}$ and $\boldsymbol{k}'$ are regarded as equal if they differ by a vector $\boldsymbol{G}$ of the reciprocal lattice).

The Floquet-Bloch Theorem
(G. Floquet, Ann. Ecole Normale (Paris) **12**, 47 (1883)
F. Bloch, Zeit. für Physik, **52**, 555 (1928)).

The Floquet-Bloch theorem is now a simple consequence of the results outlined above. A state belonging to Γ_k satisfies

$$T_n\psi(\boldsymbol{r}) = e^{-i\mathbf{k}\cdot\mathbf{n}}\psi(\boldsymbol{r}) \; . \tag{7.25}$$

There are, in general, an infinite number of states belonging to Γ_k. We will designate them by $\psi_{\nu k}(\boldsymbol{r})$ where ν is simply a way of numbering these states. They obey a Schrödinger equation of the form

$$H_0\psi_{\nu k}(\boldsymbol{r}) = E_\nu(\boldsymbol{k})\psi_{\nu k}(\boldsymbol{r}) \tag{7.26}$$

with

$$H_0 = \frac{p^2}{2m} + U(\boldsymbol{r}) \tag{7.27}$$

where $U(\boldsymbol{r})$ is the potential energy of an electron in the crystal, $U(\boldsymbol{r})$ being periodic with the period of the lattice. Thus

$$[H_0, T_n] = 0 \; . \tag{7.28}$$

We have

$$T_n\psi_{\nu k}(\boldsymbol{r}) = \psi_{\nu k}(\boldsymbol{r} - \boldsymbol{n}) = e^{-i\mathbf{k}\cdot\boldsymbol{n}}\psi_{\nu k}(\boldsymbol{r}) \; . \tag{7.29}$$

Thus

$$u_{\nu k}(\boldsymbol{r}) = e^{-i\mathbf{k}\cdot\mathbf{r}}\psi_{\nu k}(\boldsymbol{r}) \tag{7.30}$$

is a periodic function with periods $\boldsymbol{a}_1, \boldsymbol{a}_2, \boldsymbol{a}_3$. In fact

$$u_{\nu k}(\boldsymbol{r} + \boldsymbol{n}) = e^{-i\mathbf{k}\cdot(\mathbf{r}+\mathbf{n})}\psi_{\nu k}(\boldsymbol{r} + \boldsymbol{n}) = u_{\nu k}(\boldsymbol{r}) \; . \tag{7.31}$$

It is often convenient to normalize $\psi_{\nu k}(\boldsymbol{r})$ to unity over the volume $V = Nv_0$ of the crystal and $u_{\nu k}(\boldsymbol{r})$ over v_0. Under such conditions

$$\psi_{\nu k}(\boldsymbol{r}) = N^{-\frac{1}{2}}e^{i\mathbf{k}\cdot\mathbf{r}}u_{\nu k}(\boldsymbol{r}) \; ; \tag{7.32}$$

and

$$\begin{aligned}
\int\limits_{(V)} \psi^*_{\nu' k'}(\boldsymbol{r})\psi_{\nu k}(\boldsymbol{r})d\boldsymbol{r} &= \frac{1}{N}\int\limits_{(V)} e^{i(\mathbf{k}-\mathbf{k}')\cdot\mathbf{r}}u^*_{\nu' k'}(\boldsymbol{r})u_{\nu k}(\boldsymbol{r})d\boldsymbol{r} \\
&= \frac{1}{N}\int\limits_{(v_n)} e^{i(\mathbf{k}-\mathbf{k}')\cdot\mathbf{r}}u^*_{\nu' k'}(\boldsymbol{r})u_{\nu k}(\boldsymbol{r})d\boldsymbol{r} \\
&= \frac{1}{N}\sum_n e^{i(\mathbf{k}-\mathbf{k}')\cdot\mathbf{n}}\int\limits_{(v_0)} e^{i(\mathbf{k}-\mathbf{k}')\cdot\mathbf{r}}u^*_{\nu' k'}(\boldsymbol{r})u_{\nu k}(\boldsymbol{r})d\boldsymbol{r}
\end{aligned}$$

where we changed $\boldsymbol{r}$ into $\boldsymbol{r}+\boldsymbol{n}$ thus reducing all integrals to one over the volume v_0 of the primitive cell. The sum over the lattice sites yields $\delta_{k'k}$, *i.e.*, unity if $\boldsymbol{k}$ and $\boldsymbol{k}'$ are equivalent and zero otherwise. Thus

$$\int_{(V)} \psi^*_{\nu' k'}(\boldsymbol{r})\psi_{\nu k}(\boldsymbol{r})d\boldsymbol{r} = \delta_{k'k}\int_{(v_0)} u^*_{\nu' k}(\boldsymbol{r})u_{\nu k}(\boldsymbol{r})d\boldsymbol{r} \ . \tag{7.33}$$

The function $u_{\nu k}(\boldsymbol{r})$ satisfies the equation

$$H(\boldsymbol{k})u_{\nu k}(\boldsymbol{r}) = E_\nu(\boldsymbol{k})u_{\nu k}(\boldsymbol{r}) \tag{7.34}$$

where

$$H(\boldsymbol{k}) = H_0 + \frac{\hbar}{m}\boldsymbol{k}\cdot\boldsymbol{p} + \frac{\hbar^2 k^2}{2m} \tag{7.35}$$

with the boundary condition that $u_{\nu k}(\boldsymbol{r})$ be periodic with the period of the lattice. For each $\boldsymbol{k}$, Eqs. (7.34) and (7.35) constitute an eigenvalue problem. $H(\boldsymbol{k})$ is Hermitian and every solution of $H(\boldsymbol{k})u_{\nu k}(\boldsymbol{r}) = E_\nu(\boldsymbol{k})u_{\nu k}(\boldsymbol{r})$ can be expressed in terms of an orthonormal set. We suppose this has been done. It is, of course, automatic for different eigenvalues.

Perturbation Theory

So far we have only considered the translational frequency. If that were the only symmetry of the system, without taking account of time reversal symmetry, all energy levels would be non-degenerate. The Hamiltonian H_0 is real so that it obeys time reversal invariance, $[H_0, T] = 0$. Thus, the time reversed of $\psi_{\nu k}(\boldsymbol{r})$ is

$$\psi^*_{\nu k}(\boldsymbol{r}) = N^{-\frac{1}{2}}e^{-i\mathbf{k}\cdot\mathbf{r}}u^*_{\nu k}(\boldsymbol{r}) \ . \tag{7.36}$$

This state belongs to Γ_{-k} so that for every state with wave vector $\boldsymbol{k}$ there is one at $-\boldsymbol{k}$ having the same energy. Thus $E_\nu(\boldsymbol{k}) = E_\mu(-\boldsymbol{k})$ where the index μ need not be equal to ν. The existence of the point group symmetry will introduce additional degeneracies as we shall see.Now we review certain results of perturbation theory as they apply to the Bloch states $\psi_{\nu k}(\boldsymbol{r})$.

Before proceeding we investigate the form of the matrix elements of the momentum operator between Bloch states:

$$\begin{aligned}\int_{(V)} \psi^*_{\nu' k'}(\boldsymbol{r})\boldsymbol{p}\psi_{\nu k}(\boldsymbol{r})d\boldsymbol{r} &= \frac{1}{N}\int_{(V)} e^{i(\mathbf{k}-\mathbf{k}')\cdot\mathbf{r}}u^*_{\nu' k'}(\boldsymbol{r})(\boldsymbol{p}+\hbar\boldsymbol{k})u_{\nu k}(\boldsymbol{r})d\boldsymbol{r} \\ &= \frac{1}{N}\sum_n e^{i(\mathbf{k}-\mathbf{k}')\cdot\mathbf{n}}\int_{(v_0)} u^*_{\nu' k'}(\boldsymbol{r})(\boldsymbol{p}+\hbar\boldsymbol{k})u_{\nu k}(\boldsymbol{r})d\boldsymbol{r} \ , \\ &= \delta_{k'k}\int_{(v_0)} u^*_{\nu' k}(\boldsymbol{r})(\boldsymbol{p}+\hbar\boldsymbol{k})u_{\nu k}(\boldsymbol{r})d\boldsymbol{r} \ .\end{aligned}$$

We use the notation

$$\int_{(v_0)} u^*_{\nu' k}(\boldsymbol{r})\boldsymbol{p}u_{\nu k}(\boldsymbol{r})d\boldsymbol{r} = \boldsymbol{p}_{\nu'\nu}(\boldsymbol{k}) \ . \tag{7.37}$$

Then

$$\int_{(V)} \psi^*_{\nu' k'}(\boldsymbol{r})\boldsymbol{p}\psi_{\nu k}(\boldsymbol{r})d\boldsymbol{r} = \delta_{k'k}\boldsymbol{p}_{\nu'\nu}(\boldsymbol{k}) + \delta_{k'k}\delta_{\nu'\nu}\hbar\boldsymbol{k} \ . \tag{7.38}$$

The expectation value of the momentum in the state $\psi_{\nu k}(\boldsymbol{r})$ is

$$\int_{(V)} \psi^*_{\nu k}(\boldsymbol{r})\boldsymbol{p}\psi_{\nu k}(\boldsymbol{r})d\boldsymbol{r} = \boldsymbol{p}_{\nu\nu}(\boldsymbol{k}) + \hbar\boldsymbol{k} \ . \tag{7.39}$$

We note that

$$\frac{\partial H(\boldsymbol{k})}{\partial \boldsymbol{k}} = \frac{\hbar}{m}\boldsymbol{p} + \frac{\hbar^2\boldsymbol{k}}{m} \tag{7.40}$$

so that

$$\int_{(v_0)} u^*_{\nu k}\frac{\partial H(\boldsymbol{k})}{\partial \boldsymbol{k}}u_{\nu k}(\boldsymbol{r})d\boldsymbol{r} = \frac{\hbar}{m}\boldsymbol{p}_{\nu\nu}(\boldsymbol{k}) + \frac{\hbar^2\boldsymbol{k}}{m} \ . \tag{7.41}$$

By virtue of the Feynman-Hellmann theorem (actually stated first by Pauli in his Handbuch der Physik article)

$$\int_{(v_0)} u^*_{\nu k}\frac{\partial H(\boldsymbol{k})}{\partial \boldsymbol{k}}u_{\nu k}(\boldsymbol{r})d\boldsymbol{r} = \frac{\partial E_\nu(\boldsymbol{k})}{\partial \boldsymbol{k}} \tag{7.42}$$

Thus

$$\boldsymbol{p}_{\nu\nu}(\boldsymbol{k}) + \hbar\boldsymbol{k} = \frac{m}{\hbar}\frac{\partial E_\nu(\boldsymbol{k})}{\partial \boldsymbol{k}} \ . \tag{7.43}$$

The expectation value of the momentum is, thus,

$$\int_{(v)} \psi^*_{\nu k}(\boldsymbol{r})\boldsymbol{p}\psi_{\nu k}(\boldsymbol{r})d\boldsymbol{r} = \frac{m}{\hbar}\frac{\partial E_\nu(\boldsymbol{k})}{\partial \boldsymbol{k}} \ . \tag{7.44}$$

Let us consider now the point group $\{\boldsymbol{R}\}$ of the crystal. Applying the operation P_R to Eq. (7.34) we have

$$P_R H(\boldsymbol{k}) P_R^{-1} P_R u_{\nu k}(\boldsymbol{r}) = E_\nu P_R u_{\nu k}(\boldsymbol{r}) \ . \tag{7.45}$$

The Hamiltonian H_0 is invariant under P_R, *i.e.*,

$$P_R H_0 P_R^{-1} = H_0 \ , \tag{7.46}$$

but $(\hbar/m)\boldsymbol{k}\cdot\boldsymbol{p}$ is transformed according to

$$P_R\frac{\hbar}{m}\boldsymbol{k}\cdot\boldsymbol{p}P_R^{-1} = \frac{\hbar}{m}\boldsymbol{k}\cdot(R^{-1}\boldsymbol{p}) \tag{7.47}$$

so that Eq. (7.45) reads

$$\left(H_0 + \frac{\hbar}{m}\boldsymbol{k}\cdot(R^{-1}\boldsymbol{p}) + \frac{\hbar^2k^2}{2m}\right)u_{\nu k}(R^{-1}\boldsymbol{r}) = E_\nu(\boldsymbol{k})u_{\nu k}(R^{-1}\boldsymbol{r}) \ . \tag{7.48}$$

Now

$$\begin{aligned} \boldsymbol{k}\cdot(R^{-1}\boldsymbol{p}) &= \sum_i k_i(R^{-1}\boldsymbol{p})_i = \sum_{i,j} k_i(R^{-1})_{ij}p_j \\ &= \sum_{i,j} p_j R_{ji} k_i = \boldsymbol{p}\cdot(R\boldsymbol{k}) \,. \end{aligned}$$

Thus Eq. (7.48) becomes

$$\left(H_0 + \frac{\hbar}{m}(R\boldsymbol{k})\cdot\boldsymbol{p} + \frac{\hbar^2}{2m}(R\boldsymbol{k})^2\right) u_{\nu k}(R^{-1}\boldsymbol{r}) = E_\nu(\boldsymbol{k}) u_{\nu k}(R^{-1}\boldsymbol{r}) \,. \tag{7.49}$$

The function $u_{\nu k}(R^{-1}\boldsymbol{r})$ is periodic with the period of the lattice since $R^{-1}\boldsymbol{n}$ is, itself, a vector of the lattice. Thus, $u_{\nu k}(R^{-1}\boldsymbol{r})$ is an eigenvector of the effective Schrödinger equation for $R\boldsymbol{k}$ instead of $\boldsymbol{k}$. We can write

$$u_{\nu k}(R^{-1}\boldsymbol{r}) = u_{\nu', R\boldsymbol{k}}(\boldsymbol{r}) \,.$$

The function $u_{\nu', R\boldsymbol{k}}(\boldsymbol{r})$ has the same eigenvalue as $u_{\nu k}(\boldsymbol{r})$. Two situations may occur. To describe them we consider the figure formed by the totality of vectors $R\boldsymbol{k}$ as R ranges over the elements of the point group. The figure thus formed is called the star of $\boldsymbol{k}$. For a general point $\boldsymbol{k}$ in the Brillouin zone the set $\{R\boldsymbol{k}\}$ has as many vectors as there are elements in the point group $\{R\}$. In this case the state $u_{\nu k}(\boldsymbol{r})$ for fixed $\boldsymbol{k}$ is non-degenerate. It may be that the star of $\boldsymbol{k}$ is a smaller set, *i.e.*, some elements of $\{R\}$ may leave $\boldsymbol{k}$ invariant. The subgroup of $\{R\}$ leaving $\boldsymbol{k}$ invariant is called the group of $\boldsymbol{k}$. This group may have irreducible representations of dimension larger than one in which case there are degenerate levels occurring at the same value of $\boldsymbol{k}$. We say that two or more bands touch at those values of $\boldsymbol{k}$.

It has become traditional to adopt a nomenclature due to Bouckaert, Smoluchowski and Wigner to represent the different points in the Brillouin zone. This nomenclature can be found in their paper (Phys. Rev. **50**, 58 (1936)).

We shall discuss the zinc-blende structure as an example The structure is face-centered-cubic (fcc) with a basis of two distinct atoms at (0,0,0) and (a/4) (1,1,1). The point group if T_d and the space group is symorphic. The space group if T_d^2 ($F\bar{4}3m$). The fundamental Brillouin zone (BZ) is shown in figure 7.3.

The primitive vectors of the structure can be chosen as

$$\boldsymbol{a}_1 = \frac{a}{2}(\boldsymbol{e}_2 + \boldsymbol{e}_3), \boldsymbol{a}_2 = \frac{a}{2}(\boldsymbol{e}_3 + \boldsymbol{e}_1), \boldsymbol{a}_3 = \frac{a}{2}(\boldsymbol{e}_1 + \boldsymbol{e}_2),$$

where $\boldsymbol{e}_1, \boldsymbol{e}_2$ and $\boldsymbol{e}_3$ are unit vectors along the cubic axes. The volume of the primitive cell is

$$v_0 = \boldsymbol{a}_1 \cdot (\boldsymbol{a}_2 \times \boldsymbol{a}_3) = \frac{a^3}{4} \,.$$

The translation vectors of the structure are

$$\begin{aligned} \boldsymbol{n} &= n_1\boldsymbol{a}_1 + n_2\boldsymbol{a}_2 + n_3\boldsymbol{a}_3 = \frac{a}{2}[(n_2+n_3)\boldsymbol{e}_1 + (n_3+n_1)\boldsymbol{e}_2 + (n_1+n_2)\boldsymbol{e}_3] \\ &= \frac{a}{2}(\ell_1\boldsymbol{e}_1 + \ell_2\boldsymbol{e}_2 + \ell_3\boldsymbol{e}_3) \,. \end{aligned}$$

Here

$$\ell_1 = n_1 + n_3, \ell_2 = n_3 + n_1, \ell_3 = n_1 + n_2$$

and

$$\ell_1 + \ell_2 + \ell_3$$

is an even integer.

The reciprocal vectors, defined by $\boldsymbol{a}_i \cdot \boldsymbol{b}_j = 2\pi\delta_{ij}$ are

$$\boldsymbol{b}_1 = \frac{2\pi}{a}(-\boldsymbol{e}_1 + \boldsymbol{e}_2 + \boldsymbol{e}_3), \boldsymbol{b}_2 = \frac{2\pi}{a}(\boldsymbol{e}_1 - \boldsymbol{e}_2 + \boldsymbol{e}_3), \boldsymbol{b}_3 = \frac{2\pi}{a}(\boldsymbol{e}_1 + \boldsymbol{e}_2 - \boldsymbol{e}_3).$$

They form a bcc reciprocal lattice with a conventional cubic cell of side $(4\pi/a)$. In the extended zone scheme we shall show that $E_\nu(\boldsymbol{k})$ is a continuous function of $\boldsymbol{k}$.

First we describe a few points in the BZ of the fcc structure.

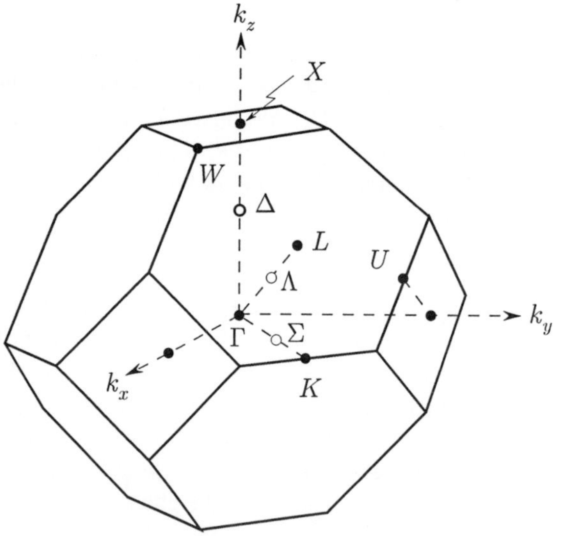

Figure 7.3

(i) Point Γ : $\boldsymbol{k} = 0$.

(ii) Point X : $k = (2\pi/a)$ (1,0,0) and its equivalents $(2\pi/a)(0,1,0)$, $(2\pi/a)$ (0,0,1). These are $\frac{1}{2}(\boldsymbol{b}_2 + \boldsymbol{b}_3)$, $\frac{1}{2}(\boldsymbol{b}_3 + \boldsymbol{b}_1)$, $\frac{1}{2}(\boldsymbol{b}_1 + \boldsymbol{b}_2)$. The points $(2\pi/a)$ $(-1,0,0)$, $(2\pi/a)$ $(0,-1,0)$, $(2\pi/a)$ $(0,0,-1)$ are equivalent to the above since the differ from them by $\boldsymbol{b}_2 + \boldsymbol{b}_3, \boldsymbol{b}_3 + \boldsymbol{b}_1$ and $\boldsymbol{b}_1 + \boldsymbol{b}_2$, respectively.

(iii) Point L : $\boldsymbol{k} = (\pi/a)\ (\hat{e}_1 + \hat{e}_2 + \hat{e}_3) = \frac{1}{2}(\boldsymbol{b}_1 + \boldsymbol{b}_2 + \boldsymbol{b}_3)$. The point $\boldsymbol{k}' = (\pi/a)(-\hat{e}_1 - \hat{e}_2 - \hat{e}_3)$ differs from $\boldsymbol{k}$ by a vector of the reciprocal lattice. The points $(\pi/a)\ (1,1,1)$, $(\pi/a)\ (1\bar{1}\bar{1})$, $(\pi/a)\ (\bar{1}1\bar{1})$, $(\pi/a)\ (\bar{1}\bar{1}1)$ are also L points not equivalent to $(\pi/a)(111)$. In fact, $(\pi/a)\ (111)$ - $(\pi/a)(1\bar{1}\bar{1}) = (2\pi/a)\ (011) = \boldsymbol{b}_1 + \frac{1}{2}(\boldsymbol{b}_2 + \boldsymbol{b}_3)$

(iv) Point K : $(3\pi/2a)(110)$,
There are 12 K points:
$(3\pi/2a)(\pm1, \pm1, 0), (3\pi/2a)(\pm1, 0, \pm1), (3\pi/2a)(0, \pm1, \pm1)$

The groups of these vectors are:

Group of Γ : The complete point group T_d

Group of X : $\{E, 2S_4, C_2, 2C_2', 2\sigma_d\} = D_{2d}$
$2S_4$ and C_2 are about k_z (see figure), $2C_2'$ are about k_x and k_y and the $2\sigma_d$ are dihedral planes bisecting the (k_x, k_y) plane.

Group of L : $\{E, 2C_3, 3\sigma_v\} = C_{3v}$
The $2C_3$'s are rotations by $\pm 2\pi/3$ about [111] and the $3\sigma_v$'s are dihedral mirror planes containing ΓL and k_x, ΓL and k_y, ΓL and k_z.

Group of K : $\{E, \sigma_d\} = C_s$

The groups of the points on the lines Λ and Σ are the same as those for L and K, respectively.

Group of Δ : $\{E, C_2, \sigma_v, \sigma_v'\} = C_{2v}$. C_2 is about k_z (see figure) and σ_v and σ_v' are the mirror planes containing k_z and bisecting the first (and third) and second (and fourth) quadrants of the (k_x, k_y) plane, respectively.

The character tables are given in the tables in Appendix A (p. 221).

In going from a point $\boldsymbol{k}_0$ whose group of $\boldsymbol{k}$ is G_{k_0} to a neighboring $\boldsymbol{k}$ whose group is G_k, in general, if $\boldsymbol{k}_0$ has higher symmetry than $\boldsymbol{k}$, G_k is a subgroup of G_{k_0}. If a level belongs to an irreducible representation Γ_{k_0} of G_{k_0}, the perturbation $H(\boldsymbol{k}) - H(\boldsymbol{k}_0)$ has the symmetry of G_k and a representation generated by the states of Γ_{k_0} is often reducible as a representation of G_k. Thus there are, in those cases, splittings of these levels. Their form is ascertained by performing the reduction of the states in Γ_{k_0} with respect to G_k. The results for the zinc-blende structure are shown in the form of the compatibility tables in Appendix A (p. 222).

For the zinc-blende structure there are several notations used which we now show in tabular form.

Koster	BSW	Chemical	Basis functions
Γ_1	Γ_1	A_1	$x^2+y^2+z^2, xyz$
Γ_2	Γ_2	A_2	
Γ_3	Γ_{12}	E	(u_1, u_2)
Γ_4	Γ_{15}	F_1	(S_1, S_2, S_3)
Γ_5	Γ_{25}	F_2	$(x, y, z), (yz, zx, xy)$

$u_1 = z^2 + \omega^2 x^2 + \omega y^2 = \frac{1}{2}(2z^2 - x^2 - y^2) - \frac{i}{2}\sqrt{3}(x^2 - y^2) = \frac{1}{2}(\gamma_1 - i\gamma_2)$
$u_2 = z^2 + \omega x^2 + \omega^2 y^2 = \frac{1}{2}(\gamma_1 + i\gamma_2)$.
S_1, S_2, S_3 are the components of a pseudo-vector. The functions $x(y^2 - z^2), y(z^2 - x^2), z(x^2 - y^2)$ generate the representation Γ_4

The notation for the group $O_h = O \times i$ is as follows

Koster	BSW	Chemical	Basis functions
Γ_1^+	Γ_1	A_{1g}	$x^2+y^2+z^2$
Γ_2^+	Γ_2	A_{2g}	$x^4(y^2-z^2)+y^4(z^2-x^2)+z^4(x^2-y^2)$
Γ_3^+	Γ_{12}	E_g	$2z^2-x^2-y^2$, $\sqrt{3}(x^2-y^2)$
Γ_4^+	Γ'_{15}	F_{1g}	$yz(y^2-z^2)$, $zx(z^2-x^2)$, $xy(x^2-y^2)$
Γ_5^+	Γ'_{25}	F_{2g}	yz , zx , xy
Γ_1^-	Γ'_1	A_{1u}	xyz
Γ_2^-	Γ'_2	A_{2u}	$(xyz)(\Gamma_2^+)$
Γ_3^-	Γ'_{12}	E_u	$(xyz)(\Gamma_3^+)$
Γ_4^-	Γ_{15}	F_{1u}	x, y, z
Γ_5^-	Γ_{25}	F_{2u}	$x(y^2-z^2), y(z^2-x^2), z(x^2-y^2)$

Perturbation Theory

(i) Let $\boldsymbol{k}_0$ be a point in the BZ and $E_\nu(\boldsymbol{k}_0)$ a non-degenerate level. For a point $\boldsymbol{k}$ near $\boldsymbol{k}_0$ we have the following Schrödinger equation for $E_\nu(\boldsymbol{k})$:

$$H(\boldsymbol{k})u_{\nu k} = E_\nu(k)u_{\nu k} \tag{7.50}$$

if we suppose

$$H(\boldsymbol{k}_0)u_{\nu k_0} = E_\nu(\boldsymbol{k}_0)u_{\nu k_0} \tag{7.51}$$

solved, then, to second order in

$$H' = H(\boldsymbol{k}) - H(\boldsymbol{k}_0) = \frac{\hbar}{m}(\boldsymbol{k} - \boldsymbol{k}_0)\cdot\boldsymbol{p} + \frac{\hbar^2}{2m}(k^2 - k_0^2) \tag{7.52}$$

we have

$$E_\nu(\boldsymbol{k}) = E_\nu(\boldsymbol{k}_0) + <\nu\boldsymbol{k}_0|H'|\nu\boldsymbol{k}_0> + \sum_{\nu'}{}' \frac{|<\nu'\boldsymbol{k}_0|H'|\nu\boldsymbol{k}_0>|^2}{E_\nu(\boldsymbol{k}_0) - E_{\nu'}(\boldsymbol{k}_0)} \tag{7.53}$$

and

$$u_{\nu k} = u_{\nu k_0} + \sum_{\nu'}{}' \frac{u_{\nu' k_0} <\nu'\boldsymbol{k}_0|H'|\nu\boldsymbol{k}_0>}{E_\nu(\boldsymbol{k}_0) - E_{\nu'}(\boldsymbol{k}_0)} , \tag{7.54}$$

where the sum over ν' is over all $|\nu'\boldsymbol{k}_0>$ states except $|\nu\boldsymbol{k}_0>$. This restriction is indicated by a prime in the summation symbol. We obtain

$$\begin{aligned} E_\nu(\boldsymbol{k}) &= E_\nu(\boldsymbol{k}_0) + \frac{\hbar^2}{2m}(k^2 - k_0^2) + \frac{\hbar}{m}(\boldsymbol{k}-\boldsymbol{k}_0)\cdot\boldsymbol{p}_{\nu\nu}(\boldsymbol{k}_0) \\ &+ \frac{\hbar^2}{m^2}\sum_{\nu'}{}' \frac{(\boldsymbol{k}-\boldsymbol{k}_0)\cdot\boldsymbol{p}_{\nu\nu'}(\boldsymbol{k}_0)\boldsymbol{p}_{\nu'\nu}(\boldsymbol{k}_0)\cdot(\boldsymbol{k}-\boldsymbol{k}_0)}{E_\nu(\boldsymbol{k}_0)-E_{\nu'}(\boldsymbol{k}_0)} \end{aligned} \tag{7.55}$$

Use of Eq. (7.43) gives

$$\begin{aligned} E_\nu(\boldsymbol{k}) &= E_\nu(\boldsymbol{k}_0) + \frac{\hbar^2}{2m}(\boldsymbol{k}-\boldsymbol{k}_0)^2 + (\boldsymbol{k}-\boldsymbol{k}_0)\cdot\left(\frac{\partial E_\nu(\boldsymbol{k})}{\partial \boldsymbol{k}}\right)_{k=k_0} \\ &+ \frac{\hbar^2}{m^2}(\boldsymbol{k}-\boldsymbol{k}_0)\cdot\left(\sum_{n'}{}' \frac{\boldsymbol{p}_{\nu\nu'}(\boldsymbol{k}_0)\boldsymbol{p}_{\nu'\nu}(\mathbf{k}_0)}{E_\nu(\boldsymbol{k}_0)-E_{\nu'}(\boldsymbol{k}_0)}\right)\cdot(\boldsymbol{k}-\boldsymbol{k}_0)\,. \end{aligned} \tag{7.56}$$

(ii) $E_\nu(\boldsymbol{k}_0)$ is degenerate. In this case, second order perturbation reduces to the diagonalization of an $\ell\times\ell$ matrix where ℓ is the dimension of the degeneracy. This matrix is given by its elements:

$$\begin{aligned} K_{\nu'\nu} &= E(\boldsymbol{k}_0)\delta_{\nu'\nu} + <\nu'\boldsymbol{k}_0|H'|\nu\boldsymbol{k}_0> \\ &+ \sum_n{}' \frac{<\nu'\boldsymbol{k}_0|H'|n\boldsymbol{k}_0><n\boldsymbol{k}_0|H'|\nu\boldsymbol{k}_0>}{E(\boldsymbol{k}_0)-E_n(\boldsymbol{k}_0)} + \cdots\,. \end{aligned} \tag{7.57}$$

Here ν and ν' run over the ℓ states forming the representation of the level and the sum over n includes all states belonging to $\boldsymbol{k}_0$ except the ℓ states of energy $E(\boldsymbol{k}_0)$. We obtain

$$\begin{aligned} K_{\nu'\nu} &= \left(E(\boldsymbol{k}_0) + \frac{\hbar^2}{2m}(k^2-k_0^2)\right)\delta_{\nu'\nu} + \frac{\hbar}{m}(\boldsymbol{k}-\boldsymbol{k}_0)\cdot\boldsymbol{p}_{\nu'\nu}(\boldsymbol{k}_0) \\ &+ \frac{\hbar^2}{m^2}(\boldsymbol{k}-\boldsymbol{k}_0)\cdot\left(\sum_n{}' \frac{\boldsymbol{p}_{\nu'n}(\boldsymbol{k}_0)\boldsymbol{p}_{n\nu}(\boldsymbol{k}_0)}{E(\boldsymbol{k}_0)-E_n(\boldsymbol{k}_0)}\right)\cdot(\boldsymbol{k}-\boldsymbol{k}_0) + \cdots\,. \end{aligned} \tag{7.58}$$

As an example we consider the shape of the bands for a crystal with the zinc-blende structure in the vicinity of the Γ-point. Here $\boldsymbol{k}_0 = 0$. We write $E(\boldsymbol{k}_0) = E_0$ and take $E_0 = 0$ by shifting the origin of measurements of the energy. Then

$$K_{\nu'\nu} = \frac{\hbar^2 k^2}{2m}\delta_{\nu'\nu} + \frac{\hbar}{m}\boldsymbol{k}\cdot\boldsymbol{p}_{\nu'\nu} - \frac{\hbar^2}{m^2}\boldsymbol{k}\cdot\left(\sum_n{}' \frac{\boldsymbol{p}_{\nu'n}\boldsymbol{p}_{n\nu}}{E_n}\right)\cdot\boldsymbol{k} + \cdots \tag{7.59}$$

We define the operator

$$\underset{\sim}{\Omega} = -\frac{2}{m}\sum_n{}' \frac{\boldsymbol{p}|n><n|\boldsymbol{p}}{E_n}\,. \tag{7.60}$$

We note that the sum

$$\sum_n{}' \frac{|n><n|}{E_n}$$

belongs to the totally symmetric representation by virtue of Unsöld's theorem. This means that $\underset{\sim}{\Omega}$ belong to

$$\Gamma_5 \times \Gamma_5 = \Gamma_{25} \times \Gamma_{25} = \Gamma_1 + \Gamma_3 + \Gamma_4 + \Gamma_5 = \Gamma_1 + \Gamma_{12} + \Gamma_{15} + \Gamma_{25} \ .$$

We require the matrix elements of

$$\boldsymbol{k} \cdot \underset{\sim}{\Omega} \cdot \boldsymbol{k} = \sum_{i,j} \Omega_{ij} k_i k_j \tag{7.61}$$

so that since $k_i k_j$ is symmetric only the symmetric part of $\underset{\sim}{\Omega}$ is of interest:

$$\boldsymbol{k} \cdot \underset{\sim}{\Omega} \cdot \boldsymbol{k} \in \Gamma_1 + \Gamma_3 + \Gamma_5$$

We can thus write

$$\begin{aligned} \boldsymbol{k} \cdot \underset{\sim}{\Omega} \cdot \boldsymbol{k} = & \frac{1}{3} k^2 (\Omega_{xx} + \Omega_{yy} + \Omega_{zz}) + \frac{1}{3} (k_z^2 + \omega k_x^2 + \omega^2 k_y^2)(\Omega_{zz} + \omega^2 \Omega_{xx} + \omega \Omega_{yy}) \\ & + \frac{1}{3} (k_z^2 + \omega^2 k_x^2 + \omega k_y^2)(\Omega_{zz} + \omega \Omega_{xx} + \omega^2 \Omega_{yy}) \\ & + k_y k_z (\Omega_{yz} + \Omega_{zy}) + k_z k_x (\Omega_{zx} + \Omega_{xz}) + k_x k_y (\Omega_{xy} + \Omega_{yx}) \ , \end{aligned} \tag{7.62}$$

where $\omega = e^{2\pi i/3}$ and x, y, z are along the cubic axes.

$$\Omega_{zz} + \omega^2 \Omega_{xx} + \omega \Omega_{yy} \qquad \text{and} \qquad \Omega_z + \omega \Omega_{xx} + \omega^2 \Omega_{yy}$$

transform as $z^2 + \omega^2 x^2 + \omega y^2 = u_1$ and $z^2 + \omega x^2 + \omega^2 y^2 = u_2$ under the operations of T_d. As we have seen before, the functions u_1 and u_2 generate the Γ_3 irreducible representation of T_d. Equation (7.62) can also be written as

$$\begin{aligned} \boldsymbol{k} \cdot \underset{\sim}{\Omega} \cdot \boldsymbol{k} = & \frac{1}{3} k^2 (\Omega_{xx} + \Omega_{yy} + \Omega_{zz}) \\ & + \frac{1}{6} [(2\Omega_{zz} - \Omega_{xx} - \Omega_{yy})(2k_z^2 - k_x^2 - k_y^2) + 3(\Omega_{xx} - \Omega_{yy})(k_x^2 - k_y^2)] \\ & + k_y k_z (\Omega_{yz} + \Omega_{zy}) + k_z k_x (\Omega_{zx} + \Omega_{xz}) + k_x k_y (\Omega_{xy} + \Omega_{yx}) \end{aligned} \tag{7.63}$$

which is useful in some cases.

We consider a level of symmetry Γ_3 and denote by u_1 and u_2 the orthonormal wave functions associated with it. Since $\boldsymbol{p} \in \Gamma_5$ and $\Gamma_5 \times \Gamma_3 = \Gamma_4 + \Gamma_5$, the matrix elements of the momentum in the states u_1 and u_2 vanish. Similarly the quantities $\Omega_{ij} + \Omega_{ji}$ $(i \neq j)$ belong to Γ_5 and again the matrix elements of these operators between u_1 and u_2 vanish.

In order to determine the remaining matrix elements of $\underset{\sim}{\Omega}$ we need the Clebsch-Gordan coefficients for $\Gamma_3 \times \Gamma_3 = \Gamma_1 + \Gamma_2 + \Gamma_3$. They are given in the following table.

	u_1v_1	u_1v_2	u_2v_1	u_2v_2
Γ_1	0	$\frac{1}{\sqrt{2}}$	$\frac{1}{\sqrt{2}}$	0
Γ_2	0	$\frac{1}{\sqrt{2}}$	$-\frac{1}{\sqrt{2}}$	0
$\Gamma_3(u_1)$	0	0	0	1
$\Gamma_3(u_2)$	1	0	0	0

We find

$$\begin{aligned} < u_1|\Omega_{xx} &+ \Omega_{yy} + \Omega_{zz}|u_1 > \neq 0 \\ < u_1|\Omega_{zz} &+ \omega^2\Omega_{xx} + \omega\Omega_{yy}|u_1 >= 0 \\ < u_1|\Omega_{zz} &+ \omega\Omega_{xx} + \omega^2\Omega_{yy}|u_1 >= 0 \ . \end{aligned}$$

These equations lead to

$$< u_1|\Omega_{xx}|u_1 >=< u_1|\Omega_{yy}|u_1 >=< u_1|\Omega_{zz}|u_1 >$$

Since $< u_2|\sum_i \Omega_{ii}|u_2 >=< u_1|\Sigma_i\Omega_{ii}|u_1 >$ we find that $< u_2|\Omega_{xx}|u_2 >= < u_2|\Omega_{yy}|u_2 >=< u_2|\Omega_{zz}|u_2 >=< u_1|\Omega_{zz}|u_1 >$. The off diagonal elements of $\Omega_{xx} + \Omega_{yy} + \Omega_{zz}$ are equal to zero.

$$\begin{aligned} < u_1|\Omega_{xx} &+ \Omega_{yy} + \Omega_{zz}|u_2 >= 0 \\ < u_1|\Omega_{zz} &+ \omega^2\Omega_{xx} + \omega\Omega_{yy}|u_2 >= 0 \\ < u_1|\Omega_{zz} &+ \omega\Omega_{xx} + \omega^2\Omega_{yy}|u_2 > \neq 0 \end{aligned}$$

From these equations we obtain

$$\begin{aligned} < u_1|\Omega_{xx}|u_2 > &= \omega^2 < u_1|\Omega_{zz}|u_2 > \\ < u_1|\Omega_{yy}|u_2 &= \omega < u_1|\Omega_{zz}|u_2 > \end{aligned}$$

It is possible to prove that $< u_1|\Omega_{zz}|u_2 >$ is real. In fact consider the operation S_4 about the z-axis. This operation interchanges u_1 and u_2;

$$\begin{aligned} < u_1|\Omega_{zz}|u_2 > &= < P_{S_4}u_2|\Omega_{zz}|P_{S_4}u_1 > \\ &= < u_2|P_{S_4}^{\dagger}\Omega_{zz}P_{S_4}|u_1 > \end{aligned}$$

but P_{S_4} leaves Ω_{zz} invariant so that

$$< u_2|\Omega_{zz}|u_1 >=< u_1|\Omega_{zz}|u_2 > \ .$$

We now define

$$\gamma_1 = -1+ < u_1|\Omega_{zz}|u_1 >= -1 - \frac{2}{m}\sum_n{}' \frac{|< u_1|p_z|n >|^2}{E_n} \qquad (7.64)$$

and

$$\gamma_2 =< u_1|\Omega_{zz}|u_2 >= -\frac{2}{m}\sum_n{}' \frac{< u_1|p_z|n >< n|p_z|u_2 >}{E_n} \tag{7.65}$$

The matrix $K_{\nu'\nu}$ $(\nu,\nu'=1,2)$ is

$$\underset{\sim}{K} = \gamma_1\frac{\hbar^2k^2}{2m}\underset{\sim}{1} + \frac{\hbar^2}{2m}\gamma_2 \begin{pmatrix} 0 & k_z^2+\omega^2k_x^2+\omega k_y^2 \\ k_z^2+\omega k_x^2+\omega^2k_y^2 & 0 \end{pmatrix} \tag{7.66}$$

The energy eigenvalues are

$$E_\pm = \frac{\hbar^2}{2m}(\gamma_1k^2 \pm \gamma_2\{k^4 - 3(k_y^2k_z^2+k_z^2k_x^2+k_x^2k_y^2\}^{\frac{1}{2}}) \tag{7.67}$$

Note that the quantity under the radical is non-negative. In fact

$$\begin{aligned} k^4 - 3(k_y^2k_z^2+k_z^2k_x^2+k_x^2k_y^2) &= k^4+k_x^4+k_y^4-(k_y^2k_z^2+k_z^2k_x^2+k_x^2k_y^2) \\ &= \frac{1}{2}[(k_y^2-k_z^2)^2+(k_z^2-k_x^2)^2+(k_x^2-k_y^2)^2] \geq 0 \end{aligned}$$

Note further that it is zero only if $k_x^2=k_y^2=k_z^2$, *i.e.*, along $< 111 >$ directions. Thus

$$k_y^2k_z^2+k_z^2k_x^2+k_x^2k_y^2 \leq \frac{1}{3}k^4 \ .$$

For $\boldsymbol{k} \,\| < 111 >$, there is no splitting of the Γ_{12} states in agreement with the compatibility of Γ_{12} and Λ_3. For $\boldsymbol{k} \,\| < 100 >$ we have

$$E_\pm = \frac{\hbar^2}{2m}(\gamma_1 \pm \gamma_2)k^2 \ ,$$

i.e., there is the expected splitting into Δ_1 and Δ_2 levels. We remark that we can set $\gamma_2 > 0$ without loss of generality.

We consider now the variation of the energy eigenvalue with the wave vector in the vicinity of a Γ_{25} level at Γ. We label the Γ_{25} states by ϵ_1, ϵ_2 and ϵ_3 behaving as yz, zx and xy. Each of these states is its own time-reversed conjugate. From the Clebsch-Gordan coefficients for $\Gamma_5 \times \Gamma_5$ $(\Gamma_1+\Gamma_3+\Gamma_4+\Gamma_5)$ we deduce that the matrix elements of $p_1 = p_x, p_2 = p_y$ and $p_3 = p_z$ are proportional to the matrices.

$$\begin{pmatrix} 0&0&0\\0&0&1\\0&1&0 \end{pmatrix} = -\{I_2,I_3\}, \begin{pmatrix} 0&0&1\\0&0&0\\1&0&0 \end{pmatrix} = -\{I_3,I_1\}, \begin{pmatrix} 0&1&0\\1&0&0\\0&0&0 \end{pmatrix} = -\{I_1,I_2\} \ ,$$

respectively. Here

$$I_1 = \begin{pmatrix} 0&0&0\\0&0&-i\\0&i&0 \end{pmatrix} \ , \quad I_2 = \begin{pmatrix} 0&0&i\\0&0&0\\-i&0&0 \end{pmatrix} \ , \quad I_3 = \begin{pmatrix} 0&-i&0\\i&0&0\\0&0&0 \end{pmatrix} .$$

The same holds true for the matrix elements of

$$\Omega_{23}+\Omega_{32} \quad , \quad \Omega_{31}+\Omega_{13} \quad , \quad \Omega_{12}+\Omega_{21} \quad ,$$

since they transform as Γ_5.

However, the matrix elements of $\boldsymbol{p}$ vanishes. In fact, the matrix element

$$\begin{aligned} <\epsilon_1|p_3|\epsilon_2> &= <T\epsilon_1|p_3|T\epsilon_2>= -<T\epsilon_1|Tp_3T^{-1}|T\epsilon_2> \\ &= -<\epsilon_1|p_3|\epsilon_2>^* = -<\epsilon_2|p_3|\epsilon_1> \ . \end{aligned}$$

Consider now the mirror plane σ_d through the 3-axis and bisecting the first and third quadrants of the (1,2) plane. This operation interchanges ϵ_1 and ϵ_2 so that

$$<\epsilon_2|p_3|\epsilon_1>=<P_{\sigma_d}\epsilon_1|p_3|P_{\sigma_d}\epsilon_2>= \quad = \quad <\epsilon_1|p_3|\epsilon_2>$$

There

$$<\epsilon_1|p_3|\epsilon_2>=0 \ ,$$

i.e., the linear term in $\boldsymbol{k}$ in the K-matrix vanishes. We shall see later that, if due account is taken of the spin-orbit interaction, there will exist such a contribution.

The matrix elements of the Γ_5 terms in $\boldsymbol{k}\cdot\underset{\sim}{\Omega}\cdot\boldsymbol{k}$ are all expressed in terms of the single constant

$$\begin{aligned} <\epsilon_1|\Omega_{xy}+\Omega_{yx}|\epsilon_2> &= -\frac{2}{m}\sum_n{}' \frac{1}{E_n}\big(<\epsilon_1|p_x|n><n|p_y|\epsilon_2> \\ &\quad +<\epsilon_1|p_y|n><n|p_x|\epsilon_2>\big) \ . \end{aligned} \tag{7.68}$$

The relation

$$[x,H_0]=\frac{i\hbar}{m}p_x$$

allows us to transform this result as follow: We first evaluate

$$<\epsilon_1|p_x|n>=\frac{m}{i\hbar}<\epsilon_1|[x,H_0]|n>=\frac{mE_n}{i\hbar}<\epsilon_1|x|n>$$

so that

$$\begin{aligned} &-\frac{2}{m}\sum_n{}'\frac{1}{E_n}(<\epsilon_1|p_x|n><n|p_y|\epsilon_2>+<\epsilon_1|p_y|n><n|p_x|\epsilon_2> \\ &= \frac{2i}{\hbar}\sum_n{}'(<\epsilon_1|x|n><n|p_y|\epsilon_2>+<\epsilon_1|y|n><n|p_x|\epsilon_2> \\ &= \frac{2i}{\hbar}<\epsilon_1|xp_y+yp_x|\epsilon_2> \end{aligned} \tag{7.69}$$

where we used the fact that $<\epsilon_i|p_k|\epsilon_j>=0$ and the completeness of the eigenvectors of H_0. Luttinger uses the notation

$$\gamma_3=-\frac{i}{3\hbar}<\epsilon_1|xp_y+yp_x|\epsilon_2> \tag{7.70}$$

so that

$$< \epsilon_1|\Omega_{xy} + \Omega_{yx}|\epsilon_2 >= -6\gamma_3 \ . \tag{7.71}$$

There remains the calculation of the matrix elements of the Γ_1 and Γ_3 terms in $\boldsymbol{k} \cdot \underset{\sim}{\Omega} \cdot \boldsymbol{k}$. For this purpose we prefer to use Eq. (7.63) rather than Eq. (7.62).

The matrix elements of $\Omega_{xx} + \Omega_{yy} + \Omega_{zz}$ are equal to zero off the main diagonal and equal to

$$< \epsilon_3|\Omega_{xx} + \Omega_{yy} + \Omega_{zz}|\epsilon_3 >=< \epsilon_3|\Omega_{zz}|\epsilon_3 > +2 < \epsilon_3|\Omega_{xx}|\epsilon_3 > \tag{7.72}$$

on that diagonal. To calculate the Γ_3 terms we use the Clebsch-Gordan coefficients for $\Gamma_3 \times \Gamma_5$. The matrix elements can be read off the table. We obtain for

$$2\Omega_{zz} - \Omega_{xx} - \Omega_{yy}$$

a matrix proportional to

$$\begin{pmatrix} -\frac{1}{2} & 0 & 0 \\ 0 & -\frac{1}{2} & 0 \\ 0 & 0 & 1 \end{pmatrix} = -\frac{1}{2}(2I_3^2 - I_1^2 - I_2^2) \tag{7.73}$$

while for $\sqrt{3}(\Omega_{xx} - \Omega_{yy})$ we find

$$\begin{pmatrix} \frac{1}{2}\sqrt{3} & 0 & 0 \\ 0 & -\frac{1}{2}\sqrt{3} & 0 \\ 0 & 0 & 0 \end{pmatrix} = -\frac{1}{2}\sqrt{3}(I_1^2 - I_2^2) \ . \tag{7.74}$$

Thus, the matrix associated with the Γ_3 terms in Eq. (7.63) is

$$\begin{aligned} -\frac{1}{6}(< \epsilon_3|\Omega_{zz}|\epsilon_3 > \ &- \ < \epsilon_3|\Omega_{xx}|\epsilon_3 >)[(2k_z^2 - k_x^2 - k_y^2)(2I_z^2 - I_x^2 - I_y^2) \\ &+ \ 2(k_x^2 - k_y^2)(I_x^2 - I_y^2)] \\ &= -(< \epsilon_3|\Omega_{zz}|\epsilon_3 > - < \epsilon_3|\Omega_{xx}|\epsilon_3 >)\sum_i k_i^2(I_i^2 - \frac{1}{3}I^2) \end{aligned} \tag{7.75}$$

Defining

$$\gamma_1 = -1 + \frac{1}{3}(< \epsilon_3|\Omega_{zz}|\epsilon_3 > +2 < \epsilon_3|\Omega_{xx}|\epsilon_3 >) \tag{7.76}$$

and

$$\gamma_2 = -\frac{1}{6}(< \epsilon_3|\Omega_{zz}|\epsilon_3 > - < \epsilon_3|\Omega_{xx}|\epsilon_3 >) \tag{7.77}$$

we obtain

$$\underset{\sim}{K} = \frac{\hbar^2}{m}\left(-\frac{1}{2}\gamma_1 k^2 + 3\gamma_2\sum_i k_i^2(I_i^2 - \frac{1}{3}I^2) + 3\gamma_3\sum_{i<j} k_i k_j\{I_i, I_j\}\right) . \tag{7.78}$$

We can transform the matrix elements in Eqs. (7.76) and (7.77) in the same manner as we transformed that for $\Omega_{xy} + \Omega_{yx}$ in Eq. (7.79) to obtain

$$\gamma_1 + 4\gamma_2 = -1 - \frac{2i}{\hbar} < \epsilon_3|zp_z|\epsilon_3 > \tag{7.79}$$

and

$$\gamma_1 - 2\gamma_2 = -1 - \frac{2i}{\hbar} < \epsilon_3|xp_x|\epsilon_3 > . \tag{7.80}$$

As examples, we consider the splitting of the Γ_{25} bands near $\boldsymbol{k} = 0$ for a few high symmetry directions. For $\boldsymbol{k} \parallel [001]$ we find

$$\underset{\sim}{K} = \frac{\hbar^2 k^2}{m} \left(-\frac{1}{2}\gamma_1 - 2\gamma_2 + 3\gamma_2 I_z^2 \right) . \tag{7.81}$$

This shows that the Γ_{25} level splits into a singlet and a doublet in this approximation. According to the compatibility tables it should be into $\Delta_1 + \Delta_3 + \Delta_4$ where Δ_3 and Δ_4 have approximately the same energy. For $\boldsymbol{k} \parallel [111]$

$$\underset{\sim}{K} = \frac{\hbar^2 k^2}{m} \left(-\frac{1}{2}\gamma_1 - 2\gamma_3 + 3\gamma_3 I_n^2 \right)$$

with

$$I_n = \frac{1}{\sqrt{3}}(I_1 + I_2 + I_3) .$$

The eigenvalues of I_n^2 are 0 and 1, the latter being a double root. Thus, the Γ_{25} states split into a singlet of energy $-(\hbar^2 k^2/m)\left(\frac{1}{2}\gamma_1 + 2\gamma_3\right)$ and a doublet of energy $(\hbar^2 k^2/2m)\left(-\frac{1}{2}\gamma_1 + \gamma_3\right)$. Thus, it splits into a singlet Λ_1 and a doublet Λ_3. For $\boldsymbol{k} \parallel [110]$, we find

$$K = \frac{\hbar^2 k^2}{m} \left(-\frac{1}{2}\gamma_1 + \gamma_2 - \frac{3}{2}\gamma_2 I_3^2 + \frac{3}{2}\gamma_3\{I_1, I_2\} \right) .$$

The matrix

$$\gamma_2 I_3^2 - \gamma_3\{I_1, I_2\} = \begin{pmatrix} \gamma_2 & \gamma_3 & 0 \\ \gamma_3 & \gamma_2 & 0 \\ 0 & 0 & 0 \end{pmatrix}$$

has eigenvalues $\gamma_2 \pm \gamma_3$ and 0. Thus, along [110], Γ_{25} splits into three singlets of energies

$$\frac{\hbar^2 k^2}{m} \left(-\frac{1}{2}\gamma_1 + \gamma_2 \right) \qquad \text{and} \qquad \frac{\hbar^2 k^2}{m} \left(-\frac{1}{2}\gamma_1 - \frac{1}{2}\gamma_2 \pm \frac{3}{2}\gamma_3 \right) ,$$

corresponding to $2\Sigma_1 + \Sigma_2$.

In many articles and books the constants

$$A = -\frac{\hbar^2 \gamma_1}{2m} \quad , \quad B = -\frac{\hbar^2 \gamma_2}{m} \quad \text{and} \quad D = -\frac{\sqrt{3}\hbar^2 \gamma_3}{m}$$

are often used instead of the Luttinger parameters. Then $\underset{\sim}{K}$ reads

$$\underset{\sim}{K} = Ak^2 - 3B\sum_i k_i^2(I_i^2 - \frac{1}{3}I^2) - D\sqrt{3}\sum_{i<j} k_ik_j\{I_i, I_j\}$$

The form of $\underset{\sim}{K}$ can be deduced in the following alternative way. The matrix $\underset{\sim}{K}$ is a Hermitian 3×3 matrix so that it is determined by nine real numbers. The nine matrices $I_1, I_2, I_3, I_1^2, I_2^2, I_3^2, \{I_2, I_3\}, \{I_3, I_1\}$ and $\{I_1, I_2\}$ are linearly independent. The matrices I_1, I_2, I_3 behave as pseudo-vectors so that they belong to the Γ_4 irreducible representation of T_d. Thus

$$I_1^2 + I_2^2 + I_3^2 \quad \in \Gamma_1$$

$$2I_3^2 - I_1^2 - I_2^2 \quad , \sqrt{3}(I_1^2 - I_2^2) \quad \in \Gamma_3$$

and

$$\{I_2, I_3\} \quad , \quad \{I_3, I_1\} \quad , \quad \{I_1, I_2\} \quad \in \Gamma_5 \; .$$

The only invariants we can form with the components of the wave vector $\boldsymbol{k}$ are

$$(I_1^2 + I_2^2 + I_3^2)k^2 \; ,$$

$$(2I_3^2 - I_1^2 - I_2^2)(2k_3^2 - k_1^2 - k_2^2) + 3(I_1^2 - I_2^2)(k_1^2 - k_2^2) = 6\sum_i k_i^2\left(I_i^2 - \frac{1}{3}I^2\right)$$

and

$$\sum_{i<j}\{I_i, I_j\}k_ik_j \; .$$

Thus, if we transform $\boldsymbol{k}$ as well as $\{I_i, I_j\}$ the invariant Hamiltonian is a linear combination of these three invariant.

We note that

$$\sum_i I_ik_i = \boldsymbol{I} \cdot \boldsymbol{k}$$

belongs to Γ_2 so that it cannot appear in the Hamiltonian matrix $\underset{\sim}{K}$. We further remark that in the group O this is an allowed invariant because O consists only of rotations so that vectors and pseudo-vectors behave in identical ways.

7.8 Effect of the spin-orbit interaction on electronic energy levels in crystals

If we wish to include the additional degree of freedom of an electron due to its intrinsic spin, each eigenvalue of the Hamiltonian

$$H_0 = \frac{p^2}{2m} + U(\boldsymbol{r}) \tag{8.1}$$

considered in the previous section has its degeneracy doubled. We recall that in this case the energy eigenstates of an electron are classified according to the irreducible representations of the double group $\bar{G} = G \times \bar{E}$ where G is the symmetry group of H_0 and $\bar{E}$, representing a rotation by 2π about an arbitrary axis, reverses the sign of spinors. The spin-orbit interaction

$$H_S = \frac{\hbar}{4m^2c^2}\boldsymbol{\sigma} \cdot (\nabla U \times \boldsymbol{p}) \tag{8.2}$$

where $\boldsymbol{\sigma}$ the Pauli operator may remove the degeneracy but in no case do we obtain singlet states in the absence of an externally applied magnetic field. All degeneracies are, of course, of even order. Under these circumstances the Hamiltonian operator is not that in Eq. (8.1) but rather

$$H_o = \frac{p^2}{2m} + U(\boldsymbol{r}) + \frac{\hbar}{4m^2c^2}\boldsymbol{\sigma} \cdot (\nabla U \times \boldsymbol{p})\ . \tag{8.3}$$

We note that since $U(\boldsymbol{r})$ is periodic with the periods of the lattice, so is H_o. Furthermore H_0 is time reversal invariant since $T\boldsymbol{p}T^{-1} = -\boldsymbol{p}$ and $T\boldsymbol{\sigma}T^{-1} = -\boldsymbol{\sigma}$, *i.e.*,

$$[H_0, T] = 0 \tag{8.4}$$

where T is the time reversal operator studied in Chapter 7, Sec. 3. For a single electron we recall that $T^2 = -1$. The eigenfunctions of H_0 in Eq. (8.3) are of the Bloch form

$$\psi_{ks}(\boldsymbol{r}) = N^{-\frac{1}{2}}e^{i\mathbf{k}\cdot\mathbf{r}}u_{ks}(\boldsymbol{r}) \tag{8.5}$$

where s labels one of the two orthogonal spinors belonging to the $\boldsymbol{k}$-vector $\boldsymbol{k}$. The states (5) are two-component wave functions. We have omitted the band index for convenience but it must be restored when there is danger of confusion. The energy eigenvalue corresponding to $\psi_{ks}(\boldsymbol{r})$ is denoted by $E_s(\boldsymbol{k})$. For any two spinor eigenfunctions of this form, belonging to different bands we have the orthogonality condition

$$\int_{(V)} \psi^{\dagger}_{n'k's'}\psi_{nks}d\boldsymbol{r} = \delta_{n'n}\delta_{k'k}\delta_{s's} \tag{8.6}$$

For a single band,

$$\int_{(v_0)} u^{\dagger}_{ks'}u_{ks}d\boldsymbol{r} = \delta_{s's}\ . \tag{8.7}$$

Recalling that $T = -i\sigma_y K$,

$$T\psi_{ks}(\boldsymbol{r}) = -iN^{-\frac{1}{2}}e^{-i\mathbf{k}\cdot\mathbf{r}}\sigma_y u^*_{kr}(\boldsymbol{r}) \tag{8.8}$$

so that $T\psi_{ks}(\boldsymbol{r})$ belongs to $-\boldsymbol{k}$. Since H_0 is time-reversal invariant $T\psi_{ks}(\boldsymbol{r})$ is an eigenvector of H_0 with the eigenvalue $E_s(\boldsymbol{k})$. If $\widehat{\zeta}$ is an arbitrary direction

$$\begin{aligned} < T\psi_{ks}|\sigma_\zeta|T\psi_{ks} > &= - < T\psi_{ks}|T\sigma_\zeta T^{-1}|T\psi_{ks} > \\ &= - < \psi_{ks}|\sigma_\zeta|\psi_{ks} > \ . \end{aligned} \tag{8.9}$$

This means that the expectation values of σ_ζ, if different from zero, in ψ_{ks} and $T\psi_{ks}$ are of opposite signs. This observation justifies labeling the spinor states with $s = \pm$, *i.e.*, we denote them by ψ_{k+} and ψ_{k-}. This notation is, admittedly, somewhat arbitrary but is convenient for discussion. It only implies that the spinors ψ_{k+} and $T\psi_{k+} = \psi_{-k,-}$ are orthogonal and, hence, linearly independent. Thus we write

$$T\psi_{k,\pm} = \psi_{-k,\mp} \tag{8.10}$$

and

$$E_\pm(\boldsymbol{k}) = E_\mp(-\boldsymbol{k}) \ . \tag{8.11}$$

If the crystal has a center of inversion ($\boldsymbol{r} = 0$),

$$[P_i, H_o] = 0 \tag{8.12}$$

and

$$P_i\psi_{k,\pm}(\boldsymbol{r}) = \psi_{k,\pm}(-\boldsymbol{r}) = \psi_{-k,\pm}(\boldsymbol{r}) \tag{8.13}$$

so that

$$E_\pm(-\boldsymbol{k}) = E_\pm(\boldsymbol{k}) \ . \tag{8.14}$$

Combining Eqs. (8.11) and (8.14) when H_0 is both time-reversal and inversion invariant we deduce that

$$E_\pm(\boldsymbol{k}) = E_\mp(\boldsymbol{k}) \tag{8.15}$$

so that at each point $\boldsymbol{k}$ in the fundamental Brillouin zone there is a double degeneracy due to the intrinsic spin of the electron.

The effective Schrödinger equation for u_{ks} is

$$\left(\frac{p^2}{2m} + U(\boldsymbol{r}) + \frac{\hbar}{4m^2c^2}\boldsymbol{\sigma}\cdot(\nabla U \times \boldsymbol{p}) + \frac{\hbar}{m}\boldsymbol{k}\cdot\boldsymbol{\pi} + \frac{\hbar^2k^2}{2m}\right) u_{ks}(\boldsymbol{r}) = E_s(\boldsymbol{k})u_{ks}(\boldsymbol{r}) \, , \tag{8.16}$$

where

$$\boldsymbol{\pi} = \boldsymbol{p} + \frac{\hbar}{4mc^2}\boldsymbol{\sigma} \times \nabla U \tag{8.17}$$

We can now proceed as in the previous section. We consider the states of H_0 prior to including the spin-orbit interaction. Let us start with the solutions for $\boldsymbol{k} = 0$. We can always start from an arbitrary point $\boldsymbol{k}_0$ but to simplify the writing we shall take $\boldsymbol{k}_0 = 0$. The reader can provide the steps necessary for the appropriate generalization. We consider states $\nu(\nu = 1, 2 \ldots \ell)$ of H_0 (Eq. (8.1) and take their energy as the origin of energies. The perturbation is

$$H' = \frac{\hbar}{m}\boldsymbol{k}\cdot\boldsymbol{\pi} + \frac{\hbar}{4m^2c^2}\boldsymbol{\sigma}\cdot(\nabla U \times \boldsymbol{p}) + \frac{\hbar^2k^2}{2m} \ . \tag{8.18}$$

Taking into account the spin the states of H_0 (Eq. (8.1)) are $|\nu s>$ (ν = 1,2,...ℓ ; $s = \pm$). We shall denote all other manifolds by $|ns>$. In second order perturbation we form the matrix $\underset{\sim}{K}$ whose components $K_{\nu's',\nu s}$ are given by

$$K_{\nu's',\nu s} =< \nu's'|H'|\nu s> - \sum_{ns''}{}' \frac{< \nu's'|H'|ns''>< ns''|H'|\nu s>}{E_n} \ . \tag{8.19}$$

For convenience we define the pseudovector $\boldsymbol{W}$ by

$$\boldsymbol{W} = \frac{1}{4mc^2} \nabla U \times \boldsymbol{p} \tag{8.20}$$

so that

$$H_S = \frac{\hbar}{m} \boldsymbol{\sigma} \cdot \boldsymbol{W} \ . \tag{8.21}$$

The matrix elements of H' in the $\{|\nu s>\}$ manifold are

$$\frac{\hbar}{m} \boldsymbol{\sigma}_{s's} \cdot \boldsymbol{W}_{\nu'\nu} \ . \tag{8.22}$$

The commutator of $\boldsymbol{p}$ and H_0 allows us to determine the matrix elements of ∇u:

$$\nabla U = \frac{i}{\hbar} [\boldsymbol{p}, H_0] \tag{8.23}$$

so that

$$< \nu'|\nabla U|\nu >= 0 \tag{8.24}$$

and

$$< n|\nabla U|\nu >= -\frac{i}{\hbar} E_n \boldsymbol{p}_{n\nu} \tag{8.25}$$

where we must remember that we took $E_\nu = 0$. Thus

$$< \nu's'|H'|\nu s >= \frac{\hbar^2 k^2}{2m} \delta_{\nu'\nu} \delta_{s's} + \frac{\hbar}{m} \boldsymbol{k} \cdot \boldsymbol{p}_{\nu'\nu} \delta_{s's} + \frac{\hbar}{m} \boldsymbol{W}_{\nu'\nu} \cdot \boldsymbol{\sigma}_{s's} \ , \tag{8.26}$$

and

$$< ns''|H'|\nu s >= \frac{\hbar}{m} \boldsymbol{k} \cdot \boldsymbol{p}_{n\nu} \delta_{s''s} - \frac{i\hbar E_n}{4m^2c^2} (\boldsymbol{k} \times \boldsymbol{\sigma}_{s''s}) \cdot \boldsymbol{p}_{n\nu} + \frac{\hbar}{m} \boldsymbol{\sigma}_{s''s} \cdot \boldsymbol{W}_{n\nu} \ . \tag{8.27}$$

The second term in the right hand side of Eq. (8.27) can be neglected in comparison with the first. Thus

$$< ns'|H'|\nu s >\approx \frac{\hbar}{m} (\boldsymbol{k} \cdot \boldsymbol{p}_{n\nu} \delta_{s''s} + \boldsymbol{\sigma}_{s''s} \cdot \boldsymbol{W}_{n\nu}) \ . \tag{8.28}$$

The second order part of the matrix $\underset{\sim}{K}$ is, thus,

$$\begin{aligned} -\frac{\hbar^2}{m^2} \sum_n{}' \frac{1}{E_n} (&\boldsymbol{k} \cdot \boldsymbol{p}_{\nu'n} \boldsymbol{p}_{n\nu} \cdot \boldsymbol{k} \delta_{s's} + \boldsymbol{k} \cdot \boldsymbol{p} +_{nu'n} \boldsymbol{W}_{n\nu} \cdot \boldsymbol{\sigma}_{s's} \\ &+ \boldsymbol{\sigma}_{s's} \cdot \boldsymbol{W}_{\nu'n} \boldsymbol{p}_{n\nu} \cdot \boldsymbol{k} + \sum_{s''} \boldsymbol{\sigma}_{s's''} \cdot \boldsymbol{W}_{nu'n} \boldsymbol{W}_{n\nu} \cdot \boldsymbol{\sigma}_{s''s}) \ . \end{aligned} \tag{8.29}$$

The first term of expression (8.29) is identical to $(\hbar^2/2m)$ multiplied by $< \nu'|\boldsymbol{k}\cdot \underset{\sim}{\Omega}\cdot\boldsymbol{k}|\nu>$ and $\delta_{s's}$ where $\underset{\sim}{\Omega}$ is the same tensor used in the previous section. We define

$$\Phi_{ij} = -\frac{2}{m}\sum_n{}' \frac{1}{E_n}(p_i|n>< n|W_j + W_j|n>< n|p_i) \tag{8.30}$$

and

$$\Lambda_{ij} = -\frac{2}{m}\sum_n{}' \frac{1}{E_n} W_i|n>< n|w_j \; . \tag{8.31}$$

We finally write

$$\begin{aligned} K_{\nu's',\nu s} &= \frac{\hbar^2k^2}{2m}\delta_{\nu'\nu}\delta_{s's} + \frac{\hbar}{m}\boldsymbol{k}\cdot\boldsymbol{p}_{nu'nu}\delta_{s's} + \frac{\hbar}{m}\boldsymbol{\sigma}_{s's}\cdot\boldsymbol{W}_{\nu'\nu} \\ &+ \frac{\hbar^2}{2m}\sum_{i,j}(\delta_{s's} < \nu'|k_ik_j\Omega_{ij}|\nu> + k_i\sigma^{(j)}_{s's} < \nu'|\Phi_{ij}|nu> \\ &+ < s'|\sigma^{(i)}\sigma^{(j)}|s>< \nu'|\Lambda_{ij}|\nu> \;) \end{aligned} \tag{8.32}$$

We now apply these results to establish the $\boldsymbol{k}$-dependence of the energy eigenvalue for states in the vicinity of a three-fold degenerate $\Gamma_5(=\Gamma_{25})$ level at the Γ point of a zinc-blende crystal. We recall that we proved that the matrix elements of the momentum operator $< \epsilon_1|p_3|\epsilon_2 >$, since $\boldsymbol{W}$ is a pseudovector operator.Thus, the term $(\hbar/m)\boldsymbol{\sigma}_{s's}\cdot W_{\nu'\nu}$ is simply of the form $\boldsymbol{\sigma}\cdot\boldsymbol{I}$. We write it as

$$\frac{1}{3}\Delta\boldsymbol{I}\cdot\boldsymbol{\sigma} \tag{8.33}$$

where Δ is given by

$$\Delta = \frac{3i\hbar}{4m^2c^2} < \epsilon_1\left|\frac{\partial U}{\partial x}p_y - \frac{\partial U}{\partial y}p_x\right|\epsilon_2 > \; . \tag{8.34}$$

The terms involving $\underset{\sim}{\Omega}$ in Eq. (8.32) are identical to those studied in the previous section. The terms in $\underset{\sim}{\phi}$ give a linear contribution in $\boldsymbol{k}$ to the energy eigenvalues. Finally the terms in $\underset{\sim}{\Lambda}$ can be rewritten as matrix elements of the operator

$$\sum_{i,j}\sigma_i\sigma_j\Lambda_{ij} = \sum_i\Lambda_{ii} + \sum_{i<j}\sigma_i\sigma_j(\Lambda_{ij}-\Lambda_{ji}) \; . \tag{8.35}$$

The first term on the right hand side of Eq. (8.35) belongs to the totally symmetric representation so that it shifts all levels by the same amount. The second term is proportional to $\boldsymbol{\sigma}\cdot\boldsymbol{I}$ since $\Lambda_{ij}-\Lambda_{ji}$ is an antisymmetric second rank tensor and $\sigma_i\sigma_j = i\epsilon_{ijk}\sigma_k$. Thus, it gives a correction to the first order term $(\Delta/3)\boldsymbol{\sigma}\cdot\boldsymbol{I}$ and can be incorporated in it. Thus, we need not concern ourselves with it further.

At $\boldsymbol{k} - 0$, the $\Gamma_5 \times \Gamma_6$ level splits into a quadruplet Γ_8 and a doublet Γ_7 by virtue of the spin orbit interaction. In fact, setting

$$\widehat{\boldsymbol{J}} = \widehat{\boldsymbol{I}} + \frac{1}{2}\hat{\boldsymbol{\sigma}} \tag{8.36}$$

the quantity

$$\frac{\Delta}{3}\boldsymbol{I}\cdot\boldsymbol{\sigma} = \frac{\Delta}{3}\left(\hat{j}^2 - \hat{I}^2 - \frac{1}{4}\sigma^2\right) = \frac{\Delta}{3}\left(J(J+1) - 2 - \frac{3}{4}\right) \tag{8.37}$$

where J can take the values $j = \frac{3}{2}$ and $j = \frac{1}{2}$. The values of $\frac{\Delta}{3}\boldsymbol{I}\cdot\boldsymbol{\sigma}$ are $\frac{\Delta}{3}$ for $j = \frac{3}{2}$ and $-\frac{2}{3}\Delta$ for $j = \frac{1}{2}$.

We have

$$\begin{aligned} \underset{\sim}{K} &= \frac{\hbar^2}{m}\Bigg(-\frac{1}{2}\gamma_1 k^2 + 3\gamma_2 \sum_i k_i^2 \left(I_i^2 - \frac{1}{3}\right) + 3\gamma_3 \sum_{i<j} k_i k_j \{I_i, I_j\} \\ &\quad + \frac{1}{2}\sum_i k_i S_i \Bigg) + \frac{1}{3}\Delta \boldsymbol{I}\cdot\boldsymbol{\sigma} \,. \end{aligned} \tag{8.38}$$

To transform $\underset{\sim}{K}$ into separate submatrices for $j = \frac{3}{2}$ and $j = \frac{1}{2}$ we simply calculate the elements of $\underset{\sim}{K}$ in the representation in which $\widehat{J}^2$ and $\widehat{J}_z$ are diagonal. This is accomplished with the use of the Wigner-Eckart theorem. We write the components of $\{I_i, I_j\}$ in irreducible form as discussed in chapter 6, Sec. 9, *i.e.*,

$$\begin{aligned} I_0^{(2)} &= \left(\frac{2}{3}\right)^{\frac{1}{2}} (2I_3^2 - I_1^2 - I_2^2) \\ I_{\pm 1}^{(2)} &= \mp\{I_3, I_\pm\} \\ I_\pm^{(2)} &= I_\pm^2 \end{aligned} \tag{8.39}$$

where $I_\pm = I_1 \pm i I_2$. Then in the representation $|jm>$

$$< jm'|J_\kappa^{(2)}|jm >= F(j) < jm'|I_\kappa^{(2)}|jm > \quad , \quad \kappa = 0, \pm 1, \pm 2 \,, \tag{8.40}$$

where $F(j)$ is a constant depending of j only and the $J_\kappa^{(2)}$ are defined in a way similar to that in Eqs. (8.39). To find $F(j)$ we calculate a particular case, say for $\kappa = 0$ and $m' = m = j$. We obtain

$$< jj|2J_3^2 - J_1^2|jj >= F(j) < jj|2I_3^2 - I_1^2 - I_2^2|jj > \ .$$

The left hand side of this equation is

$$< jj|3J_3^2 - J^2|jj >= j(2j-1) \ .$$

For $j = \frac{3}{2}$, $m = \frac{3}{2}$ we have $I_3 = 1$, $\sigma_3 = 1$ so that

$$< \frac{3}{2}, \frac{3}{2}|3I_3^2 - I^2|\frac{3}{2}, \frac{3}{2} >=< 1, \frac{1}{2}|3I_3^2 - I^2|1, \frac{1}{2} >= 1 \ .$$

Thus, for $j = \frac{3}{2}$, $f\left(\frac{3}{2}\right) = 3$. For $j = \frac{1}{2}$, $I_3 = 1$, $\sigma_3 = -1$, $F\left(\frac{1}{2}\right) = 0$. Thus, when Δ is much larger than $(\hbar^2 k^2/2m)\bar{\gamma}$ where $\bar{\gamma}$ is a typical average Luttinger parameter we can express $\underset{\sim}{K}$ in block form for the values of j equal to 3/2 and 1/2. They are

$$\begin{aligned}\underset{\sim}{K}_{3/2} &= \frac{\hbar^2}{m}\left(-\frac{1}{2}\gamma_1 k^2 + \gamma_2 \sum_i k_i^2 \left(J_i^2 - \frac{1}{3}J^2\right) + \gamma_3 \sum_{i<j} k_i k_j \{J_i, J_j\}\right) \\ &+ \frac{C}{\sqrt{3}}(k_1\{J_1, J_2^2 - J_3^2\} + k_2\{J_2, J_3^2 - J_1^2\} + k_3\{J_3, J_1^2 - J_2^2\}) \\ &+ \frac{1}{3}\Delta\,,\end{aligned} \tag{8.41}$$

and

$$\underset{\sim}{K}_{\frac{1}{2}} = -\frac{2}{3}\Delta - \frac{\hbar^2}{2m}\gamma_1 k^2\,. \tag{8.42}$$

Chapter 8

Problems

1. What is the order of the group H generated by two distinct, non-unit elements a and b subject to the conditions

$$a^3 = b^2 = (ab)^2 = e \ ?$$

2. Let

$$u = \begin{pmatrix} 0 & 1 \\ -1 & 0 \end{pmatrix} \ , \quad v = \begin{pmatrix} 0 & i \\ i & 0 \end{pmatrix} \ ,$$

and construct the group generated by these elements under matrix multiplication. Show that this group is a non-Abelian group of order 8 isomorphic to a group generated by a, b and c subject to

$$a^2 = b^2 = c^2 = d \ , d^2 = e, bc = a, ca = b, ab = c \ .$$

3. Let H be a group containing a unique element $a \neq e$ such that $a^2 = e$. Show that $xa = ax$ for all x in H.

4. Let a, b and c be elements of a finite group and form the cyclic subgroups generated by abc, bca and cab. Show that they are of the same order.

5. D_4 is the group of rotations about the center of a square leaving it invariant.

   (i) Enumerate the elements of D_4 and construct its multiplication table.

   (ii) Construct the character table of D_4 and basis functions proportional to polynomials in the coordinates x, y, z of a point (take the origin of coordinates at the center of the square and the x and y axes parallel to its sides).

6. Construct the character table of $D_{3d} = D_3 \times C_i$ where $C_i = \{E_i i\}$, i being the inversion operation $\boldsymbol{r} \rightarrow -\boldsymbol{r}$. Express the basis functions in terms of the components of a vector (x, y, z), of a symmetric second rank tensor $\underset{\sim}{\alpha}$ and a pseudo-vector $S_x, S_y S_z$ (2nd rank antisymmetric tensor).

7. The diamond structure consists of a fcc lattice with a basis of two carbon atoms at (0,0,0) and $(a/4)(1,1,1)$. Taking $\hat{e}_1, \hat{e}_2$ and $\hat{e}_3$ as unit vectors parallel to the cubic axes show that the lattice points are at

$$\boldsymbol{n} = \frac{a}{2}(\ell_1\hat{e}_1 + \ell_2\hat{e}_2 + \ell_3\hat{e}_3)$$

where $\ell_1\ell_2$ and ℓ_3 are integers whose sum is even. The primitive vectors can be selected to be $\boldsymbol{a}_1 = (a/2)(\hat{e}_2 + \hat{e}_3)$, $\boldsymbol{a}_2 = (a/2)(\hat{e}_3 + \hat{e}_1)$ and $\boldsymbol{a}_3 = (a/2)(\hat{e}_1 + \hat{e}_2)$. Find all centers of inversion of the structure. How many of these are independent? We say that two points are independent if their position vectors do not differ by a translation of the lattice.

8. Prove or disprove the following statement:

> "If a finite group H has an invariant proper subgroup N, then the irreducible representations of H are direct products of those of N and those of the factor group H/N".

9. Consider the methane molecule CH_4 shown in the diagram.

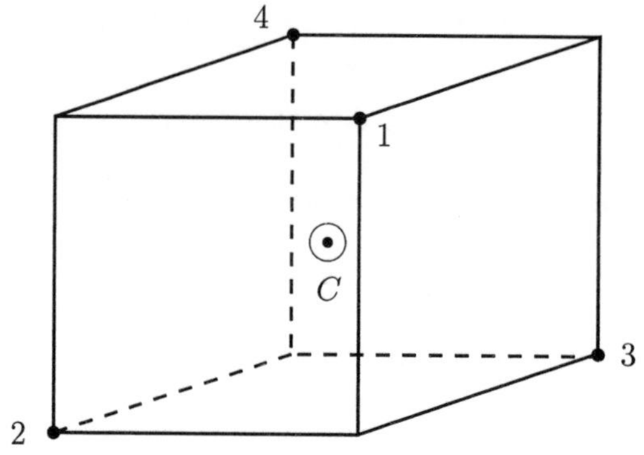

Take the origin at the center of the cube at the location of the carbon atom C. Hydrogen atoms are located at 1, 2, 3 and 4. Enumerate the symmetry elements of the CH_4 molecule and find the symmetry classification of the normal modes of vibration. Denote the displacement of the C-atom by (X,Y,Z) where X,Y and Z are its components parallel to the edges of the cube in the figure. The displacements of hydrogen atoms with respect to the C-atoms are denoted by

$$(u_i, v_i, w_i) \quad , \quad i = 1, 2, 3, 4 \; .$$

The w_i component is parallel to the bond CH_4, the u_i component is in the mirror plane of the molecule containing the bond CH_i. Finally v_i is normal to that mirror plane.

Show that

$$\begin{aligned}
\alpha &= \frac{1}{2}(w_1 + w_2 + w_3 + w_4) , \\
\gamma_1 &= \frac{1}{2}(u_1 - u_2 - u_3 - u_4) \quad , \quad \gamma_2 = \frac{1}{2}(v_1 - v_2 - v_3 + v_4) \\
\epsilon_{r1} &= \frac{1}{2}(w_1 + w_2 - w_3 - w_4) \quad , \quad \epsilon_{r2} = \frac{1}{2}(w_1 - w_2 + w_3 - w_4) , \\
\epsilon_{r3} &= \frac{1}{2}(w_1 - w_2 - w_3 + w_4) , \\
\epsilon_{\theta 1} &= \frac{1}{4}(-u_1 + u_2 - u_3 + u_4) + \frac{1}{4}\sqrt{3}(v_1 - v_2 + v_3 - v_4) , \\
\epsilon_{\theta 2} &= \frac{1}{4}(-u_1 - u_2 + u_3 + u_4) + \frac{1}{4}\sqrt{3}(-v_1 - v_2 + v_3 + v_4) , \\
\epsilon_{\theta 3} &= \frac{1}{2}(u_1 + u_2 + u_3 + u_4) , \\
\delta_1 &= \frac{1}{4}\sqrt{3}(-u_1 + u_2 - u_3 + u_4) - \frac{1}{4}(v_1 - v_2 + v_3 - v_4) , \\
\delta_2 &= -\frac{1}{4}\sqrt{3}(-u_1 - u_2 + u_3 + u_4) + \frac{1}{4}(-v_1 - v_2 + v_3 + v_4) ,
\end{aligned}$$

and

$$\delta_3 = \frac{1}{2}(v_1 + v_2 + v_3 + v_4)$$

generate the representation

$$\Gamma_1 + \Gamma_3 + 2\Gamma_5 + \Gamma_4$$

with

Γ_1	α	
Γ_3	(γ_1, γ_2)	
$2\Gamma_5$	$(\epsilon_{r1}, \epsilon_{r2}, \epsilon_{r3})$;	$(\epsilon_{\theta 1}, \epsilon_{\theta 2}, \epsilon_{\theta 3})$
Γ_4	$(\delta_1, \delta_2, \delta_3)$	

(Note: select v_i so that it represents a rigid rotation of the molecule).

Let Δr_i be the change in length of the bond CH_i and $\Delta\theta_{ij}$ the change in the angle of the bonds CH_i and CH_j and suppose that the energy associated with such a deformation of the molecule is

$$V = \frac{1}{2}k_r \sum_{i=1}^{4}(\Delta r_i)^2 + \frac{1}{2}k_\theta \sum_{i<j}(R\Delta\theta_{ij})^2$$

where R is the equilibrium bond length.

Find the Hamiltonian for vibrations of the system in terms of the symmetry coordinates above and describe the different modes of vibration.

Show that the breathing mode α has an angular frequency $(k_r/m)^{/half}$ where m is the mass of H (in this mode C does not move). Show that the Γ_3 (α's) modes have angular frequency

$$\omega_\gamma = (3k_\theta/m)^{\frac{1}{2}}$$

The ϵ modes have frequencies ω^2 obtained solving

$$\omega^4 - (\omega_\theta^2 + \omega_r^2)\omega^2 + 2k_r k_\theta \frac{M+4m}{m^2 M} = 0$$

Here M is the mass of C and

$$\omega_\theta = \left[\frac{2k_\theta}{m}\left(1 + \frac{8m}{3M}\right)\right]^{\frac{1}{2}}$$

and

$$\omega_r = \left[\frac{k_r}{m}\left(1 + \frac{4m}{3M}\right)\right]^{\frac{1}{2}} .$$

Draw diagrams showing the normal modes.

10. Find the symmetry classification of the vibrational modes of the benzene molecule. Determine the classification also according to infrared and Raman active modes. (the benzene molecule C_6H_6 is a planar molecule with its carbon and hydrogen atoms arranged in an hexagonal ring:

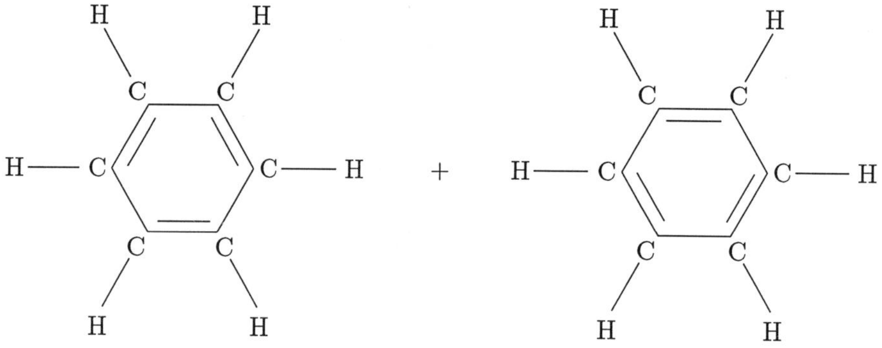

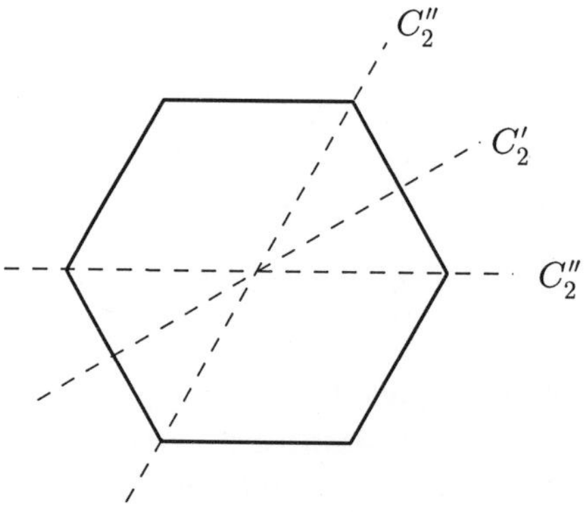

The symmetry group is D_{6h}, *i.e.*, D_6 with an additional horizontal mirror plane.

$$D_{6h} = \{E, C_2, 2C_3, 2C_6, 3C_2', 3C_2'', i, iC_2, 2iC_3, 2iC_6, 3iC_2', 3iC_2''\}$$

Chapter 9

Solutions to the problems

1. Since $a^3 = e$, the elements e, a, a^2 are in H. The element b cannot be any of these because $b \neq a$ and $b \neq e$. Now b cannot be equal to a^2 because this would imply that $b^2 = a^4$ so that a^4 would be e, but since $a^3 = e$ it would follow that $a = e$. The product $ab \neq e$ because if $ab - e$, since b is its own inverse we would conclude that $a = b$, contrary to the statement of conditions. In the same way $ba \neq e$. Furthermore ab cannot be equal to a, a^2 or b. Also $ab \neq a, a^2$ or b. The inverse of ab is

$$(ab)^{-1} = b^{-1}a^{-1} = \boldsymbol{a}^2 \, .$$

But from $(ab)^2 = e$ we have that

$$(ab)^{-1} = ab \, .$$

Thus $ba^2 = ab$ and $a^2b = ba$. Thus

$$H = \{e, a, a^2, b, ab, ba\}$$

2. Let G be the group under matrix multiplication generated by u and v. Then

$$\begin{aligned}
u^2 &= \begin{pmatrix} -1 & 0 \\ 0 & -1 \end{pmatrix} = -e \quad , \quad e = \begin{pmatrix} 1 & 0 \\ 0 & 1 \end{pmatrix} \\
u^3 &= -u \\
u^4 &= e \\
v^2 &= -e, v^3 = -v, v^4 = e \\
w &= uv = \begin{pmatrix} i & 0 \\ 0 & -i \end{pmatrix} \\
w^2 &= -e \\
w^3 &= -w \\
w^4 &= e \\
vu &= -uv \\
uw &= u^2v = -v \\
wu &= uvu = -u^2v = v \\
vw &= -wv = u
\end{aligned}$$

Thus

$$G = \{e, u, v, w, -e, -u, -v, -w\}$$

G is isomorphic to $H = \{e, a, b, c, d, da, db, dc\}$ under the isomorphism

$$\begin{array}{cccccccc} e & u & v & w & -e & -u & -v & -w \\ \updownarrow & \updownarrow & \updownarrow & \updownarrow & \updownarrow & \updownarrow & \updownarrow & \uparrow \\ e & a & b & c & d & da & db & dc \end{array}$$

3. Clearly a commutes with e and with itself. Let x be any other element of H. Since a is the only element whose square is e, $x^{-1} \neq x$. The square of $x^{-1}ax$ is equal to e. Thus,

$$x^{-1}ax = a \Rightarrow ax = xa \qquad Q.E.D.$$

4. The order of the cyclic subgroup generated by abc is the smallest positive integer for which

$$(abc)^n = e\ .$$

Then

$$(abc)^{n-1} = (abc)^{-1} = c^{-1}b^{-1}a^{-1}\ .$$

but

$$(bca)^n = bc(abc)^{n-1}a = e\ .$$

In a similar way

$$(cab)^n = e\ .$$

Suppose now that an integer m exists such that $0 < m < n$ and

$$(bca)^m = e\ .$$

Then $(abc)^m = a(bca)^{m-1}bc = e$ contrary to the assumption that n is the smallest positive integer for which $(abc)^n = e$. We conclude that abc, bca and cab are of the same order.

5.

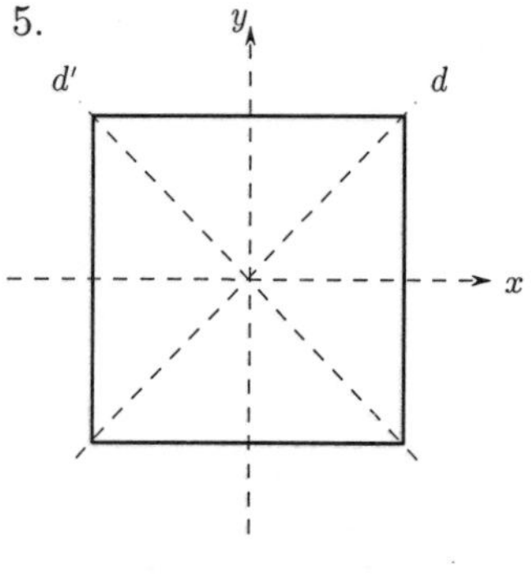

Symmetry elements (rotations only)

(i) E the identity

(ii) C_2; rotation by π about z

(iii) C_4, C_4^{-1}; rotations by $\frac{\pi}{2}$ and $-\frac{\pi}{2}$ about z

(iv) C_{2x}, C_{2y}; rotations by π about x and y

(v) C_{2d}, C'_{2d}; rotations by π about d and d'.

Multiplication table:

D_4	E	C_4	C_4^{-1}	C_2	C_{2x}	C_{2y}	C_{2d}	$C_{2d'}$
E	E	C_4	C_4^{-1}	C_2	C_{2x}	C_{2y}	C_{2d}	$C_{2d'}$
C_4	C_4	C_2	E	C_4^{-1}	C_{2d}	$C_{2d'}$	C_{2y}	C_{2x}
C_4^{-1}	C_4^{-1}	E	C_2	C_4	$C_{2d'}$	C_{2d}	C_{2x}	C_{2y}
C_2	C_2	C_4^{-1}	C_4	E	C_{2y}	C_{2x}	$C_{2d'}$	C_{2d}
C_{2x}	C_{2x}	$C_{2d'}$	C_{2d}	C_{2y}	E	C_2	C_4^{-1}	C_4
C_{2y}	C_{2y}	C_{2d}	$C_{2d'}$	C_{2x}	C_2	E	C_4	C_4^{-1}
C_{2d}	C_{2d}	C_{2x}	C_{2y}	$C_{2d'}$	C_4	C_4^{-1}	E	C_2
$C_{2d'}$	$C_{2d'}$	C_{2y}	C_{2x}	C_{2d}	C_4^{-1}	C_4	C_2	E

Example:

$$C_{2x}C_{2y}\begin{bmatrix} 3 & 2 \\ 4 & 1 \end{bmatrix} = C_{2x}\begin{bmatrix} 2 & 3 \\ 1 & 4 \end{bmatrix} = \begin{bmatrix} 1 & 4 \\ 2 & 3 \end{bmatrix} = C_2\begin{bmatrix} 3 & 2 \\ 4 & 1 \end{bmatrix}$$

Class structure

$$\begin{aligned} E \;\; ; \;\; 2C_4 &= \{C_4, C_4^{-1}\} \quad , \quad C_2(=C_4^2) \quad , \quad 2C_2' = \{C_{2x} \, , \, C_{2y}\} \\ 2C_2'' &= \{C_{2d} \, , \, C_{2d'}\} \, . \end{aligned}$$

$$N_C = 5 \Rightarrow N_\Gamma = 5$$

$$\ell_1^2 + \ell_2^2 + \ell_3^2 + \ell_4^2 + \ell_5^2 = 8 \Rightarrow \ell_1 = \ell_2 = \ell_3 = \ell_4 = 1 \, , \; \ell_5 = 2$$

(x, y, z) generate a vector representation. It is reducible since (x, y) and (z) are invariant subspaces.

D_4	E	$2C_4$	C_2	$2C_2'$	$2C_2''$	Basis functions
Γ_1	1	1	1	1	1	$x^2 + y^2; z^2$
Γ_2	1	1	1	-1	-1	z
Γ_3	1	-1	1	1	-1	$x^2 - y^2$
Γ_4	1	-1	1	-1	1	xy
Γ_5	2	0	-2	0	0	(x, y)

6. The elements of D_{3d} are those of D_3 and those obtained from them by multiplication by the inversion i.

$$iC_3^{-1} = \sigma_h C_2 C_3^{-1} = \sigma_h C_6 = S_6$$

(Note that σ_h and C_2 here are not elements of the group where C_2 is a rotation by π about the 3-axis, *i.e.*, not one of the C_2 in the x, y plane. But $\sigma_h C_2$ is, by definition i. The improper rotation S_φ is defined as $\sigma_h C_\varphi$.)

$$iC_3 = S_6^{-1}$$

The product of i and a C_2 in the plane gives a vertical plane of reflection conventionally denoted σ_d. Thus the elements of D_{3d}, classified into conjugate classes are

$$D_{3d} = \{E, 2C_3, 3C_2, i, 2S_6, 3\sigma_d\}$$

The order of the group is $h = 12$. It has six inequivalent irreducible representations of which four are one-dimensional and two are two-dimensional.

D_{3d}	E	$2C_3$	$3C_2$	i	$2S_6$	$3\sigma_d$	Basic Functions
Γ_1^+	1	1	1	1	1	1	$x^2 + y^2$; z^2
Γ_2^+	1	1	−1	1	1	−1	S_z
Γ_3^+	2	−1	0	2	−1	0	$(\alpha_{zx}, \alpha_{zy})$; $(\alpha_{xx} - \alpha_{yy}\,, -2\alpha_{xy})$, (S_x, S_y)
Γ_1^-	1	1	1	−1	−1	−1	
Γ_2^-	1	1	−1	−1	−1	1	z
Γ_3^-	2	−1	0	−2	1	0	(x, y)

the $\pm$ signs correspond to even and odd representations. In the chemical literature they use g and u (German: gerade and ungerade for even and odd).

7. The positions of the carbon atoms are at $\boldsymbol{n}$ and $\boldsymbol{n} + \boldsymbol{\delta}_0$, $\boldsymbol{n} = n_1\boldsymbol{a}_1 + n_2\boldsymbol{a}_2 + n_3\boldsymbol{a}_3 = \frac{a}{2}[(n_2 + n_3)\hat{e}_1 + (n_3 + n_1)\hat{e}_2 + (n_1 + n_2)\hat{e}_3]$. The vector $\boldsymbol{\delta}_0$ is

$$\boldsymbol{\delta}_0 = \frac{a}{4}(\hat{e}_1 + \hat{e}_2 + \hat{e}_3) = \frac{1}{4}(\boldsymbol{a}_1 + \boldsymbol{a}_2 + \boldsymbol{a}_3) \ ,$$

the displacement of one fcc sublattice with respect to the other. Let I be the sublattice $\boldsymbol{n}$ and II the sublattice $\boldsymbol{n} + \boldsymbol{\delta}_0$. No lattice position in I is a center of inversion because II does not remain invariant under inversion on any $\boldsymbol{n}$ point. The inversion operator on a center leaving the structure invariant must map I into II (and, of course, conversely). If P is a center of inversion, the inverse of O ($\boldsymbol{n} = 0$) with respect to O is $-\overrightarrow{OP}$. Thus P is a center of inversion if, and only if, $\overrightarrow{OP} - (-\overrightarrow{OP}) = 2\overrightarrow{OP}$ is a vector if II, *i.e.*,

$$\overrightarrow{OP} = \frac{\boldsymbol{n}}{2} + \frac{1}{2}\boldsymbol{\delta}_0$$

Thus $\frac{1}{2}\boldsymbol{\delta}_0$ is a center of inversion. $\frac{1}{2}\boldsymbol{\delta}_0 - \frac{1}{2}\boldsymbol{a}_1 = \frac{1}{2}\boldsymbol{\delta}_1$ where

$$\boldsymbol{\delta}_1 = \frac{a}{4}(\hat{e}_1 - \hat{e}_2 - \hat{e}_3)$$

is also but is, of course, equivalent to $\frac{1}{2}\boldsymbol{\delta}_0$. Similarly $\frac{1}{2}\boldsymbol{\delta}_0 - \frac{1}{2}\boldsymbol{a}_2 = \frac{a}{4}(-\hat{e}_1 + \hat{e}_2 - \hat{e}_3) = \frac{1}{2}\boldsymbol{\delta}_2$ and $\frac{1}{2}\boldsymbol{\delta}_0 - \frac{1}{2}\boldsymbol{a}_3 = \frac{a}{4}(-\hat{e}_1 - \hat{e}_2 + \hat{e}_3) = \frac{1}{2}\boldsymbol{\delta}_3$ are also centers of inversion equivalent to $\frac{1}{2}\boldsymbol{\delta}_0$. The directions of $\boldsymbol{\delta}_0, \boldsymbol{\delta}_1, \boldsymbol{\delta}_2, \boldsymbol{\delta}_3$ are those of the tetrahedral bonds of a carbon atom and its four nearest neighbors. A second point of inversion symmetry, not equivalent to $\frac{1}{2}\boldsymbol{\delta}_0$ is

$$\overrightarrow{OP'} = \frac{a}{2}(\hat{e}_1 + \hat{e}_2 + \hat{e}_3) + \frac{1}{2}\boldsymbol{\delta}_0 = \frac{5a}{8}(\hat{e}_1 + \hat{e}_2 + \hat{e}_3)$$

Clearly P'' given by

$$\overrightarrow{OP''} = \overrightarrow{OP'} - \boldsymbol{a}_1 - \boldsymbol{a}_2 = \frac{a}{8}(\hat{e}_1 + \hat{e}_2 - 3\hat{e}_3)$$

is a center equivalent to P'.

8. The statement is false. It is enough to show that it is false in a single example. Consider the group T_d. The group T is an invariant subgroup of T_d. The cosets of T are T itself and $\{6\sigma_d, 6S_4\} = M$. The factor group contains the two elements T and M whose multiplication table is

	T	M
T	T	M
M	M	T

i.e., the character table of T_d/T is

	T	M
Γ_1'	1	1
Γ_2'	1	−1

The character table of T is $\omega = e^{\frac{2\pi i}{3}}$.

	E	$4C_3$	$4C_3^{-1}$	$3C_2$
Γ_1	1	1	1	1
Γ_2	1	ω^2	ω	1
Γ_3	1	ω	ω^2	1
Γ_4	3	0	0	−1

Since T_d has only five non-equivalent representations direct products of the irreducible representations of T_d/T and T do not yield those of T_d.

9. The elongations Δr_i are equal to w_i and $R\Delta\theta_{ij}$ is given in the table

ij	$R\Delta\theta_{ij}$
12	$\frac{1}{2}(u_1 - u_2 - \sqrt{3}(v_1 - v_2))$
13	$\frac{1}{2}(u_1 - u_3 + \sqrt{3}(v_1 - v_3))$
14	$-(u_1 + u_4)$
23	$u_2 + u_3$
24	$-\frac{1}{2}(u_2 - u_4 + \sqrt{3}(v_2 - v_4)$
34	$-\frac{1}{2}(u_3 - u_4 - \sqrt{3}(v_3 - v_4)$

The kinetic energy is (m = mass of C, m = mass of H).

$$\begin{aligned} T &= \frac{1}{2}(M+4m)(\dot{X}^2+\dot{Y}^2+\dot{Z}^2)+\frac{1}{2}m\sum_{i=1}^{4}(\dot{u}_i^2+\dot{v}_i^2+\dot{w}_i^2) \\ &\quad +m\left[\dot{X}\left\{\frac{1}{\sqrt{6}}(-\dot{u}_1+\dot{u}_2-\dot{u}_3+\dot{u}_4)\right.\right. \\ &\quad +\frac{1}{\sqrt{2}}(\dot{v}_1-\dot{v}_2+\dot{v}_3-\dot{v}_4) \\ &\quad \left.+\frac{1}{\sqrt{3}}(\dot{w}_1+\dot{w}_2-\dot{w}_3-\dot{w}_4)\right\} \\ &\quad +\dot{Y}\left\{\frac{1}{\sqrt{6}}(-\dot{u}_1-\dot{u}_2+\dot{u}_3+\dot{u}_4)\right. \\ &\quad +\frac{1}{\sqrt{2}}(-\dot{v}_1-\dot{v}_2+\dot{v}_3+\dot{v}_4) \\ &\quad \left.+\frac{1}{\sqrt{3}}(\dot{w}_1-\dot{w}_2+\dot{w}_3-\dot{w}_4)\right\} \\ &\quad +\dot{Z}\left\{\sqrt{\frac{2}{3}}(\dot{u}_1+\dot{u}_2+\dot{u}_3+\dot{u}_4)\right. \\ &\quad \left.\left.+\frac{1}{\sqrt{3}}(\dot{w}_1-\dot{w}_2-\dot{w}_3+\dot{w}_4)\right\}\right]. \end{aligned}$$

The potential energy is

$$\begin{aligned} V &= \frac{1}{2}k_r\sum_{i=1}^{4}w_i^2+\frac{3}{4}k_\theta\left[\sum_{i=1}^{4}(u_i^2+v_i^2)-(v_1+v_4)(v_2+v_3)\right. \\ &\quad +\frac{1}{\sqrt{3}}\{(u_1-u_4)(v_2-v_3)+(u_2-u_3)(v_1-v_4)\} \\ &\quad \left.-\frac{1}{3}\{(u_1+u_4)(u_2+u_3)-4(u_1u_4+u_2u_3)\}\right]. \end{aligned}$$

Some transformations give

$$\begin{aligned} T &= \frac{1}{2}(M+4m)(\dot{X}^2+\dot{Y}^2+\dot{Z}^2)+\frac{1}{2}m(\dot{\alpha}^2+\dot{\gamma}_1^2+\dot{\gamma}_2^2+\dot{\delta}_1^2+\dot{\delta}_2^2+\dot{\delta}_3^2) \\ &\quad +\frac{1}{2}m(\dot{\epsilon}_{r1}^2+\dot{\epsilon}_{r2}^2+\dot{\epsilon}_{r3}^2+\dot{\epsilon}_{\theta1}^2+\dot{\epsilon}_{\theta2}^2+\dot{\epsilon}_{\theta3}^2) \\ &\quad +\frac{2m}{\sqrt{3}}\{\dot{X}(\sqrt{2}\dot{\epsilon}_{\theta1}+\dot{\epsilon}_{r1})+\dot{Y}(\sqrt{2}\dot{\epsilon}_{\theta2}+\dot{\epsilon}_{r2}) \\ &\quad +\dot{Z}(\sqrt{2}\dot{\epsilon}_{\theta3}+\dot{\epsilon}_{r3})\} \end{aligned}$$

$\delta_1, \delta_2, \delta_3$ are cyclic variables since they do not appear in

$$V = \frac{1}{2}k_r(\alpha^2 + \epsilon_{r1}^2 + \epsilon_{r2}^2 + \epsilon_{r3}^2) + k_\theta(\epsilon_{\theta 1}^2 + \epsilon_{\theta 2}^2 + \epsilon_{\theta 3}^2) + \frac{3}{2}k_\theta(\gamma_1^2 + \gamma_2^2)\,.$$

The Hamiltonian is

$$\begin{aligned} H &= \left(\frac{p_\alpha^2}{2M} + \frac{1}{2}k_r\alpha^2\right) + \sum_{i=1}^{2}\left(\frac{p_{\gamma i}^2}{2m} + \frac{3}{2}k_\theta\gamma_i^2\right) \\ &+ \sum_{i=1}^{3}\left[\frac{P_i^2}{2M} + \frac{p_{\theta i}^2 + p_{ri}^2}{2m} + \frac{2}{3M}(\sqrt{2}p_{\theta i} + p_{ri})^2 \right. \\ &\left. - \frac{2}{\sqrt{3}M}P_i(\sqrt{2}p_{\theta i} + p_{ri}) + \frac{1}{2}k_r\epsilon_{ri}^2 + k_\theta\epsilon_{\theta i}^2\right] \end{aligned}$$

(P_1, P_2, P_3) are canonically conjugated to X, Y, Z. The equations of motion are

$$\ddot{\alpha} = -\frac{k_r}{m}\alpha$$

$$\ddot{\gamma}_i = -\frac{3k_\theta}{m}\gamma_i$$

and

$$\begin{aligned} \ddot{\epsilon}_{\theta i} &= -\omega_\theta^2\epsilon_{\theta i} - \frac{4\sqrt{2}}{3M}k_r\epsilon_{ri} \\ \ddot{\epsilon}_{ri} &= -\frac{8\sqrt{2}}{3M}k_\theta\epsilon_{\theta i} - \omega_r^2\epsilon_{ri} \end{aligned}$$

The results follow immediately from these.

10. The character table of D_{6h} is

D_{6h}	E	C_2	$2C_3$	$2C_6$	$3C_2'$	$3C_2''$	i	iC_2	$2iC_3$	$2iC_6$	$3iC_2'$	$3iC_2''$	Basis Functions
A_{1g}	1	1	1	1	1	1	1	1	1	1	1	1	
A_{1u}	1	1	1	1	1	1	−1	−1	−1	−1	−1	−1	
A_{2g}	1	1	1	1	−1	−1	1	1	1	1	−1	−1	S_z
A_{2u}	1	1	1	1	−1	−1	−1	−1	−1	−1	1	1	z
B_{1g}	1	−1	1	−1	1	−1	1	−1	1	−1	1	−1	
B_{1u}	1	−1	1	−1	1	−1	−1	1	−1	1	−1	1	
B_{2g}	1	−1	1	−1	−1	1	1	−1	1	−1	−1	1	
B_{2u}	1	−1	1	−1	−1	1	−1	1	−1	1	1	−1	
E_{1g}	2	−2	−1	1	0	0	2	−2	−1	1	0	0	S_x, S_y
E_{1u}	2	−2	−1	1	0	0	−2	2	1	−1	0	0	x, y
E_{2g}	2	2	−1	−1	0	0	2	2	−1	−1	0	0	
E_{2u}	2	2	−1	−1	0	0	−2	−2	1	1	0	0	

	12	0	0	0	4	0	0	12	0	0	0	4	Fixed Atoms	
V	3	−1	0	2	−1	−1	−3	1	0	−2	1	1	Vector rep.	$A_{2u} + E_{1u}$
R	3	−1	0	2	−1	−1	3	−1	0	2	−1	−1	Pseudovec. rep.	$A_{2g} + E_{1g}$
Γ	36	0	0	0	−4	0	0	12	0	0	0	4	Total	

The z-axis is normal to the plane of the molecular through its center of gravity. There are six 2-fold axes in the (x, y) plane (*i.e.*, the plane of the molecule)

The reduction of the representation Γ generated by the displacements of all 12 atoms is

$$\begin{aligned}\Gamma &= 2A_{1g} + 2A_{2g} + 2A_{2u} + 2B_{1u} + 2B_{2g} + 2B_{2u} + 2E_{1g} \\ &\quad + 4E_{1u} + 4E_{2g} + 2E_{2u}\end{aligned}$$

Now $A_{2u} + E_{1u}$ account for a rigid displacement of the molecular and $E_{2g} + E_{1g}$ for a rigid rotation. Thus, the vibrational modes are classified as follows

$$\begin{aligned}\Gamma_{vib.} &= 2A_{1g} + A_{2g} + A_{2u} + 2B_{1u} + 2B_{2g} + 2B_{2u} + E_{1g} \\ &\quad + 3E_{1u} + 4E_{2g} + 2E_{2u} \ .\end{aligned}$$

Thus, we have 20 different frequencies of vibration of which only four are infrared active, namely A_{2u} and the three E_{1u}'s. The polarizability tensor generates the representation $[V \times V]$ (symmetric direct product of the vector representation V by itself). Now $[V \times V] = 2A_{1g} + E_{1g} + E_{2g}$. Thus there are seven Raman active modes, namely $2A_{1g}$, E_{1g} and $4E_{2g}$.

Appendix A

9.1 Clebsch-Gordan Coefficients and Compatibility Tables for T_d

$$\Gamma_3 \times \Gamma_3 = \Gamma_1 + \Gamma_2 + \Gamma_3$$

1. Select basis functions u_1 and u_2 behaving as

$$\begin{aligned} u_1 &\sim z^2 + \omega^2 x^2 + \omega y^2 = \frac{1}{2}(2z^2 - x^2 - y^2) - \frac{i}{2}\sqrt{3}(x^2 - y^2) \\ u_2 &\sim z^2 + \omega x^2 + \omega^2 y^2 = \frac{1}{2}(2z^2 - x^2 - y^2) + \frac{i}{2}\sqrt{3}(x^2 - y^2) \\ \omega &= e^{2\pi i/3} = -\frac{1}{2} + \frac{i}{2}\sqrt{3} \end{aligned}$$

(u_1, u_2) and (v_1, v_2) generate Γ_3 in identical form

	u_1v_1	u_1v_2	u_2u_1	u_2v_2
α'	0	$\frac{1}{\sqrt{2}}$	$\frac{1}{\sqrt{2}}$	0
β'	0	$\frac{1}{\sqrt{2}}$	$-\frac{1}{\sqrt{2}}$	0
u_1'	0	0	0	1
u_2'	1	0	0	0

Select basis functions behaving as

$$\begin{aligned} \gamma_1 &\sim (2z^2 - x^2 - y^2) \sim u_1 + u_2 \\ \gamma_2 &\sim \sqrt{3}(x^2 - y^2) \sim i(u_1 - u_2) \end{aligned}$$

	$\gamma_1\gamma_1'$	$\gamma_1\gamma_2'$	$\gamma_2\gamma_1'$	$\gamma_2\gamma_1'$
α''	$\frac{1}{\sqrt{2}}$	0	0	$\frac{1}{\sqrt{2}}$
β''	0	$\frac{1}{\sqrt{2}}$	$-\frac{1}{\sqrt{2}}$	0
γ_1''	$-\frac{1}{\sqrt{2}}$	0	0	$\frac{1}{\sqrt{2}}$
γ_2''	0	$\frac{1}{\sqrt{2}}$	$\frac{1}{\sqrt{2}}$	0

2. $\Gamma_3 \times \Gamma_5 = \Gamma_4 + \Gamma_5$; $\epsilon_1 \sim yz, \epsilon_2 \sim zx, \epsilon_3 \sim xy$

	$\gamma_1\epsilon_1$	$\gamma_1\epsilon_2$	$\gamma_1\epsilon_3$	$\gamma_2\epsilon_1$	$\gamma_2\epsilon_2$	$\gamma_2\epsilon_3$
δ_1'	$-\frac{1}{2}\sqrt{3}$	0	0	$-\frac{1}{2}$	0	0
δ_2'	0	$\frac{1}{2}\sqrt{3}$	0	0	$-\frac{1}{2}$	0
δ_3'	0	0	0	0	0	1
ϵ_1'	$-\frac{1}{2}$	0	0	$\frac{1}{2}\sqrt{3}$	0	0
ϵ_2'	0	$-\frac{1}{2}$	0	0	$-\frac{1}{2}\sqrt{3}$	0
ϵ_3'	0	0	1	0	0	0

3. $\Gamma_5 \times \Gamma_5 = \Gamma_1 + \Gamma_3 + \Gamma_4 + \Gamma_5$

	$\epsilon_1\epsilon_1'$	$\epsilon_1\epsilon_2'$	$\epsilon_1\epsilon_3'$	$\epsilon_2\epsilon_1'$	$\epsilon_2\epsilon_2'$	$\epsilon_2\epsilon_3'$	$\epsilon_3\epsilon_1'$	$\epsilon_3\epsilon_2'$	$\epsilon_3\epsilon_3'$
α''	$\frac{1}{\sqrt{3}}$	0	0	0	$\frac{1}{\sqrt{3}}$	0	0	0	$\frac{1}{\sqrt{3}}$
γ_1''	$-\frac{1}{\sqrt{6}}$	0	0	0	$-\frac{1}{\sqrt{6}}$	0	0	0	$\sqrt{\frac{2}{\sqrt{3}}}$
γ_2''	$\frac{1}{\sqrt{2}}$	0	0	0	$-\frac{1}{\sqrt{2}}$	0	0	0	0
δ_1''	0	0	0	0	0	$\frac{1}{\sqrt{2}}$	0	$-\frac{1}{\sqrt{2}}$	0
δ_2''	0	0	$-\frac{1}{\sqrt{2}}$	0	0	0	$\frac{1}{\sqrt{2}}$	0	0
δ_3''	0	$\frac{1}{\sqrt{2}}$	0	$-\frac{1}{\sqrt{2}}$	0	0	0	0	0
ϵ_1''	0	0	0	0	0	$\frac{1}{\sqrt{2}}$	0	$\frac{1}{\sqrt{2}}$	0
ϵ_2''	0	0	$\frac{1}{\sqrt{2}}$	0	0	0	$\frac{1}{\sqrt{2}}$	0	0
ϵ_3''	0	$\frac{1}{\sqrt{2}}$	0	$\frac{1}{\sqrt{2}}$	0	0	0	0	0

4. $\Gamma_4 \times \Gamma_5 = \Gamma_2 + \Gamma_3 + \Gamma_4 + \Gamma_5$

	$\delta_1\epsilon_1$	$\delta_1\epsilon_2$	$\delta_1\epsilon_3$	$\delta_2\epsilon_1$	$\delta_2\epsilon_2$	$\delta_2\epsilon_3$	$\delta_3\epsilon_1$	$\delta_3\epsilon_2$	$\delta_3\epsilon_3$
β'	$\frac{1}{\sqrt{3}}$	0	0	0	$\frac{1}{\sqrt{3}}$	0	0	0	$\frac{1}{\sqrt{3}}$
γ_1'	$\frac{1}{\sqrt{2}}$	0	0	0	$-\frac{1}{\sqrt{2}}$	0	0	0	0
γ_2'	$\frac{1}{\sqrt{6}}$	0	0	0	$\frac{1}{\sqrt{6}}$	0	0	0	$-\sqrt{\frac{2}{3}}$
δ_1'	0	0	0	0	0	$\frac{1}{\sqrt{2}}$	0	$\frac{1}{\sqrt{2}}$	0
δ_2'	0	0	$\frac{1}{\sqrt{2}}$	0	0	0	$\frac{1}{\sqrt{2}}$	0	0
δ_3'	0	$\frac{1}{\sqrt{2}}$	0	$\frac{1}{\sqrt{2}}$	0	0	0	0	0
ϵ_1'	0	0	0	0	0	$\frac{1}{\sqrt{2}}$	0	$-\frac{1}{\sqrt{2}}$	0
ϵ_2'	0	0	$-\frac{1}{\sqrt{2}}$	0	0	0	$\frac{1}{\sqrt{2}}$	0	0
ϵ_3'	0	$\frac{1}{\sqrt{2}}$	0	$-\frac{1}{\sqrt{2}}$	0	0	0	0	0

Compatibility Tables (Zinc-blende structure)

Γ $(k=0)$
BSW K

Γ , T_d	E	$8C_3$	$3C_2$	$6\sigma_d$	$6S_4$		
Γ_1 Γ_1	1	1	1	1	1	$x^2+y^2+z^2, xyz$	a
Γ_2 Γ_2	1	1	1	−1	−1		a
Γ_{12} Γ_3	2	−1	2	0	0	$2z^2-x^2-y^2, \sqrt{3}(x^2-y^2)$	a
Γ_{15} Γ_4	3	0	−1	−1	1	S_x, S_y, S_z	a
Γ_{25} Γ_5	3	0	−1	1	−1	$(x,y,z); (yz,zx,xy)$	a

X D_{2d}	E	$2S_4$	C_2	$2C_2'$	$2\sigma_d$		
X_1 Γ_1	1	1	1	1	1		a
X_4 Γ_2	1	1	1	−1	−1	S_z	a
X_2 Γ_3	1	−1	1	1	−1	x^2-y^2	a
X_3 Γ_4	1	−1	1	−1	1	x, xy	a
X_5 Γ_5	2	0	−2	0	0	(x,y) ; (S_x,S_y)	a

L, Λ, C_{3v}	E	$2C_3$	$3\sigma_v$		
Λ_1 L_1	1	1	1	$\zeta, \zeta^2; \xi^2+\eta^2$	a
Λ_2 L_2	1	1	−1	$S_\zeta, \eta^3-3\xi^2\eta$	a
Λ_3 L_3	2	−1	0	$(\xi,\eta); (S_\eta-S_\xi)$	a

Δ, C_{2v}	E	C_2	σ_v	σ_v'		
Δ_1 Γ_1	1	1	1	1	z	
Δ_2 Γ_3	1	1	−1	−1	$xy; S_z$	a
Δ_3 Γ_2	1	−1	1	−1	$x; S_y$	a
Δ_4 Γ_4	1	−1	−1	1	$y; S_x$	a

Σ C_s	E	σ
Σ_1 Γ_1	1	1
Σ_2 Γ_2	1	−1

Γ_1	Γ_2	Γ_{12}	Γ_{15}	Γ_{25}
Δ_1	Δ_2	$\Delta_1 + \Delta_2$	$\Delta_2 + \Delta_3 + \Delta_4$	$\Delta_1 + \Delta_3 + \Delta_4$
Λ_1	Λ_2	Λ_3	$\Lambda_2 + \Lambda_3$	$\Lambda_1 + \Lambda_3$
Σ_1	Σ_2	$\Sigma_1 + \Sigma_2$	$\Sigma_1 + 2\Sigma_2$	$2\Sigma_1 + \Sigma_2$

X_1	X_2	X_3	X_4	X_5
Δ_1	Δ_2	Δ_1	Δ_2	$\Delta_3 + \Delta_4$

L_1	L_2	L_3
Λ_1	Λ_2	Λ_3

Elenco dei volumi della collana
"Appunti"
pubblicati dall'Anno Accademico 1994/95

GIUSEPPE BERTIN (a cura di), *Seminario di Astrofisica,* 1995.

EDOARDO VESENTINI, *Introduction to continuous semigroups,* 1996.

LUIGI AMBROSIO, *Corso introduttivo alla Teoria Geometrica della Misura ed alle Superfici Minime,* 1997.

CARLO PETRONIO, *A Theorem of Eliashberg and Thurston on Foliations and Contact Structures,* 1997.

MARIO TOSI, *Introduction to Statistical Mechanics and Thermodynamics,* 1997.

MARIO TOSI, *Introduction to the Theory of Many-Body Systems,* 1997.

PAOLO ALUFFI (a cura di), *Quantum cohomology at the Mittag-Leffler Institute,* 1997.

GILBERTO BINI, CORRADO DE CONCINI, MARZIA POLITO, CLAUDIO PROCESI, *On the Work of Givental Relative to Mirror Symmetry,* 1998

GIUSEPPE DA PRATO, *Introduction to differential stochastic equations,* 1998

HERBERT CLEMENS, *Introduction to Hodge Theory,* 1998

HUYÊN PHAM, *Imperfections de Marchés et Méthodes d'Evaluation et Couverture d'Options,* 1998

MARCO MANETTI, *Corso introduttivo alla Geometria Algebrica,* 1998

AA.VV., *Seminari di Geometria Algebrica 1998-1999,* 1999

ALESSANDRA LUNARDI, *Interpolation Theory,* 1999

RENATA SCOGNAMILLO, *Rappresentazioni dei gruppi finiti e loro caratteri,* 1999

SERGIO RODRIGUEZ, *Symmetry in Physics,* 1999

"CompoMat" Loc. Braccone, 02040 Configni (RI), Italy
Finito di stampare dalla Nuova Grafica 86 nel dicembre 1999